Nobel Prize Women in Science

Bare-breasted woman adorns the reverse of the Nobel Prize medals for physics and chemistry. Copyright Nobel Foundation

Nobel Prize Women in Science

Their Lives, Struggles, and Momentous Discoveries

REVISED EDITION

Sharon Bertsch McGrayne

JOSEPH HENRY PRESS
Washington, D.C.

Joseph Henry Press • 2101 Constitution Avenue, N.W. • Washington, D.C. 20418

The Joseph Henry Press, an imprint of the National Academy Press, was created with the goal of making books on science, technology, and health more widely available to professionals and the public. Joseph Henry was one of the founders of the National Academy of Sciences and a leader of early American science.

Library of Congress Cataloging-in-Publication Data

McGrayne, Sharon Bertsch.
Nobel Prize women in science : their lives, struggles,
and momentous discoveries / Sharon Bertsch McGrayne.—Updated ed.
p. cm.
ISBN 0-309-07270-0 (pbk.)
1. Women scientists—Awards. 2. Science—Awards. 3. Nobel Prizes.
I. Title.
141.M358 1998
509.2′2—dc21

98-39490
CIP

Designed by Ardashes Hamparian

Printed in the United States of America

TO

Hilde Proescholdt Mangold

1898 – 1924

Hilde Proescholdt Mangold's doctoral thesis in biology won her adviser a Nobel Prize. Mangold executed the crucial experiments demonstrating the nature and location of the organizer, the chemicals that direct the embryonic development of different tissues and organs.

Unlike the other women in this book, Hilde Proescholdt Mangold did not conceive of or design her experiment. Her adviser, Hans Spemann, did so. She executed the project under his direction.

In 1924, the gas heater in Mangold's kitchen exploded. Hilde Proescholdt Mangold, the twenty-six-year-old mother of an infant son and the codiscoverer of the organizer, died of severe burns.

Eleven years after her death, Spemann won the Nobel Prize.

Frieda Robscheit-Robbins

1893 – December 18, 1973

For thirty-eight years Frieda Robscheit-Robbins was the research partner of George Hoyt Whipple. Although their joint work led to a cure for the deadly disease pernicious anemia, it was Whipple alone who won a Nobel Prize for Physiology or Medicine in 1934.

"Whipple's experiments," the Nobel Committee observed, "were planned exceedingly well and carried out very accurately, and consequently their results can lay claim to absolute reliability." Frieda Robscheit-Robbins helped to plan and carried out those experiments.

In fact, she was listed as the first author on Whipple's most important single paper, the report on which his scientific reputation rested. Generally, the first author is primarily responsible for the work summarized in the paper.

Whipple cited twenty-three scientific papers in his Nobel address. Of these, Robschiet-Robbins was the coauthor of ten. Whipple shared his prize money with Robscheit-Robbins and with two women technicians.

Frieda Robscheit-Robbins was born in Germany, educated in Chicago and California, and received her Ph.D. from the University of Rochester. She worked with Whipple from 1917 until her retirement from the University of Rochester Medical School in 1955. After thirty-eight years, she was still an associate in pathology, a junior-grade employee.

Of scientific research, she said, "You become possessed of a magnificent obsession and determination to learn the truth of your scientific theory if it takes sixteen years or many times sixteen. If you are successful, you really deserve no great credit, for by that time experiment has become the only thing in life you care to do."

Contents

Acknowledgments

SINCE THE FIRST EDITION OF THIS BOOK APPEARED, a tenth woman scientist has won a Nobel Prize. Christiane Nüsslein-Volhard won the Nobel Prize in Medicine or Physiology in 1995 for helping explain how embryos develop in flies, fish, people, and a host of other creatures. Her Nobel made a new and expanded edition of *Nobel Prize Women in Science* necessary.

The lives of the fifteen women in this book illustrate the changing patterns of discrimination against women in science, starting with legal bars to academic high schools and universities in Europe, and continuing in the United States with laws against working wives in universities.

The new chapter on Christiane Nüsslein-Volhard brings the book full circle. Many of the problems that women have faced in North America and Europe have been exacerbated by German attitudes about research science and working wives. Thus, Christiane Nüsslein-Volhard's experiences cast light on the lives of women far beyond the borders of her native Germany.

At the same time, she herself raises a difficult issue. With the decline in overt discrimination against women in science, we may be approaching an era of more equal opportunity. If that is the case, the future of women in science depends on what women want to make of it. Their future will be up to them, and not to others.

For this book Drs. Jocelyn Bell Burnell, Gertrude B. Elion, the late Dorothy Crowfoot Hodgkin, Rita Levi-Montalcini, the late Barbara McClintock, Christiane Nüsslein-Volhard, the late C. S. Wu, and Rosalyn S. Yalow graciously granted personal interviews.

Each chapter—whether its subject is alive or deceased—is based on primary and secondary sources and on extensive interviews with colleagues, students, family, friends, and experts in each field. Their cooperation is testimony to the importance that leaders of the scientific community place on attracting more woman to science.

All scientific explanations are nontechnical. Nothing is made up or fictionalized, however. All quotations come from published sources or from interviews. In many cases, the research revealed new material about these women, their lives, and their accomplishments.

Sources and recommended reading are provided at the end of the book.

I want to thank George F. Bertsch for many invaluable discussions on editorial and scientific questions. I would also like to acknowledge the helpful comments made by Ruth Ann Bertsch and Frederick M. Bertsch. The staff of the Michigan State University libraries, especially the reference department of the main library and Diane Clark of the physics library, were extremely helpful as well. Particular thanks go to my agents, Julian Bach and Carolyn Krupp, and to copy editor Connie M. Parkinson and to production supervisor Donald J. Davidson at Carol Publishing Group.

The following persons deserve special thanks because they carefully read and advised me on one or more of the chapters in manuscript: Bruce Alberts, Ruth Hogue Angeletti, George F. Bertsch, Frederick M. Bertsch, Ruth Ann Bertsch, Jacob Bigeleisen, Gloria Blatt, Verena Brink, Peter A. Brix, James Burchall, Alice Calaprice, Donald L. D. Caspar, Stanley Cohen, Mildren Cohen, Mildred S. Dresselhaus, Elizabeth E. K. Edwards, Jenny P. Glusker, James Goodman, Viktor Hamburger, Anne Harrison, William Havens, Günter Herrmann, Linda Cooke Johnson, Charles Kimberling, Aaron Klug, David M. Kipnis, Thomas A. Krenitsky, Hélène Langevin-Joliot, Joseph Larner, Leon J. Lidofsky, Neil Madsen, Joseph Meites, Ben Mottelson, Emiliana Pasca Noether, Hans Noll, Markus Noll, Ronald Oppenheim, Jeremiah P. Ostriker, Richard E. Phillips, Robert Provine, Robert G. Sachs, Anne Sayres, David Sayres, Paul Schedl, Hartmut Schultz, Trudi Schüpbach, Barbara Sears, Ruth L. Sime, Susan Simkin, Martha K. Smith, Joseph Taylor, Kenneth Trueblood, Alexander Tulinsky, Marcia Van Ness, Robert Ward, Spencer R. Weart, Alycia Weinberger, Eric Wieschaus, and Evelyn Witkin.

Author to Reader

ALFRED NOBEL, the Swedish inventor of dynamite, died in 1896, establishing with his fortune the most famous of all international awards, the Nobel Prizes. Following the dictates of his will, annual awards are given for peace and literature and for discoveries in physics, chemistry, and physiology or medicine. An economics prize was added in 1968.

When the awards were established, a single prize represented a fortune: thirty times the annual salary of a university professor, two hundred times that of a skilled construction worker. Awards were made for a discovery or improvement that had been made the preceding year and that had conferred the greatest benefit on mankind.

Prominent scientists throughout the world nominate candidates for the science awards. The physics and chemistry winners are chosen by the Swedish Academy of Sciences, and physiology or medicine winners are named by the Karolinska Institute in Stockholm.

Nobel laureates are the aristocrats of science, the elite, the cream of the crop. Approximately three hundred men have received science Nobels since 1901. In the course of almost a century, only ten women scientists have won Nobel prizes, a mere three percent.

This book examines the lives and achievements of fifteen women scientists who either won a Nobel Prize or played a crucial role in a Nobel Prize–winning project. It offers one answer to the question: Why so few women?

Nobel Prize Women in Science

1

A Passion for Discovery

"History is the essence of innumerable biographies."
Thomas Carlyle, *History*

WHY SO FEW? Why have only ten women won Nobel Prizes in science when more than three hundred men have done so? Ten out of several hundred—only three percent of all Nobel Prize scientists are women.

The fifteen women portrayed here are Nobel-class scientists. None is a typical, everyday researcher. They all either won a Nobel Prize in science or played a critical role in discoveries that won a Nobel for someone else.

Many of these women faced enormous obstacles. They were confined to basement laboratories and attic offices. They crawled behind furniture to attend science lectures. They worked in universities for decades without pay as volunteers—in the United States as late as the 1970s. Science was supposed to be tough, rigorous, and rational; women were supposed to be soft, weak, and irrational. As a consequence, women scientists were—by definition—unnatural beings. Sandra Harding, who writes about women in science from a feminist perspective, concluded that "women have been more systematically excluded from doing serious science than from performing any other social activity except, perhaps, frontline warfare."

No sooner did these women overcome one barrier than another cropped up. Pioneers like the mathematician Emmy Noether were not only legally barred from universities; they were also excluded from the academic high schools that prepared men for university educations. Until the 1920s, most European high schools for girls were finishing schools. Women who wanted university training had to hire tutors to learn mathematics, science, Latin, and Greek—all required subjects for entrance to a university. The father of physicist Lise Meitner refused to hire a tutor until she had completed teachers' training school. The dictatorial father of Rita Levi-Montalcini prevented her from obtaining an academic education until she was twenty years old; later she discovered the nerve growth factor, which may play a vital role in degenerative diseases like Alzheimer's. Both Meitner and Levi-Montalcini started their scientific careers a decade behind their male counterparts. Once in universities, women like Marie Curie, Emmy Noether, and Meitner worked for years without salaries or positions.

In the United States, the situation was different but no less difficult. American universities admitted women as students but refused to hire them as researchers. Women scientists were supposed to teach in women's colleges or in coeducational universities; they were not to do research. Expected to remain single, they needed husbands to give them access to research laboratories. Yet until the Federal Equal Opportunity Act of 1972, state laws and university rules banned hiring wives of university employees. These rules were devasting for women scientists. Even today, 70 percent of American women physicists are married to scientists. As a result, the academic landscape was littered with husband-and-wife teams in which the man had the salary, job security, and prestige, and the woman assisted him at his pleasure. Universities have dealt with the issue of married women researchers for a relatively short time.

Gerty Cori, who studied carbohydrate metabolism, enzymes, and children's diseases caused by enzyme deficiencies, did not become a professor until the year she won a Nobel Prize. Maria Goeppert Mayer, who developed the shell model of the atomic nucleus, worked for decades as a volunteer at some of North America's most prestigious universities. When Barbara McClintock was president-elect of the American Genetics Society, she quit science for a time because she could not get a university job. Gertrude Elion spent almost a decade studying to be a secretary and working in temporary, marginal jobs before she landed a position as a research chemist. Then she helped develop a new approach to drug-making.

Even the most successful women scientists faced ridicule and hostility. Rosalind Franklin—caricatured as "Rosy" in James Watson's best-seller *The Double Helix*—was a commanding leader. But Watson and Francis Crick used her experimental evidence—without her knowledge, permission, or credit—to explain the molecular structure of DNA. After her death, they won the Nobel Prize. Irène Joliot-Curie, the daughter of Marie Curie, was a teenage heroine of World War I. Yet after she won a Nobel Prize for discovering artificial radioactivity, the American press vilified her for her support of the Soviet Union following World War II.

If a woman formed a long-term scientific partnership with a man, the scientific community assumed that he was the brains of the team and she was the brawn. Medical scientists concluded that Rosalind Yalow's male collaborator was the creative force behind their discovery of the radioimmunoassay, a phenomenally sensitive test that revolutionized endocrinology and the treatment of hormonal disorders like diabetes. When her partner died, Yalow had to establish her reputation all over again.

In addition to professional discrimination, these women suffered their share of racial and religious discrimination, as well as poverty, war, substance abuse, physical handicaps, and illness. Marie Curie, Irène Joliot-Curie, Dorothy Hodgkin, and Gerty Cori worked for decades despite life-threatening and crippling diseases. World War II destroyed Lise Meitner's career. Rita Levi-Montalcini began her research in her bedroom, hidden from the Nazis. Gertrude Elion worked her way through school during the Depression and quit graduate school without a Ph.D. Chien-Shiung Wu, the experimental physicist who overturned the fundamental law of parity, could not get a research job during World War II because of discrimination against Asians—even though her country was allied with the United States. Jocelyn Bell Burnell, a graduate student when she discovered pulsars, later worked part-time while raising a family.

In the face of such obstacles, what sustained these women?

What prevented them from giving up, as many other women scientists did?

First, they adored science. They triumphed because they were having a wonderful time. Their hobbies ranged from music to mountain climbing, books, gourmet cooking, church, and child-rearing. But it was science that illuminated their lives. Words like "pleasure," "joy," and "satisfaction" permeate their speech. They survived in science because they were passionately determined and in love with their work.

Science thrilled them because they were making some of the most important scientific breakthroughs of the twentieth century. Two of the greatest intellectual achievements of the century occurred in evolution and in atomic and subatomic physics. These women helped explain how individual characteristics are passed down through generations of organisms and how atoms and their constituent particles behave. They opened up new fields of science in mathematics, biology, chemistry, astronomy, physics, and medicine.

Before Emmy Noether escaped from Nazi Germany to the United States, she created abstract algebra, a major new field of mathematics known in its elementary school version as "The New

Math." Barbara McClintock revolutionized genetics several times over as a young woman, yet molecular biologists ignored her discovery of transposable genetic elements for decades. Dorothy Hodgkin, an English physicial chemist, pioneered the use of molecular structure to explain biological functions by deciphering the atomic structures of penicillin, vitamin B_{12}, and insulin.

Marie Curie, winner of two Nobel Prizes in science, focused scientific attention on radioactivity, the key to the atomic nucleus, and discovered radium, the first real hope in cancer therapy. Lise Meitner, officially retired after her escape from the Nazis, deciphered the experiment of the century by explaining that the atomic nucleus can split and release enormous amounts of energy. For the fission project that she initiated and explained, her German partner received the Nobel Prize.

Sympathetic parents and relatives were particularly influential. All of these women, with the exception of Rosalyn Yalow, came from professional or academic families: their fathers were architects, engineers, physicians, dentists, lawyers, and university professors. Yalow's father owned a small paper and twine shop in one of New York's immigrant neighborhoods. Emmy Noether's father, on the other hand, was a prominent mathematician who nurtured his daughter's talent. Maria Goeppert Mayer's father urged her to have a career; she wanted to become the seventh-generation professor in his family. Chien-Shiung Wu's father was one of China's leading feminists. Dorothy Hodgkin and Rosalind Franklin received financial assistance from mothers and aunts. Marie Curie and her sister forged a pact to support each other through the university; Marie in turn then helped her daughter Irène Joliot-Curie. In contrast, the fathers of Levi-Montalcini and Rosalind Franklin vehemently opposed their daughters' aspirations. In Barbara McClintock's family, it was her mother who disapproved of women professors.

Religious values stressing education were critical. Jocelyn Bell Burnell is a Quaker, a member of the Society of Friends, a small denomination that has produced a disproportionate number of the world's great scientists. Half of the women have Jewish backgrounds. The Jews' commitment to learning and abstract thinking has helped men as well as women in science; Jews number only three percent of the United States population, but they account for approximately twenty-seven percent of the Nobelists brought up in the United States. Being Jewish was particularly helpful to women. Of the three Nobel winners who were born and educated in the United States, two are Jews. Conversely, Catholic and Protestant America has produced only one woman Nobel Prize winner in science: Barbara McClintock.

Behind many of these successful women stood a man. More than half of the women married and raised children. All but one of the husbands supported their wives' science, sometimes at considerable sacrifice. Pierre Curie and Carl Cori refused prestigious job offers in leading laboratories to further their wives' careers. Wu and her husband had a commuter marriage. Three prominent male physicists encouraged a generation of English women in crystallography, including Dorothy Hodgkin. The great mathematician David Hilbert was Emmy Noether's mentor. Joseph Mayer may have been more of a feminist than his wife, Maria Goeppert Mayer. Gertrude Elion's research partnership with George Hitchings endured for decades, as did Yalow's with Solomon Berson. Unfortunately for Jocelyn Bell Burnell's career, her thesis advisor failed to become her mentor, and she received little or no career counseling.

The importance of institutional support for women scientists is highlighted by one remarkable fact: Two schools account for the majority of Nobel Prizes received by American women scientists. Of the six American women who won science Nobels, four were associated with either Hunter College in New York City or Washington University in St. Louis. Gertrude Elion and Rosalyn Yalow were undergraduates at Hunter College during its heyday as a free municipal college for New York's brightest women. Gerty Cori and Rita Levi-Montalcini won Nobels for research conducted at Washington University in St. Louis, Missouri. At the time, Washington University was notably liberal in its treatment of working women. How many more women might have succeeded had they enjoyed such support! Girls' schools played a role too. Barbara McClintock is the only one in the group who never attended a girls-only school.

Finally, good luck and good timing were vitally important. Pioneers like Marie Curie, Lise Meitner, and Emmy Noether came of age just as European universities opened their doors to women. Most of the women—eight out of fifteen—were born within fifteen years of each other. Eleven of the fifteen were born within a single generation: from 1896 to 1921. Their formative years spanned the first women's movement, when suffrage campaigns swept North America and Europe, World War I, when women took over men's jobs, and the 1920s, when the social constraints on women's behavior moderated. Four of the women slipped into jobs vacated by men during World War II. Will the second women's movement of the 1960s and 1970s have a similar effect on the Nobel Prize?

Given the enormousness of the problems they faced and the importance of the discoveries they made, the real question to be asked about these women is not "Why so few?" A better question is

"Why so many?" As Chien-Shiung Wu noted about women physicists, "Never before have so few contributed so much under such trying circumstances."

* * *

FIRST GENERATION PIONEERS

2

Marie Skłodowska Curie

November 7, 1867–July 4, 1934

PHYSICIST AND RADIOCHEMIST

Nobel Prize in Physics 1903
Nobel Prize in Chemistry 1911

"HUSBAND STEALER! Get the foreign woman out!" The sound of shouts and catcalls filled Marie Curie's home. A stone struck the house. The crowd was growing larger and more hostile. Inside, Marie Curie and her seven-year-old daughter Ève huddled white-faced and silent.

For days, the Parisian tabloids had been filled with the Polish woman's scandal: "The Vestal Virgin of Radium" had stolen a French mother's husband. Curie had been maintained like "a concubine in a conjugal domicile.... The fire of radium had lit a flame in the heart of a scientist, and the scientist's wife and children [were] now in tears."

Reporters, editors, and her accused lover fought duels; the French Cabinet deliberated; and the University of Paris tried to get Curie to resign her professorship and return to Poland. In the midst of the crisis, Curie collapsed, contemplated suicide—and won her second Nobel Prize.

At the time, Marie Curie was the world's most famous scientist, a household word, a saintly icon, and an easy target. The first woman professor in France, she had discovered radium, the world's first hope for cancer patients. She was the physicist and chemist who turned the attention of the scientific world to radioactivity and proved that the atom was not inert, indivisible, or solid. For sixty-one years, she was the only person who had won two Nobel Prizes for science.

An intensely private woman, she hid deep emotions behind a composed façade. Longing for a simple family life and freedom to work in her laboratory, she disliked distractions and disturbances. When fame and feelings intruded, they threatened the settled pattern of her existence. Only later did she learn to contain the passions that had almost destroyed her and to use her unsought celebrity to serve science.

She was born Marya Skłodowska in Warsaw, Poland, in 1867, just after the American Civil War. Called Manya as a child, she changed her name to its French form, "Marie," as a student in France. Her childhood was often tense and unhappy. Russia and Germany had divided Poland between them and were trying to destroy every vestige of Polish nationalism and culture. As Chancellor Otto von Bismarck of Prussia wrote, "Hit the Poles till they despair of their very lives. I have every sympathy for their position, but if we are to survive, our only course is to exterminate them." In Tsarist Poland, Marie Curie learned early that expressing her emotions could be dangerous.

Both her parents, Valdislav and Bronislawa Skłodowski, were teachers. Vladislav taught high school physics and mathematics, and although Bronislawa gave birth to five children in eight years and had tuberculosis, she was the full-time director of a private school for girls. She quit her job only after the birth of Marie, her last child. Bad investments and medical bills had cost the family their savings, so they lived in Bronislawa's boarding school. Her babies were born literally within earshot of her paying customers. Life without privacy required silence and self-control, and Marie grew up hating loud voices, commotion, theatrics, and any display of emotion.

Although they held themselves tightly in check, the Skłodowskis were a passionate family of strong convictions. Marie's mother was an extremely devout Roman Catholic, the entire family was intensely patriotic, and they all believed fervently in education. "My father and mother worshiped their profession in the highest degree," Marie recalled later. The children were expected to learn to read before they entered school. When Marie was four years old, she quizzed her older sister Bronya on her letters. Impatiently Marie grabbed the book and began reading it aloud. The family became so excited that Marie burst into tears and apologized for being able to read.

By the time Marie was eleven years old, her oldest sister had died of typhus and her mother of tuberculosis. Hoping to prevent the spread of her disease, Madame Skłodowska had never kissed or fondled her youngest child. Marie, in turn, had idolized her mother without understanding her aloofness. When she died, Marie fell into a profound depression and concluded that God did not exist.

At school, Marie studied in an atmosphere of political intimidation and oppression. The Russians ran Polish schools like police states. Polish teachers were fired and children punished for speaking their own language. As Russians took over Polish teaching positions, Marie's father moved from job to job and apartment to apartment,

Marie Curie, 1913.

Marie Curie with her daughter, Ève (left), and Irène (right), in 1908.

Pierre and Marie Curie in front of their house at Sceaux, 1895.

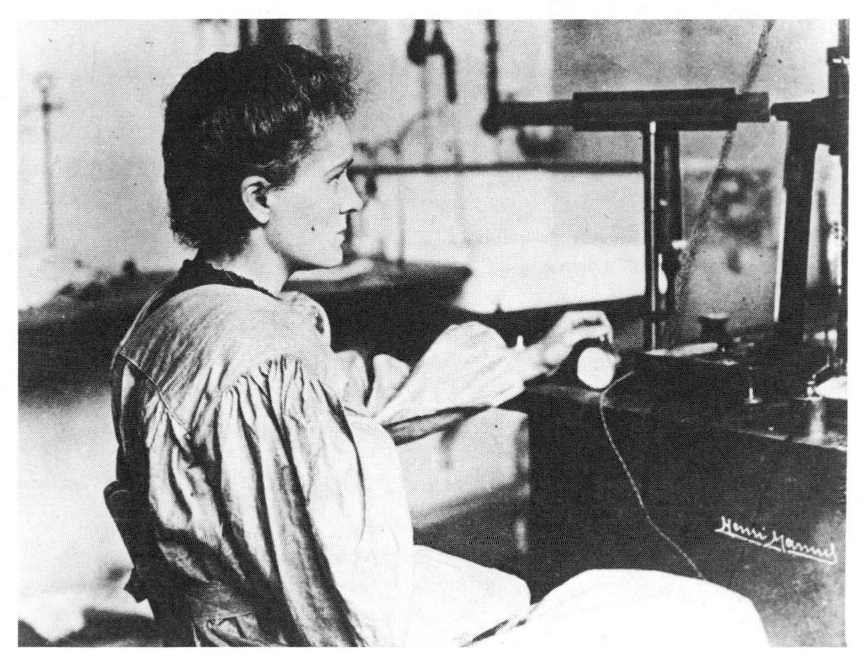

Marie Skłodowska-Curie in her laboratory.

each smaller than the last. They became so poor that her father took in student boarders for tutoring. Eventually, the family became so crowded that Marie slept on a sofa in the dining room and rose early each morning to clear the room for the boarders' breakfast.

Despite such distractions, Marie was the brightest student in her class, the one chosen to recite in Russian whenever the government inspector made his rounds. Only ten years old, she was responsible for fooling the government into thinking that her school taught Russian culture in Russian, instead of Polish culture in Polish. Marie performed perfectly, but she was so tense that, when the inspector left, she burst into tears. She remained nervous about public speaking for the rest of her life.

In secondary school, Russian professors treated their Polish pupils like enemies. "The moral atmosphere was altogether unbearable," she protested years later. Students were constantly spied upon. One of her brother's friends was hanged for political activities. It is not surprising that schools became centers of Polish nationalism and organized resistance. Education became a patriotic duty and a moral imperative.

Responsibility for preserving Polish culture fell—not on Polish soldiers—but on young middle-class Polish women like Marie. Following the collapse of the last Polish uprising in 1863, a group of Warsaw intellectuals called the Positivists argued that force alone could not defeat Russia. The Positivists campaigned for women's emancipation and education; science; toleration of the Jews; abolition of class distinctions; reform of the Polish Roman Catholic Church; and education for the peasants. Women formed the backbone of the Positivist movement and founded the clandestine Flying University, where anyone could attend secret lectures in return for teaching one. As Marie put it, "You cannot hope to build a better world without improving the individual." A Positivist poem, "Forward Through Work," declared:

> The strains of the harp are not for you....
> Neither saber, nor spear, nor arrow.
> What you need is unremitting toil,
> The food of the mind, the bread of the soul.

Marie Skłodowska's passionate commitment to hard work, learning, and science had its roots in Poland's dream for nationhood.

At the age of fifteen, Marie graduated first in her class in every subject. The strain had been enormous. In 1883, she collapsed in the first of several physical breakdowns she suffered throughout her life. Her father arranged for her to take the year off, visiting relatives in the countryside and enjoying herself. After dancing all night, she declared, "All I can say is that maybe never again, never in my whole life, will I have such fun."

After a year of pleasure, Marie had to start supporting herself—and her older sister Bronya. Their brother was a student at the University of Warsaw, but the Russian government prohibited women from attending any university within its empire. So Bronya and Marie made a pact: Marie would help support Bronya through medical school in Paris, and then Bronya would help pay Marie's way.

Few jobs were open to middle-class women, so Marie spent the next six years (from September 1885 to September 1891) as a governess. She was little more than a servant in a wealthy home. For three and a half of those years, she lived with the Zorawski family sixty miles from Warsaw. She tutored two of the Zorawski children seven hours a day and was at the family's beck and call the remainder of the time. With the father's approval, she also taught local peasant children to read and write, an illegal activity for which she could have been severely punished. She used every moment of her precious free time for study. To help, her father tutored her in mathematics in his letters, Zorawski permitted her to use the technical library in his sugar beet factory, and a factory chemist gave her twenty chemistry lessons.

That summer, the Zorawski's eldest son vacationed at home. Kazimierz was not only handsome and dashing, he was also studying mathematics at the University of Warsaw. Predictably, he and Marie fell in love. Appalled that their son and heir would consider marrying a mere governess, the Zorawskis forbade the engagement. Trapped, Marie stayed at the Zorawskis for two and a half more years because Bronya needed money in Paris. Writing of her anguish, Marie confessed, "Creatures who feel as keenly as I do...have to dissimulate it as much as possible.... I would give half my life to be independent again, to have my own home." Marie did not give up hope of marrying Kazimierz until she was almost twenty-four years old.

In 1891, Marie finally left for Paris with forty rubles in her purse, a trunk, and a camp chair to sit on in the fourth-class train car.

Although she would live the rest of her life in France and became famous under her French name "Marie Curie," she remained passionately involved in Polish affairs. She donated money to the Russian uprising against the hated Tsars in 1905 and hired Polish nursemaids to teach her children Polish. After World War I, she raised money for a Polish radium institute and trained Polish scientists.

Decades later, the French were still not reconciled to their heroine's Polish origins. For the one-hundredth-anniversary celebration of Marie Curie's birth in 1967, French government officials tried to find a photograph that would not make Curie look "so Polish." They also wanted to use only her French name. At the insistence of the Polish government and her French grandchildren, anniversary posters called her Marie Curie-Skłodowska.

For Poles, France represented the glories of cultural and political freedom. The Latin Quarter, the student section of Paris, was the intellectual heart of Europe in the 1890s. At its center was the University of Paris, one of the few universities in Europe that admitted women. Twelve thousand men and a handful of women attended lectures there. French cultural life thrived with post-Impressionist painters, Verdi's operas, the Eiffel Tower, broad boulevards designed by Baron Haussmann, electric street lighting, and automobiles. Politically, the Republic occupied the center between the right-wing nobility and the left-wing workers. Underneath the glittering veneer, however, unemployment was high, anti-Semitism was rife, and Protestants were unpopular. And scientifically, France was conservative and behind the times.

When Marie Curie enrolled in the University of Paris on November 5, 1891, she had been out of school for eight years. Her French was inadequate, and she had studied less math and science than any French high school graduate. Although she could have lived with her sister Bronya in the midst of a large community of Polish political émigrés, she chose to stay by herself. Poor like most other students, she rented a sixth-floor garret room and subsisted on bread with occasional eggs, fruit, and cups of hot chocolate.

She delighted in what her family jokingly called her "heroic period." She had space of her own, privacy, independence, and as much time as she wanted to study. She was enjoying science the way other people love music—for the delight and profound enjoyment that it gave her. As she recalled later, "This life, painful from certain points of view, had, for all that, a real charm for me. It gave me a very precious sense of liberty and independence."

Her heroic period lasted only two years. In 1893, she won a Polish fellowship of six hundred rubles a year. She also earned the equivalent of a master's degree in physics and scored first in her class. The following year, she earned a similar degree in mathematics and placed second.

One of the world's great love stories began in 1894 when Marie Skłodowska met Pierre Curie. Curie was the laboratory director of the Municipal School of Industrial Physics and Chemistry in Paris. At thirty-five, he was already an important physicist specializing in crystals and magnetic materials. He and his brother Jacques had discovered piezoelectricity, the electricity generated by squeezing certain crystals, like quartz. Such crystals are used today in microphones, broadcasting electronics, stereo systems, and wristwatches.

When Marie Skłodowska first saw Pierre Curie, he was silhouetted romantically against the light of a french door. His auburn crew cut, limpid eyes, grave smile, and simple manners instantly reassured her. His conversation about physics and social issues fascinated her. On his part, Pierre Curie saw a Slavic beauty, much in vogue at the time, with curly blond hair, gray eyes, wide cheekbones, and broad forehead. More important, he sensed her brilliance and passion for physics.

Ignoring the proper etiquette for a young lady, Marie Skłodowska gave Pierre Curie her address the very day they met. As a foreigner and student, she lived somewhat outside the bourgeois conventions of the day, and when they interfered with her life or work, she was prepared to overlook them. Later, she had dinner with Pierre and invited him to her room for tea. In return, he gave her Émile Zola's new book, just banned by the Catholic Church.

Pierre Curie was an idealistic dreamer. He had been raised in the republican, anticlerical tradition of the French Revolution. Unable to adapt to school life, he had been tutored by his father and older brother. Pierre Curie, however, was certainly not part of the French scientific establishment. He had not attended the elite École Normale, and he taught at a new technical school for talented, working-class Parisians. There he earned approximately as much as a day laborer while conducting his famous magnetic experiments in a hallway between his lab and a staircase. Totally uninterested in awards for himself, Pierre Curie would see that his wife got her full share of credit for her scientific discoveries.

Maria Skłodowska still dreamed of returning to Poland to teach physics. Pierre Curie, on the other hand, talked about their sharing a life consecrated to scientific research. He argued, correctly, that she could do more and better research in France than in impoverished Poland. "It is necessary to make a dream of life, and to make a dream a reality," he wrote her. He asked her to marry him and, failing that, to share an apartment with him. Eventually, he made the ultimate sacrifice: he offered to give up his research career and live in Poland with her. At that, she gave in and agreed to marry him. Marie and his father, however, convinced Pierre to finish his doctoral dissertation in order to qualify for a professorship and a laboratory. His thesis on the relationship between temperature and magnetism became known as Curie's law. With his doctorate and a strong recommendation from the great English physicist Lord William Thomson Kelvin, Pierre Curie was promoted to professor and was paid the comfortable income of six thousand francs a year.

Marie and Pierre Curie were married in 1895 without rings, blessings, or priests. A relative gave them enough money to buy two bicycles, and wearing a split skirt and a black straw hat, Marie took off with Pierre on the first of many bicycle tours.

Marie Curie was studying for a teaching certificate in order to qualify as a professor in girls' high schools. In August 1896, she placed first in the teachers' examination and secured financial support from the metallurgical industry to study the magnetic properties of steel.

Managing a three-room apartment with little household help, she stripped her life of unnecessary details like curtains, rugs, and excess furniture. She cared little about houses but a lot about their gardens and the views from their windows. She also had a strong sense of her worth and superiority and did not like to waste her time on anything that bored her, like housekeeping.

Marie and Pierre paid none of the formal social calls that occupied so much time for others of their class. Instead, they visited relatives and held a Sunday afternoon open house for friends and students. Professors were government employees, so the Curie circle was intensely interested in national politics. Their closest friends were a chemist, André Debierne; an eminent physicist, Jean Perrin, and his wife, Aline; and another physicist, Paul Langevin. Marie Curie focused her activities exclusively on science and her family—in that order—but she was knowledgeable and opinionated about political issues too. The Curies and their friends were leftist and anticlerical republicans who wanted the government to spend more for education and scientific research. While Catholic groups viewed science as morally bankrupt, the Curie circle thought in terms of moral and material progress, disease prevention, better agricultural methods, electric tramways, street lighting, and the like.

Two years after Marie and Pierre Curie were married and a month before Marie's thirtieth birthday, their daughter Irène was born. Pierre's widowed father moved in with them to care for the child while Marie worked. Even in liberal France, working mothers were almost unheard of, but both Curie men supported her desire to continue studying. She was one of only two women working for doctorates in Europe. A graduate student mother, she kept three sets of books: a baby's notebook, her lab reports, and her household accounts. She recorded Irène's first step as carefully as she recorded her scientific data and the cost of meat.

When Irène was three months old, her mother began looking for a research topic for a doctoral dissertation. The decision was not easy. At the beginning of the twentieth century, physicists believed that they had already discovered everything about the physical universe. As a prominent German physicist announced, "Nothing else has to be done in physics except make better measurements." Marie Curie's thesis, however, would blast physics wide open. More than anything else, her discoveries pointed to the powerful forces inside the atom, forces that physicists have spent the twentieth century exploring.

Few scientists paid great attention when Henri Becquerel discovered radioactivity in uranium in 1896. X rays, discovered the year before, had already grabbed the headlines and the glamour. Becquerel's radiation occurs when the heavy, unstable nucleus of an atom breaks apart and ejects its excess energy as clusters of protons and neutrons (called alpha particles), as superfast electrons, or as gamma rays of pure energy. Ironically, the more fashionable X rays are much less powerful; they originate in the clouds of electrons surrounding each atom. Marie Curie did not know the cause of radioactivity at the time, but she wanted to work in a totally new field where she could do laboratory research instead of library reading. Once she had decided on Becquerel's radiation for her thesis topic, Pierre Curie secured space at his college for her work.

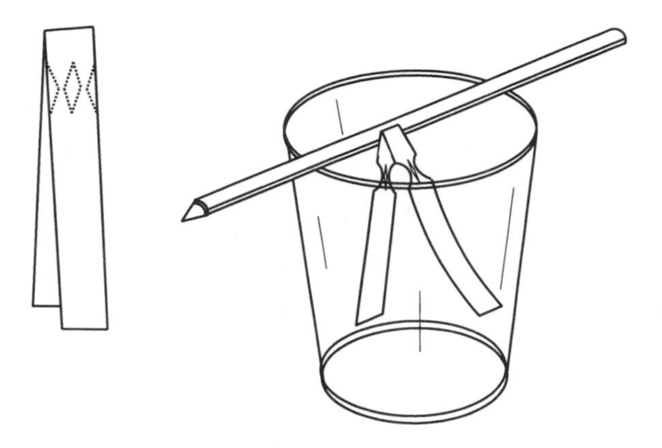

Fig. 2.1 An Electroscope.

For a simplified version of the instrument that Marie Curie used to detect radioactive substances, cut and fold a piece of gum foil. Tape it to a pencil in a glass to avoid drafts. Rub a metal pen or a balloon on wool to create static electricity and bring it close to one side of the foil. The strips separate and, as the electricity leaks away, close.

Becquerel had shown that the radiation emanating from the element uranium could blacken photographic plates, even those wrapped in heavy cardboard. He had also discovered that uranium makes the air around it conduct electricity. Marie Curie realized that this phenomenon, called ionization, could be used to detect radioactivity in other substances. So she decided to conduct a systematic search of all the known elements. She used Pierre Curie's invention, the piezoelectric quartz balance, to measure the weak electrical charges. Within days of starting her project, she discovered that another element, thorium, produces the same powerful effects as uranium.

Next, she decided to measure the strength of the electric current produced by different uranium and thorium compounds—that is, substances composed of uranium or thorium in combination with other elements. She discovered a simple, but totally unexpected, phenomenon: the strength of the radiation depends only on the amount of uranium or thorium in the compounds.

Normally, different compounds of the same element share few common chemical or physical properties, such as hardness, color, or solubility. For example, pure uranium is a heavy, shiny metal; one compound is a black powder; and another compound is a transparent, yellow crystal that glows green. Yet all three emit radiation according to the amount of uranium they contain. Thus, she deduced, radioactivity does not depend on how atoms are arranged into molecules. Instead, radioactivity originates within the atoms themselves. This simple but breathtaking discovery is Marie Curie's most important scientific contribution.

Next, she had what the Nobel Prize-winning physicist Emilio Segrè considered a stroke of genius. She extended her research beyond uranium and thorium and their simple compounds to their natural ores. Testing museum samples, she discovered that two uranium ores—pitchblende and chalcolite—were three and four times more radioactive than could be predicted from the amount of uranium or thorium they contained. What could be producing the extra amount of radioactivity? She hypothesized the presence of an unknown, highly radioactive element in the ore. She also coined a new word, *radioactivity*.

In April 1898, Marie Curie wrote an article announcing her discovery that thorium was radioactive and her hypothesis that a new radioactive element existed within the ore. Unfortunately, only two months before, a German had also discovered thorium's radioactivity. However, he had not reached the all-important conclusion that radioactivity emanates from within the atom. Nor had he realized that the ores contained a new element.

Intrigued, Pierre Curie dropped his own beloved crystal research and joined Marie's radioactivity project. He never returned to crystals.

The Curies' only clue to the new element was its radioactivity. Breaking down pitchblende into its constituent chemicals, Marie and Pierre discovered that only two components—those mixed with either bismuth or barium—were strongly radioactive. Clearly, the barium and bismuth compounds must have something radioactive in them. So they began separating those compounds into their components as well. When the bismuth compound was heated red-hot in a vacuum, the radioactive element was deposited on the cooler part of the test tube. The new substance was four hundred times more active than uranium. Marie Curie named it polonium in honor of her native land. Not only had she discovered a new element, she had also opened a new field of physics; radioactivity became the primary technique for exploring the interior of the atom.

While separating out polonium at the end of 1898, Marie Curie also discovered a second, even more active element, which she named radium. A third radioactive element, actinium, was found in the pitchblende by André Debierne, the shy but devoted chemist who was reportedly deeply in love with Marie Curie.

The Curies announced the discovery of their two new elements, polonium and radium, in articles published in July and December 1898. For her work, Marie won a thirty-eight-hundred-franc prize from the French Academy of Sciences.

Logically, the next step was to isolate the elements. Based on their calculations, the Curies knew they would have to purify tons of ore in order to extract a few grains of radium salts. The only space large enough at the school was an abandoned dissection shed. The shack was stifling hot in summer and freezing cold in winter. It had no ventilation system for removing poisonous fumes, and its roof leaked. A chemist accustomed to Germany's modern laboratories called it "a cross between a stable and a potato cellar and, if I had not seen the work table with the chemical apparatus, I would have thought it a practical joke." This ramshackle shed became the symbol of the Marie Curie legend.

Marie and Pierre shared the physics and chemistry work, moving back and forth between the disciplines as they proceeded, according to their granddaughter Hélène Langevin-Joliot, a nuclear physicist who has studied their notebooks in detail. Marie Curie was the group's leader and its driving force, directing lab discussions and keeping the operation moving. Pierre added scientific concepts. Marie also did the heavy physical labor, well aware that she was doing menial work that any technician could have accomplished. "Sometimes I had to spend a whole day mixing a boiling mass with a heavy iron rod nearly as large as myself. I would be broken with fatigue at that day's end," she complained. Every winter, she was sick with pneumonia or some other serious illness, often for several months at a time. As she remarked later, "If we had had a fine laboratory, we should have made more discoveries and our health would have suffered less."

"And yet," she concluded, "it was in this miserable old shed that the best and happiest years of our life were spent, entirely consecrated to work." She had achieved her goal—a simple family life and some interesting work. As she confided to her relatives about Pierre, "He was as much and much more than all I had dreamed at the time of our union. My admiration of his unusual qualities grew continually."

Isolating radium salts from tons of pitchblende devastated their health. Besides handling radioactive material, they were breathing the radon gas emitted by radium. They spent hours in the shed, even eating meals that Marie cooked there. Almost a century later, their notebooks are still dangerously radioactive. Marie Curie later estimated that in a properly equipped laboratory the four years could have been compressed into one, minimizing their exposure to radiation. Marie was spared the worst, because she commuted twice weekly to Sèvres to lecture at a teachers' training college for women. She suffered a late miscarriage in 1903 after a long bicycling trip, but had a second healthy child, Ève Denise Curie, in 1904. Nevertheless, Pierre told a friend, Marie was "always tired without being exactly ill."

Pierre suffered more. While Marie liked to keep some radium salts glowing by her bedside, Pierre carried a test tube of it in his pocket to show friends. He performed medical experiments on his arm and showed that radium burns took months to heal. As he liked to point out, radium is one million times more radioactive than uranium. He and Marie loved to visit the shed at night to see the luminous test tubes, glowing like faint fairy lights.

Pierre, with Marie Curie's acquiescence, made the momentous decision not to apply for patents on the industrial processes they devised for extracting radium salts from ore. Like many scientists at the time, they believed that research should be "disinterested"—conducted for its own sake, not for any material rewards. The Curies were bitter because they did not have a proper laboratory, but they rejected an opportunity to make millions that could have been used for research. By September 1902, the Curies had isolated one-tenth of a gram of radium chloride from several tons of pitchblende. Marie also ascertained the atomic weight of radium at 226.

Radium was the most important element to be discovered since oxygen. Glowing with a bluish light, it emits electrically charged particles. A piece of radium roughly the size of a penny produces approximately five hundred calories of heat every day for a thousand years. Clearly, something important was happening inside the radium atoms. As late as 1897 when the electron was discovered, physicists had assumed that atoms were solid, indivisible, stable, and immutable. Yet radium was proof that some powerful force existed inside the atom, a force powerful enough to emit heat and give off light for years on end. In his lab, Ernest Rutherford showed that atoms of radioactive substances even change from one element to another. More than any other scientist, Marie Curie forced her colleagues to pay attention to the invisibly small world inside the atom. Scientists were forced to alter their definition of an element and to recognize a new force of nature.

The honors started pouring in at once. The British had always been Pierre Curie's greatest fans, and they asked him to speak to the eminent Royal Institution in London in June 1903. By then, he had violent joint pains and his legs sometimes trembled so badly that he could not get out of bed. The evening of his Royal Institution talk, his fingers were so covered with sores that he could barely dress himself. Observers thought he looked sick and feeble. As Pierre demonstrated the properties of radium and radioactivity, he fumbled, spilling some of Marie's precious radium; fifty years later, areas of the hall had to be decontaminated.

Marie was so busy working in her shed that she did not have time to present her doctoral thesis until June 1903. Her examining committee declared that her thesis, "Researches on Radioactive Substances," was the greatest scientific contribution ever made by a doctoral dissertation. That evening, the Curies celebrated with their friends, Paul Langevin, Jean Perrin and his wife, and Ernest Rutherford. One of the greatest experimental physicists of all time, Rutherford endeared himself to Marie when he confided that radioactivity is "a splendid subject to work on.... You know, it must be dreadful not to have a laboratory to play around in."

The June evening was warm, and at eleven P.M., the group moved outdoors into the garden. In the darkness, Pierre reached into his pocket and pulled out a tube of radium. Watching it glow, the group fell silent. Rutherford could see that Pierre's hands were so raw and inflamed that he could barely hold the tube. A doctor diagnosed rheumatism and prescribed strychnine.

In 1903, the French Academy of Sciences nominated Henri Becquerel and Pierre Curie to share the Nobel Prize for physics for their work on radioactivity. Marie Curie was not included. Luckily, one of the most powerful Swedish physicists on the nominating committee, Magnus Gösta Mittag-Leffler, was a great supporter of women scientists. The Swede wrote Pierre Curie that he and he alone was being considered for the prize.

Pierre was clearly Nobel Prize material for his magnetism research, but radioactivity was another matter. He cared little about prizes himself, but he wanted his wife to get credit for her work. So he wrote back to Mittag-Leffler, "If it is true that one is seriously thinking about me [for the prize], I very much wish to be considered together with Madame Curie with respect to our research on radioactive bodies." Pointing to her role in the discovery of radium and polonium, he added with Gallic logic: "Don't you think that it would be more artistically satisfying [*plus joli d'un point de vue artistique*] if we were to be associated in this manner?"

The problem was that Marie Curie had not been nominated for the 1903 prize. She had, however, received two nominations the previous year. With some bureaucratic sleight-of-hand, one of those nominations was declared valid for 1903, and she was nominated with her husband and Becquerel.

When the Royal Swedish Academy of Sciences met to discuss the Curies' nomination, it changed their prize in a small but crucial way. As originally conceived, the Curies would have won the physics prize for "their discovery of the spontaneously radioactive elements." Chemists objected, however. They wanted to leave a way open to award the Curies a *second* Nobel Prize for the discovery of radium. "The discovery of such a singularly remarkable element as radium might eventually be considered for a Nobel Prize in chemistry," they argued. So the Academy decided to give the Curies the 1903 prize in physics for "their joint researches on the radiation phenomena discovered by Professor Henri Becquerel." Radium could wait for another year. Until the 1980s, when the Nobel archives opened its records of the 1903 debate, many scientists considered Marie Curie's second prize in 1911 undeserved because they assumed that it was awarded for her later, less crucial research.

The Curies were too sick to collect their prize in December 1903. It was eighteen months later before Pierre could give the obligatory lecture and pick up their prize money.

Marie Curie's Nobel Prize for Physics created two stars: Curie herself and the science Nobels. Until then, the press had paid no attention to the science prizes. The literature and peace awards received broad coverage, but the physics, chemistry, and medicine or physiology prizes were considered too esoteric for the mass media. Marie Curie made the science prizes so popular that the press never again ignored them.

Ironically, the public was enthralled with Marie Curie for her discovery of radium—the element that she had not yet won a prize for. Radioactivity was complicated, but radium was glamorous. It was expensive; it was a possible cure against cancer; and it was magical, changing one element into another and producing a seemingly inexhaustible supply of energy. Marie Curie symbolized both the selfless pursuit of science and its humanitarian benefits. She also personified the triumph of the lone individual against impossible odds. In France, her shed stood for governmental neglect of scientific research. The Curies' refusal to patent their discovery and their neglect by the French scientific establishment only fueled the legend.

By the end of 1903, Marie Curie was a household word, the world's most famous scientist. Her discovery of radium followed soon after the invention of the Linotype, telegraph, and telephone, which in turn created the mass media with their scandal sheets, popular magazines, and large-circulation newspapers. Reporters parked on the Curies' front step, hoping for exclusive interviews. Without receptionists, secretaries, or public relations officials to explain even the rudiments of radioactivity, the Curies were overwhelmed. "For a year now I have done no work, and I haven't had a moment to myself," Pierre Curie complained in July 1905. "Obviously, I haven't yet found a way to protect us from frittering away our time, and yet I must. It's a matter of life and death intellectually."

Both felt utterly fatigued. "We can no longer dream of the great work days of times gone by," Pierre wrote. "As to research, I am doing nothing at present.... My wife, on the contrary, leads a very active life, between her children, the School at Sèvres, and the laboratory. She does not lose a minute and occupies herself more regularly than I can with the direction of the laboratory in which she passes the greater part of the day."

She and Pierre still resented the inadequacy of their laboratory. The gateway to research facilities in France was a professorship in a major university. Between 1898 and 1904, the Curies had published thirty-six papers, won a Nobel Prize, and turned down an enticing offer for them both from the University of Geneva in Switzerland. Pierre, however, did not become a professor at a prestigious French university until 1904, when the Sorbonne, a part of the University of Paris, promised him a laboratory. At the same time, Marie became a professor at the women teachers' college at Sèvres and was told she would become superintendent of Pierre's laboratory when it was built at the Sorbonne. As late as 1906, however, the university had not begun construction. When Pierre was offered a prestigious award from the Legion of Honor, he refused it, saying "I do not in the least feel the need of a decoration, but I do feel the greatest need for a laboratory." He was never to have a "real" laboratory.

On Thursday, April 19, 1906, Pierre Curie attended a physics luncheon in Paris on his way to visit his publisher. Crossing a busy street in the rain while trying to open his umbrella, he tottered into a horsedrawn wagon. The driver frantically pulled on the reins, but a wheel rolled over Curie's head. His skull was crushed. Pierre Curie, the dreamer who had made Marie Curie's research possible, was dead at age forty-eight. They had been married eleven years.

Containing her anguish in public, Marie Curie calmly received visitors, told Irène that her father was dead, and helped lay Pierre's body out in the front room. In private, she filled a little gray notebook with heartfelt grief. In a series of love letters to Pierre, she moaned, "What a terrible shock your poor head has felt, your poor head that I have so often caressed in my two hands.... We put you into the coffin Saturday morning, and I held your head up for this move. We kissed your cold face for the last time. Then a few periwinkles from the garden on the coffin and the little picture of me that you called 'the good little student' and that you loved. It is the picture that must go with you into the grave."

Marie Curie appeared close to a breakdown but, within two

weeks, she was writing business letters about the laboratory. When friends suggested taking up a collection and securing a widow's pension to help her, she vehemently refused. She insisted on being considered a scientist, not a helpless widow. After some hesitation, the University of Paris offered her a position as an assistant lecturer at ten thousand francs yearly, starting May 1, 1906. It was her first university salary, and she would be the first woman professor in the Sorbonne's six-hundred-and-fifty-year history.

On November 5, 1906, the fifteenth anniversary of her enrollment as a student at the Sorbonne, Marie Curie entered its physics auditorium to give her first lecture. As usual, when speaking in public, she was nervous. The auditorium looked more like a theater than a classroom. Society women in formal dresses and large elaborate hats filled the front rows. A stenographer sat ready to save every word for posterity. Newspaper editors—no mere reporters—took notes. What they understood of her lecture is unclear. As one wrote, "The magnificent forehead won notice first. It was not merely a woman who stood before us, but a brain—a living thought."

After five minutes of fervent applause, Marie Curie quietly began where her husband had left off, "When we consider the progress made in physics in the past ten years, one is surprised by the..." If the audience had hoped for histrionics and melodrama, she did not intend to oblige. Instead, she built a shield of impassivity around her to mask her feelings. She rarely spoke about Pierre Curie to anyone; even late in life, she found it difficult to talk about him with her daughters. She reserved her private thoughts for the little gray exercise book. There she wrote, "My little Pierre, I want to tell you that the laburnum is in flower, the wisteria, the hawthorn and the iris are beginning. You would have loved it all. I would also like to tell you that they've given me your chair, and some imbeciles even congratulated me on it." In the fall of 1908, while editing Pierre Curie's papers, she became a full professor of general physics.

A few months after Pierre's death, Marie Curie faced a scientific crisis. In August 1906, Lord William Thomson Kelvin wrote the editor of the *Times* of London to announce, incorrectly, that radium was not an element after all. It is, he said, merely a compound of lead and helium. Kelvin's letter precipitated a controversy and threatened Marie Curie's position in science. Lord Kelvin, one of the most important scientists of the day, had been Pierre Curie's biggest supporter; it was Kelvin who had urged the Municipal School to make him a professor. For Kelvin to cast doubt on the element radium was devastating.

Marie Curie, however, could not prove him wrong. Although she had isolated radium chloride in 1902, she had not produced pure radium. To defend her discovery, she began another arduous four-year project to purify the radium salts. In the end, with enormous persistence and determination, she produced a few grains of pure radium. She had proved that radium is indeed an element.

While struggling to purify radium, she was also building a new life as a single parent. Pierre's father remained with her to care fondly for Irène and Ève until his death in 1910. To spare herself memories of Pierre, she moved the family first to a Parisian suburb and finally to a large apartment in central Paris. Because she disliked the rigid French school system, she organized a cooperative school for little Irène and her friends; she and other Sorbonne parents taught the classes.

Concerned that the children might be exposed to tuberculosis, she insisted on outdoor exercise for the girls. She took them backpacking and saw that they learned gymnastics, swimming, boating, horseback riding, and skiing, as well as sewing and cooking. During the 1920s, she built a summer house for August vacations in the Breton fishing village of L'Arcouest where many Sorbonne professors summered. And since she enjoyed swimming in warm water, she built a house for herself on the Riviera.

Despite Marie Curie's formidable determination as a scientist, she often appeared timid and vulnerable to her daughters. By the time Ève was a young girl, her mother had pared their social lives to the minimum. When she had to face a large group, she suffered physically. On days when she taught classes, she seemed like a different person.

She still did not understand the ramifications of being a twentieth-century media celebrity, however. Four years after Pierre Curie's death, she made two disastrous mistakes that engulfed her in publicity and scandal in 1911.

First, she assumed that she could seek election to the prestigious French Academy of Sciences like any other ordinary professor. She had forgotten that she was world famous in part because she was the Sorbonne's first *woman* professor. As a member of the Academy, she would be able to present research at the Academy's weekly meetings and publish it free of charge a few days later in the Academy's journal, the most prestigious scientific publication in France. To help her research group, she decided to announce her candidacy for the Academy. A sixty-six-year-old devout Catholic man, close to retirement, was already in the race.

With Curie in the running, the character of the contest abruptly changed. It was no longer a sedate competition between eminent academics. Instead, it became a sensational press campaign between, on the one hand, liberals, feminists, and anticlerical groups and, on the other hand, nationalistic Catholics and anti-Semites opposed to the election of a foreign woman. On January 23, 1911, Curie lost by one vote. She received the news by telephone in her laboratory. Hiding her dismay, she returned to work without saying a word. Her assistants, who had hidden a congratulatory bouquet under a workbench, said nothing either. She never again sought membership in the Academy, and for the next ten years she refused to publish in its journal. As for the Academy, it refused to admit women until 1979—sixty-eight years later.

Curie's second mistake was catastrophic. Once again, she underestimated the hatred that a woman in her position could generate. She assumed that a forty-three-year-old widowed professor could have a private relationship with a married man.

Paul Langevin was a gifted and influential French physicist who had been a student and friend of Pierre Curie. Langevin and Marie Curie were two of the first French scientists to recognize the importance of quantum theory and Einstein's relativity theories. Most French scientists regarded the theories as Germanic and anti-French.

In 1911, Langevin was a handsome and charming ladies' man, five years younger than Marie Curie. With beautiful chestnut brown eyes and a flowing mustache, Langevin enjoyed being taken for an army officer. His marriage had been a battlefield for years. His wife and in-laws wanted him to quit research and take a highly paid job in industry. Curie and her friends wanted to "save him for research."

When Langevin rented two rooms on the Rue du Banquier, a ten-minute walk from Curie's laboratory, she began visiting him daily. Buying food in nearby shops, she prepared their lunch in the flat. Neighbors later told the newspapers that they "behaved like lovers."

At some point that year, Paul Langevin's desk was broken into and a drawer was forced. Mme. Langevin and her brother-in-law came into possession of some letters. Curiously, all of them were undated. Even stranger, some were letters from Marie Curie while others were from Langevin himself to Marie Curie. Why his letters were in his own desk instead of in hers has never been explained. But then again, the letters may not have been genuine. No one will ever know because, following the death of Langevin's wife, her son burned them. According to Curie's friends, the letters were clever pastiches assembled from genuine letters in order to incriminate the pair.

Rumors circulated all summer. On October 26, Jeanne Langevin began proceedings to secure a legal separation. A few days later, while Curie and Langevin were attending a physics conference in Brussels, *Le Journal* broke the story: "A Story of Love. Mme. Curie and Professor Langevin." In late November, a rabidly xenophobic, anti-Semitic tabloid published excerpts from letters filed in court during the legal proceedings.

Today these letters—urging Langevin to leave his wife before he got her pregnant again—barely qualify as love letters. They seem more indiscreet than scandalous. A modern audience might even be delighted that two unhappy people had found romance and companionship. Had Marie Curie been a male professor, turn-of-the-century French readers might have agreed. In 1900, affairs were commonplace in France, abortionists ran classified advertisements in the newspapers, and one-quarter of all French births were illegitimate. The Sorbonne had overlooked affairs conducted by its male faculty members.

But to the right-wing press, Marie was a Pole who had stolen a Frenchwoman's husband. Curie and science were regarded as immoral, anti-French, and probably Jewish. The affair shook the University of Paris and the French government at the highest levels. Had it not been for a handful of close friends, Marie Curie would have left France and returned to Poland. It was not until Jeanne and Paul Langevin reached an out-of-court settlement on December 9, 1911—with no mention of Marie Curie—that the scandal subsided.

At the height of the sensationalism, on November 4, 1911, Marie Curie received a letter from the Nobel Foundation notifying her that she would be awarded a second Nobel Prize, this time in chemistry for her discovery of radium. It was the prize that the Nobel Committee had paved the way for in 1903 by excluding radium from her first Nobel. With Pierre dead, she shared this award with no one.

The timing of the second award may not have been coincidental, however. It had been engineered by Svante Arrhenius, an eminent Swedish physicist and a powerful figure on the Nobel committees. Arrhenius may have heard rumors about the impending scandal and arranged a show of support. Curie had received only two nominations in 1911, one of them from Arrhenius himself.

Leaving Paris behind, Marie and Irène Curie traveled to Stockholm. In her speech, Marie Curie clarified precisely what she had contributed to the discovery and what Pierre had accomplished. "The study of this phenomenon was extended to other substances, first by me, and then by Pierre Curie and myself....I have turned....I was struck by..." and so on.

On her return to Paris, Marie Curie collapsed and was taken on a stretcher to a nursing home, where she was admitted under an assumed name. Convinced that she had besmirched the Curie name, she insisted on using a pseudonym even to correspond with Irène. It was a year before she resumed work.

Given the social climate of the day, Curie and Langevin had to end whatever romantic relationship had existed. Two years later, Langevin returned to his wife. Surprisingly, Marie Curie's granddaughter and Paul Langevin's grandson later married without any inkling of their grandparents' famous scandal. Jeanne Langevin attended the wedding and held her tongue. Today Hélène Langevin-Joliot is "more than eighty percent sure an affair occurred, but things were so different then, and people were so different. Conceivably, Langevin's apartment might have been for his work, to get out of the house, and to have an office." Although many assume that there was an affair, no one will ever know for sure—precisely what Marie Curie would have wanted.

As for Marie Curie, she retreated once more into her shell, hiding her emotions. She had learned that, for the Vestal Virgin of Radium, romance was simply too dangerous. It was the summer of 1913 before she had recovered enough to supervise the construction of her physics institute and go hiking with her daughters, her old friend Albert Einstein, and his young son.

By the time World War I broke out in August 1914, Marie Curie was ready for action again. Unlike the French medical corps, she realized immediately that X rays would be invaluable for treating shrapnel and fracture wounds in front-line hospitals. Hiding her priceless radium supply in a Bordeaux bank vault, she returned to Paris and began organizing a mobile X-ray service. Ten days after the war started, she filed a formal request with the Minister of War for approval to begin work. All through August and September, she visited Parisian laboratories and wealthy women to ask for equipment, cars, and money. Her first vehicle—called a "petite curie"—was ready for the Battle of the Marne in November 1914. By the end of the war, she had opened two hundred X-ray stations in French and Belgian battle zones and trained one hundred and fifty women technicians, including her daughter Irène. Altogether, her X-ray units examined more than one million soldiers.

In addition to running the X-ray service, Curie began collecting the radon gas that emanated from her radium. Enclosing the gas in tiny glass tubes, she shipped it to hospitals around the world for use against cancerous tumors. For forty-eight hours after every radon session, she felt utterly exhausted. By now, she had been exposed to more radiation than any other human being.

The French government never recognized Marie Curie for her war work. Not even her patriotism could wipe out the memory of the 1911 scandals. Nevertheless, she had learned valuable lessons. She now knew that she could raise money, negotiate with government officials, and administer a large operation. Above all, she realized that it was possible to use the Curie name and fame to promote a cause.

Marie Curie dedicated the rest of her life to building a French research institute for the study of radioactivity. Germany, Great Britain, and Denmark had already established physics institutes where groups of scientists could work together on common problems. The group approach had proven extremely successful. Although French biologists could study at the Pasteur Institute in Paris, French physics professors still worked alone or with two or three students at most. Before World War I Curie had begun talking with representatives of the University of Paris and the Pasteur Institute about a research center where scientists could do basic research in radioactivity and where physicians could apply that research to medical problems. A building for the institute was completed just before the war broke out.

When the Radium Institute opened after World War I, Marie Curie was more than fifty years old, physically ill, and emotionally exhausted. Her institute had a building but little more, not even a typewriter. With the French economy devastated by postwar inflation, she was unlikely to get financial support from government officials, industrialists, or philanthropists. None of them had supported scientific research even when times were good.

The Curie circle tried to convince the French public that a nation that does not invest in research is a nation on the decline. Unfortunately, France preferred the image of poor self-sacrificing scientists like Louis Pasteur and Marie Curie toiling in their attics and sheds. Supporting her colleagues' efforts to build French science, Curie helped lobby government officials. Recognizing that scientific discoveries could be used to finance research, she even changed her mind about scientists' patenting their discoveries. These were all long-term campaigns, however, and the institute needed money immediately.

In 1920, a friend asked her to see a leading American journalist named Missy Meloney. Meloney was tiny and frail; she limped from a childhood accident and suffered from tuberculosis. The first woman with a seat in the press gallery of the United States Senate, she was editor of the *Delineator*, one of the largest women's magazines in the United States at that time. As she described herself, she was "busy as a switch engine.... Life for me has become like a highly charged electric wire, and I cannot let go." Her cable address was IDEALISM.

At the institute, Meloney waited to meet Curie in a bare little office filled with cheap furniture. "The door opened and I saw a pale, timid little woman in a black cotton dress, with the saddest face I had ever looked upon. Her well-formed hands were rough. I noticed a characteristic, nervous little habit of rubbing the tips of her fingers over the pad of her thumb in quick succession. I learned later that working with radium had made them numb." Face to face with the saintly Curie, the fast-talking Meloney was speechless. To make conversation, Curie explained that U.S. researchers had approximately fifty grams of radium. "And in France?" Meloney asked.

"My laboratory has hardly more than a gram."

According to Meloney's story, she promptly declared, "You ought to have everything in the world you need to go on with your work. Someone must undertake this."

"Who will?" Curie asked rather hopelessly.

With a flourish, Meloney promised, "The women of America." Curie did not think Meloney would do anything, but she did think the American was sincere.

Meloney pulled it off. She arranged one of the largest fundraising campaigns the world had ever seen. First she secured promises from New York City's newspaper editors not to print a word about the Langevin scandal. Incredibly, the editor of the sensationalist Hearst newspapers even gave her its Langevin file. Then she appointed fundraising committees of wealthy women and physicians and raised one hundred thousand dollars to buy Curie a gram of radium. She arranged a gala tour of the United States, honorary degrees from twenty universities, and a White House reception with President Warren Harding.

The event was so spectacular that even the French contributed money, though most of it came from Jewish philanthropists. Meloney knew few American women scientists, however, so the campaign benefited Marie Curie without helping other women in science. In fact, by creating an almost impossible standard for women scientists to live up to, Marie Curie may have made their professional progress more difficult. Although universities did not expect every male scientist to be an Albert Einstein, women scientists were continually measured against Marie Curie—and naturally found wanting.

When Marie, Irène, and Ève Curie arrived at the dock in New York City, they were greeted by cheering crowds, bands, banners, confetti, and frenzied newspaper reporters. Ève and Irène had had no idea their mother was so famous. They were stunned by the assault of journalists shouting at her, "Turn your head left," "Turn your head right." Irène and Ève knew their mother as a quiet professor. This celebrity was no one they knew.

Marie Curie proved to be a public-relations dream. Her black-garbed, self-effacing appearance was the embodiment of the selfless scientist. As the *Scientific American* gushed, she was "unassuming, plainly but neatly dressed, womanly and motherly in appearance.... She remains just plain Madame Curie, working for the good of humanity and for the expansion of scientific knowledge."

The three-week trip exhausted Curie, but she returned in 1929 for a second Meloney extravaganza and collected enough money for a gram of radium for Poland. Meloney had become Curie's close friend and confidante. Their second tour ended just three days before the stock market crash of 1929.

The glamour of the Curie name financed an international research center in France. Although she continued to do research, helping to untangle the process by which radioactive elements decay into other elements, she spent the rest of her life organizing, administering, and fundraising for the institute. Thanks to her, the Radium Institute became one of the world's leading centers for nuclear research. In it, Marguerite Perey discovered francium, a new radioactive element; Salomon Rosenblum analyzed alpha rays; and Irène Curie and her husband, Frédéric Joliot-Curie, discovered artificial radioactivity. Curie reserved a certain number of positions for foreigners and women each year. In 1933, out of approximately forty scientists at the institute, seventeen were foreigners.

In addition to the institute, Curie became a major force behind the internationalization of physics. She helped Poland develop a radium institute, worked with a League of Nations committee on international publishing standards and student fellowships, helped establish the international standard for radium units, and donated radium and radon to researchers around the world.

Between her two American tours, Curie underwent four cataract operations. (Cataracts are among the first symptoms of radiation sickness.) At one point, she was so blind that her lecture notes were printed two-and-a-half inches high and Irène had to guide her to and from work. Marie tried to hide her condition, visiting her ophthalmologist as Madame Carr.

By the mid-1920s, evidence was accumulating about the dangers of radium. In 1924, a New York dentist discovered that the young women who painted radium on watch faces were dying of cancer caused by licking their paintbrushes into points. Several of Curie's lab workers died of anemia and leukemia during the 1920s, and she herself had a raft of symptoms, including anemia, tinnitus in her ears, and chronic exhaustion. The institute, however, did not conduct research on the health problems associated with radium. Institute rules were to economize on water, gas, and electricity; not to smoke in the labs; and to change lab coats and breathe fresh air deeply whenever outside. Part of the problem was that Marie Curie herself was so strong. Fresh air, ocean swimming, and mountain climbing improved her health. When the employees died, Curie blamed them for not taking enough fresh air.

Although she ignored health hazards, she remained absolutely up to date on new developments in physics. The year before she died, she attended a conference and was often heard correcting her scientist daughter and son-in-law on facts.

Toward the end of her life, she made one last effort to insure her privacy. Sorting through her files, she destroyed almost all her personal papers, leaving only her love letters from Pierre, letters from a beau she had known during her student days, and the diary she had written after Pierre's death. They were testaments to her desirability and to the strength of the love that she and Pierre had shared. She saved nothing from the Langevin years.

She also organized an orderly transition at the institute, arranging for her aide Debierne and then Irène to take over. The year before her death, she witnessed the discovery of artificial radioactivity by Irène and her husband, Frédéric Joliot. Their work would make the institute's stockpile of natural radioactive elements obsolete, but it ensured a Nobel Prize for Irène.

At sixty-seven, Marie Curie was still consumed with curiosity and a spirit of scientific adventure. As she wrote, "I am one of those who think that science has great beauty. A scholar in his laboratory is not just a technician; he is also a child face to face with natural phenomena that impress him like a fairy tale." A few weeks before her death, she hiked alone partway up Mont Blanc to watch the sun set over the mountains.

On July 4, 1934, she died of leukemia in a nursing home in the French Alps. Death came slowly, her powerful heart continuing to beat long after all hope had gone. During her last two days, she said nothing about her family. Instead, her mind passed among little shreds of an experiment. For Marie Skłodowska Curie, all but science had disappeared.

* * *

3

Lise Meitner

November 7, 1878–October 27, 1968

NUCLEAR PHYSICIST

USING A PRIVATE ENTRANCE, Lise Meitner entered her basement laboratory—and stayed there. A former carpentry shop, it was the only room in Berlin's chemistry institute that she was permitted to enter. No females—except, of course, cleaning women—were allowed upstairs with the men. Prohibited even from using a rest room in the chemistry building, she had to use facilities in a hotel up the street.

For two years, from 1907 to 1909, Meitner performed radiation experiments in the cellar, careful never to be seen upstairs. Normally shy and timid, she sometimes longed so desperately to hear a chemistry lecture that she sneaked into the amphitheater upstairs and hid under the tiers of seats to listen.

Ten years later, Lise Meitner was the director of a center for radiation physics in Berlin. For twenty years, she reigned supreme there, creating one of the glories of the golden age of physics in the 1920s and 1930s. Later, fleeing Nazi persecution, she left Germany illegally and went into exile. At the age of sixty, she deciphered the experiment of the century by explaining that, incredibly, the nucleus of an atom could split and release enormous amounts of energy. For the fission experiment that she initiated and explained, her German partner received the Nobel prize.

Lise—pronounced *lee-zet* without the "t"—was born in 1878, the third of eight children in a gifted and liberal Viennese family. Her grandparents were Jewish, but her lawyer father, Philipp Meitner, was an agnostic. He encouraged his eight children to learn about science, and two of his daughters and a son earned university degrees. Lise's mother, Hedwig Skovran Meitner, was a talented pianist who taught her children music. Lise's oldest sister became a concert pianist and composer. Music and physics became Lise's two grand passions.

As a child, Lise noticed a beautiful, iridescent puddle of water with a bit of oil in it. Why was the puddle filled with colors? she asked. The answer entranced her; she had not imagined that such marvelous things could be learned about nature. She became convinced that, if she worked hard enough, she could understand its laws.

Later, as a teenager dreaming about her future, Lise always made the same wish: "Life need not be easy, provided only that it is not empty." It was a wish that was granted many times over.

For nine years, Meitner struggled to get an education. She called 1892 to 1901 her "lost years" and believed that they handicapped her for life. The Viennese system of education ended for girls when they were fourteen. By then, she had learned a bit of arithmetic, French, religion, and pedagogy—enough to run a household, raise children, and converse charmingly with a husband. According to Austrian law, that was enough for a girl. Females were banned from the academic high schools that prepared boys to enter universities.

Even scientists, she discovered, opposed the education of females. She was particularly irritated by a famous physiologist who wrote a popular book called *The Physiological Feeble-Mindedness of Woman,* and by a prominent physician who claimed that feminists were destroying the family. By the time girls were admitted to Vienna's academic high schools, it was 1899—too late for Meitner.

When Meitner showed no interest in marriage, her worried father inquired how she planned to support herself without a husband. Lise replied that she wanted to study physics. X rays and radioactivity had been discovered in 1895 and 1896, respectively, and she had always been interested in mathematics and science. Her dream must have seemed pathetically impractical. There were practically no jobs for physicists of either sex. Industry did not hire them, and universities considered physics a dead subject. In Germany, the president of the national bureau of standards had announced, "Nothing else has to be done in physics than just make better measurements." In all of Hungary, for example, there were only three or four physics professors. The few jobs that were available would hardly go to a woman. As the Nobel Prize–winning physicist James Franck explained, "Everyone who went into physics at that time went into physics because he had to...because he felt he could not be happy in any other way."

To make sure that Lise could support herself, her father insisted that she spend three years earning a certificate to teach French in girls' finishing schools. Only then would he hire a private tutor to prepare her for university entrance examinations. By then, Lise Meiter was twenty-one years old.

Lise Meitner.

Lise Meitner.

(Below) Lise Meitner and Otto Hahn in the laboratory of the Kaiser Wilhelm Institute for Chemistry, 1913.

In two years, Meitner completed eight years of school work, including eight of Latin and six of Greek. Whenever she took a break, her younger brothers and sisters teased her, "Lise, you're going to flunk. You have just walked through the room without studying!"

Luckily, her tutor had a gift for making mathematics and physics extraordinarily stimulating. He even showed his students a real physics laboratory; most tutors taught only from diagrams of experimental apparatus. When Lise saw the lab, she was astonished. Some of the equipment looked very different from what she had imagined.

Of the fourteen women who took the university entrance examination, only four passed. Meitner was one of them. Austria had recently opened its universities to women, and Meitner enrolled in Vienna in 1901, just a few months short of her twenty-third birthday. "The lost years" were over.

Women university students were widely regarded as freaks, and Meitner was extremely shy, but music and her physics theory professor kept her spirits up. The cheapest seats in the Vienna opera house, high up under the ceiling, were her idea of "musical heaven." She liked to follow the performances with the scores. Professor Ludwig Boltzmann was her other salvation. His theory class was located in a rundown building that reminded her of a henhouse. "If a fire breaks out here, very few of us will get out alive," she thought. Boltzmann was an emotional, enthusiastic lecturer who talked about science in the most personal terms. It was Boltzmann who gave her his vision of physics as the ultimate battle for truth. In 1905, Meitner became the second woman to earn a physics doctorate in Vienna; her doctoral dissertation concerned conduction in nonhomogeneous materials.

To be sure that she could always support herself financially as a teacher, Meitner spent the next year practice teaching at a girls' high school. In her spare time, she studied more physics at the university. Fascinated by newspaper accounts of the Curies' discovery of radium in 1898 she began investigating radioactivity.

The discoveries of natural radioactivity and of radium's spontaneous change into polonium and then into lead provided scientists with the first evidence that atoms of one element can break apart and become atoms of another element. Until then, physicists had thought of atoms as, in Meitner's words, "solid unsplittable little lumps." Scientists could chip off a few outside electrons from atoms, but no one dreamed of splitting the heart of the atom, the nucleus itself. While she had no plans to specialize in the subject, she wrote two papers about it in two years. It is often said that physicists do their best work in their youth, but Meitner did not start studying radioactivity until she was twenty-seven years old.

Eventually, a number of other women physicists specialized in radioactivity, too. Besides Meitner and Marie Curie, there were Ellen Gleditsch in Oslo; Elisabeth Rona and Berta Karlik in Vienna; Irène Joliot-Curie in Paris; and Marguerite Perey in Paris and Strasbourg. Until the early 1930s, radioactivity was a sideline of physics, and women faced less competition from male physicists there. In 1932, when the discovery of the neutron made the nucleus the most exciting field in physics, these women emerged as experts.

As her first postdoctoral project in Vienna, Meitner devised a simple but ingenious method to show that the alpha particles streaming out of naturally radioactive materials are deflected slightly while traveling through matter. She did the experiment with a piece of radium donated to the university by Marie and Pierre Curie, in thanks for Austria-Hungary's gift of pitchblende. (Ernest Rutherford and Hans Geiger made similar measurements in Manchester at approximately the same time in preparation for their 1909 gold foil experiment, which proved that atoms have nuclei.) Although Meitner was a beginner working alone, she had been on the track of an important discovery. She also explained an optics experiment conducted by John W. Strutt, Lord Rayleigh, an Englishman who won the fourth Nobel Prize for physics, and made several predictions based on his experiment.

With these successes behind her, Meitner felt brave enough to ask her parents to let her go abroad for further study. Although her first choice, Marie Curie, rejected her, Max Planck agreed to admit her to the university in Berlin. Prussian universities did not give women credit toward degrees yet, but she was permitted to audit classes. Amazed at her courage, Meitner's father agreed to give her a modest living allowance. She left home at the age of twenty-nine, expecting to be gone only a few terms. She stayed in Berlin for thirty-one years.

Meitner always believed that Germany had rescued her from scientific "oblivion" in Austria. From the mid-nineteenth century until the Nazi takeover in 1933, Germany was the scientific center of the world. The country had transformed itself from a destitute and backward region into a great, technologically advanced power by investing in universities and technical schools. Its skilled and educated workforce was its only natural resource. Berlin had famous Nobel Prize–winning physicists like Max Planck, Albert Einstein, and Max von Laue. Thanks to Planck's far-sightedness, Berlin physicists accepted the existence of atoms and Einstein's special theory of relativity long before scientists elsewhere.

When Meitner arrived in Berlin in 1907, she introduced herself to Planck. Patronizingly, he asked, "But you're a doctor already! What more do you want?"

"I would like to gain some real understanding of physics," she replied.

It was obvious to her that Planck did not approve of women students. In response to a 1897 questionnaire, Planck had reported that he had allowed several women to audit his courses and had had "nothing but favorable experience.... On the other hand, I must keep to the fact that such cases must always be regarded as exceptions.... Generally, it cannot be emphasized enough that nature herself prescribed to the woman her function as mother and housewife, and that laws of nature cannot be ignored under any circumstances without grave danger—which in the case under discussion would especially manifest itself in the following generation."

Although Planck had none of Boltzmann's fire and dramatics, Meitner was extraordinarily impressed with his quantum theory. Planck had hypothesized that an atom absorbs and emits energy in particular units, which he called quanta. She thought the theory offered great possibilities for understanding the nature of matter. Once she got to know Planck, she worshiped him. He had "a rare honesty of mind and an almost naive straightforwardness," she said. He was such a wonderful man that when he went into a room, its air seemed better.

Because Planck's lectures did not keep Meitner busy enough, she searched for a laboratory where she could do some experiments. At the same time, Otto Hahn, a young German chemist who had worked on radioactivity with Ernest Rutherford, was looking for a collaborator in physics. Hahn was informal and extremely charming. People who met him for only a few moments felt as if they had known him for years. He particularly enjoyed the company of attractive women. Meitner immediately realized that, despite her shyness, she would have no hesitation in asking him all the questions she wanted to. Unfortunately for Meitner, however, Hahn worked in Professor Emil Fischer's chemistry institute, and Fischer did not allow women in his building. Only when Rutherford recommended Hahn did Fischer agree to let Meitner work in his institute—provided that she stay in a converted carpenter's shop in the basement and never enter any part of the building used by men. Meitner agreed. Although she studied radioactivity for the rest of her life, she did not remain in the basement for long.

Meitner's biggest problem was not hiding in lecture rooms or working in damp cellars. Her complaint was that she could not learn radiochemistry without seeing Hahn's experiments, which he did upstairs with the men. Despite the problem, they published three articles during the first year of their collaboration; a year later, they worked on six more. In 1908, when Prussia opened its universities to women, Fischer allowed Meitner to enter the rest of the building and installed a toilet for her. He eventually became one of Meitner's biggest backers.

Others were more prejudiced. When Meitner and Hahn walked together in the street, Fischer's young male assistants greeted him ostentatiously with "Good day, Herr Hahn," pointedly ignoring Meitner. An encyclopedia editor liked one of her articles so much that he asked "Herr Meitner" to write more. When he learned that "Herr Meitner" was actually "Fräulein Meitner," he wrote back to say that he would not dream of publishing anything written by a woman.

Meitner and Hahn were opposites. She was slim and only five feet tall; he was a head taller and bulky. She was shy; he was outgoing. She had a light, sophisticated wit while he was a joker. He liked to whistle the last movement of Beethoven's violin concerto with an odd syncopation. When questioned, he would ask innocently, "Doesn't it go like that?" They were both good storytellers but Hahn told anecdotes anytime anywhere; Meitner blossomed only with close friends. "Hahn" means rooster in German, and his stories were called "cock-tales." He liked pretty young women and paid them elaborate compliments. He was not anti-Semitic and he later disapproved of Nazism, but he was conservative and nationalistic. His hero was Kaiser Wilhelm. He was not a feminist.

Scientifically, they were opposites too. Hahn was intuitive and often did things without knowing precisely and logically why; he simply had a feeling that they might be useful sometime. Meitner was a deep, critical thinker who reasoned systematically and always asked why. Hahn wanted to discover and study new elements; Meitner was interested in understanding their radiation.

As teammates, they dispensed with gloves in order to work fast before their radioactive materials disintegrated. Although they both got small finger burns that took weeks to heal, they escaped serious injury. The chemistry department had no magnets, so they prepared their radioactive samples, and Meitner remembered their racing "like bullets shot from a gun" a kilometer up the street to the physics institute. Hahn often spent hours every day for months preparing chemically purified sources for her radiations, while Meitner made tedious calculations for him. To pass the boring hours, they sang together—German folk songs and Meitner's favorites, Brahms' *lieder.*

She was not quite an equal member of the team. Working by herself, Meitner discovered that radioactive thorium decayed into a substance she called "thorium D." An eminent professor advised her to publish the results by herself, and Hahn agreed. But she submitted the article as a team with both their names anyway. "He was much better known than I. It wasn't bad intention. It was just thoughtlessness," she explained later. When Ernest Rutherford came to visit, Hahn brought him to Meitner's carpenter shop to meet her. Rutherford was astonished. "Oh, I thought you were a man!" he declared. While the men talked radioactivity, Meitner was expected to take Mrs. Rutherford Christmas shopping. Meitner never let on how she felt except to say of Mrs. Rutherford, "She expected everyone in Berlin to speak English."

Although Rutherford was inclined to ignore Meitner's contribution to the team, Hahn insisted that she get full academic and scientific credit for their joint work. Without his continuous support and friendly encouragement during her early years as a scientist, she might not have succeeded. At colloquia, however, Hahn read their papers. Not yet an accomplished speaker, Meitner wanted him to present their work in public. They were the same age, but Meitner was shy and introverted and a decade behind Hahn in experience. She was an unpaid beginner, and he was a professor.

Friends often wondered why Hahn and Meitner did not marry. When asked, Meitner responded gracefully, "Oh, my dear, I just didn't have enough time for that." There were other reasons too. "Lise wasn't the kind of person Hahn would marry," observed Professor Günter Herrmann of the University of Mainz, who knew Hahn. Many scientists, who were poorly paid, tried to marry rich women in order to use their money for research. Hahn was not wealthy, and he indeed married a beautiful, rich socialite. Few wealthy men wanted women scientists as wives.

Meitner also saw to it that their relationship remained exceedingly proper. For the first ten years of close friendship, she called him "Herr Hahn." As Hahn explained, "There was no question of any closer relationship between us outside the laboratory. Lise Meitner had had a strict, ladylike upbringing and was very reserved, even shy. I used to lunch with my colleague Franz Fischer almost every day and go to the café with him on Wednesdays, but for many years I never had a meal with Lise Meitner except on official occasions. Nor did we ever go for a walk together. Apart from the physics colloquia that we attended, we met only in the carpenter's shop. There we generally worked until nearly eight in the evening.... One or the other of us would have to go out to buy salami or cheese before the shops shut at that hour. We never ate our cold supper together there. Lise Meitner went home alone, and so did I. And yet we were really very close friends."

Meitner worked in the basement for five years. Then German industry founded the first of several lavish scientific research institutes named for Kaiser Wilhelm. Later financed by the government, they were renamed after World War II as the Max Planck

Institutes. Thanks to Emil Fischer's influence, Hahn and Meitner moved to the radiochemistry department in the new Kaiser Wilhelm Institute for Chemistry in suburban Berlin in 1912. There powerful industrialists who helped fund the institute protected Meitner from much of the discrimination that Emmy Noether (chapter 4) faced in a government-supported university.

Meitner got her big break that same year when her old friend, Max Planck, gave her an assistantship. Until then, she had been just an unpaid, guest researcher in Berlin. As Prussia's first woman research assistant, however, she earned a small stipend for grading students' papers and organizing Planck's seminars. The post was extremely prestigious and silenced much of the overt prejudice against her. It also bolstered her self-confidence. Unfortunately, after her father's death in 1911 she no longer received an allowance, and her assistantship pay was so low that she often had only black bread and coffee to eat. Luckily, she was a frugal eater, almost a vegetarian. When Meitner received a job offer from the University of Prague to be an associate professor and, possibly, later a professor, the Kaiser Wilhelm Society finally decided to pay her a salary.

During World War I, Meitner spent two years as an X-ray nurse in an Austrian army hospital at the front, much like Marie and Irène Joliot-Curie (chapter 6) in France. Hahn had already been called up to do poison-gas research. During her leaves, Meitner worked on a complex project that she and Hahn had started before the war: a systematic search for the parent element of actinium. By the end of the war, they had discovered protactinium, a rare radioactive element found in pitchblende that slowly decays into actinium. By 1917, Meitner was the head of her own radiophysics department at the institute; for the first time, she earned enough to rent an apartment.

Curiously, Meitner's two most famous discoveries—protactinium and the explanation for fission—were conducted with Otto Hahn while one of them was abroad in wartime. In both cases, their respective contributions are amply documented in letters. Two independent studies of the protactinium project—one in the United States and the other in Germany—have concluded that Meitner did most of the protactinium work. Hahn supplied important advice on chemical procedures from afar. Yet Hahn was listed as the first author, indicating that he was predominantly responsible for the work. Their general practice may have been to use Hahn as the first author for chemistry articles and Meitner first for physics articles, regardless of their relative contributions. With fission, during and after World War II, the story would be different.

A charmed circle of physics, friendships, and music grew up around Meitner throughout the teens, 1920s, and early 1930s and nurtured her growing self-confidence. She was cut off from most men and women by her education and their prejudice, so she especially treasured her physics friends. They were the world's great physicists: Max Planck, Albert Einstein, James Franck, Niels Bohr, Max Born, Erwin Schroedinger, Max Von Laue, and others. Among her few women friends were Mrs. Planck and the plant physiologist Elisabeth Schiemann. "They were also exceptionally nice people to know," Meitner explained. "Each was ready to help the other; each welcomed the other's success. You can understand what it meant to me to be received in such a friendly manner into this circle."

The first time she met the great Danish theoretical physicist Niels Bohr, they spoke about "everything under the sun, whether grave or gay." Years later, she could still conjure up "the magic" of that day. She considered Bohr the most influential scientist of two generations of physicists. He was often difficult to understand, however, so she and several other young physicists invited him to her institute in 1920 for a "bigwig-free" conference. The "bigwigs" were senior professors, whose presence prevented juniors from asking questions. At age forty-two, Meitner was still a "junior" scientist. Each senior professor had to be visited and personally informed why he had not been invited to talk with Bohr. Food was scarce in Germany so soon after World War I, and luckily, one of the bigwigs thought to ask, "But am I permitted to invite you to dinner after the discussion?" "Ah," the juniors said. "That's another matter."

Meitner had a gift for forming and keeping friendships. According to a famous aphorism, the Germans during World War I considered that "the situation is serious but not hopeless." The Viennese, on the other hand, regarded the situation as "hopeless, but not serious." Meitner had this kind of charm—friendly, polite, but impersonal and making light of life's grave problems.

She was much loved and respected by the physics community. Emilio Segrè called her "the gentle Lise." She was interested in everything. She enjoyed a good joke and didn't mind being teased. Niels Bohr liked skits parodying famous physicists, and Meitner always sat in the front row with the other greats, laughing and having fun. One afternoon in Berlin when Meitner and Hahn were having tea in her lab, Gustav Hertz, a Nobel Prize winner, rushed in. Refusing tea, he announced, "I am fed up with that stuff. Give me the alcohol." When a student handed him a bottle of 100-proof alcohol from the shelf, Meitner was aghast and cried, "But Hertz, you can't drink that! It's pure poison!" Paying no attention, Hertz poured a tumblerful and drank it down. He had arranged beforehand for the student to fill the bottle with water.

Music played an important role in her friendships. The institute had a choir, and Meitner and her friends met for regular musical evenings at the Plancks' or the Francks' home. Einstein played his violin, and Meitner taught her friends Brahms' songs. She was not an accomplished pianist, but in private she made her physicist nephew Otto Robert Frisch play piano duets with her. In return, she introduced him to Berlin's concerts, Brahms' symphonies, and chamber music.

After Germany's defeat in World War I, the Kaiser was overthrown and Germany's first republican government was installed. Although most German academics disapproved of the Weimar Republic, it greatly improved the position of German women. For the first time, Meitner was permitted to lecture at the university in Berlin. Her 1922 inaugural talk was titled "The Significance of Radioactivity for Cosmic Processes"—which a newspaper reported as "Cosmetic Processes." In 1926, at age forty-eight, she became Germany's first woman physics professor. Meitner's promotions are particularly remarkable considering that in 1991 women still accounted for only three percent of Germany's university professors.

Hahn and Meitner were nominated frequently for the Nobel Prize during the 1920s and 1930s. Ironically, by then they were no longer collaborating. In 1920, shortly after World War I, they broke up their scientific partnership. Protactinium had required the collaboration of a physicist and a chemist, but once they had explained its decay process they were ready to pursue independent projects. With the financial support of the I. G. Farben Corporation, Meitner developed a department that rivaled the Curie Institute of Radium in Paris and Ernest Rutherford's Cavendish Laboratory in Cambridge, then the greatest experimental physics center in the world.

After the discovery of fission, a myth arose that Hahn and Meitner had worked together throughout their careers. But during the 1920s and 1930s, no one thought of them as a team. And no one thought of them as equals. Meitner was indisputably the leader. "She was a great physicist during physics' golden age. Radiochemistry wasn't an exciting field then," as Professor Emeritus Peter Brix of the University of Heidelberg noted. Günter Herrmann agreed, "She was more famous than Hahn. The glory of the Berlin institute in the 1920s came mainly from Meitner. Hahn's reputation came from his earlier work with her on the radioactive decay chains in nature and from his personality. She was among the people discussed for a Nobel Prize every year." Einstein called her "our Madame Curie" and considered the Austrian more talented than the Frenchwoman. "Hahn and Meitner were great friends, but when they talked, she was superior. For me, she's really a great scientist," added Wolfgang Paul, 1989 Nobel Prize winner from the University of Bonn and former president of the Max Planck Society. His name is only one letter different from that of Wolfgang Pauli, the Austrian Nobel Prize winner who was a giant in theoretical physics earlier in the century. Both Paul and Pauli were Meitner's friends.

When Hahn and Meitner were together, she made their relationship quite clear. No longer shy, she behaved like a slightly bossy older sister, calling him "colleague-brother" and "cockerel," a play on Hahn's name. In public, she called him "Hähnchen," the diminutive of "Hahn." When he dared to talk about physics, she dismissed him, "Be quiet, Hähnchen, you don't understand physics." *("Sei stille, Hähnchen, von Physik verstehst Du nichts.")* That part of their relationship never changed. When they were almost eighty, Meitner whispered as they walked up some steps at an awards ceremony in Mainz, "Keep your back straight, Otto. Otherwise, they'll think we're old."

Meitner ruled the institute with a small but firm hand. "She was not oppressed," Paul said, remembering 1935. "She was a very highly esteemed woman. And she wasn't shy. She was very strong. She ruled the institute. Hahn was very polite and nice and not very strong with the rules. She was an iron lady." Hallway signs were supposed to read "No smoking. (Signed) Otto Hahn, Lise Meitner." But by changing her name from "Lise" to "Lies," wags altered the meaning to imply that Meitner wore the pants in the lab: "No smoking. Signed Otto Hahn. By order of Lise Meitner." Meitner wanted everyone working hard. When a young physicist rigged an experiment so that it collected data automatically during his cigarette breaks, Meitner complained sternly, "You're not working. You should be working."

Meitner "was a distant boss. There was a wall between her and the underlings," recalled physicist Barbara Jaeckel, whose husband, Rudolf, was a physicist in Meitner's institute. When the Jaeckels became engaged, they made the ceremonial rounds of their supervisors for tea. At Meitner's apartment one Sunday afternoon, she told Barbara Jaeckel, "The wives of physicists are martyrs. Their husbands are always working." At institute parties, Meitner joined in the fun and wore a paper hat with everyone else. But she did not invite wives. Perhaps, like physicist Elisabeth Rona, she had discovered that when women were present, she was expected to join them in household gossip instead of talking shop with the men.

To keep the institute clear of radioactive contamination, Meitner placed toilet paper next to every door, to be used when touching a door handle. Lecture halls and libraries had dark and light chairs to protect employees who worked with weak radioactive materials from employees who used strong radiation. Her care paid off. She was able to keep the first floor of the laboratory free of contamination for twenty-five years, unlike the institutes in Vienna and Paris. Otherwise, the delicate effects measured during the fission experiment would have been swamped by contamination. Asked years later how they managed to work with strong radioactivity without damaging their health, Meitner replied, "Oh, we washed our hands frequently." To which Hahn added, "And I got so used to it that I still do it occasionally."

Each Wednesday, Meitner and about forty physicists from Berlin's research institutes gathered for a colloquium. The front bench was famous. It was composed almost entirely of Nobel Prize winners—plus Meitner. "She clearly belonged there," Paul recalled. "That colloquium was the greatest event in my life," the Nobel Prize–winner James Franck declared. "We could see how the great men of that time had all struggled with their problems.... The reason that many of us...tried to do something with quantum theory is that we went to that colloquium."

"All the new results which were then pouring out were presented and discussed there," Meitner recalled. "From the very first years of my stay in Berlin, I remember lectures on astronomy, physics, chemistry.... It was quite extraordinary what one could acquire there in the way of knowledge and learning." The development of physics became, she said, "a magic, musical accompaniment to my life." The first time she heard Einstein speak, he explained that energy is trapped in mass, according to his famous equation $E = mc^2$, and that every radiation has mass associated with it. Those two facts were so overwhelmingly new and surprising to Meitner that she remembered every detail of his lecture for decades.

While working apart from Hahn at the lab between the two world wars, Meitner investigated almost every experimental problem in nuclear physics. She was always at the forefront of the field. She had the bad luck, however, to choose a puzzle that was virtually unsolvable until the 1950s, when new techniques became available to C. S. Wu (chapter 11) and others.

The puzzle concerned the streams of electrons emitted by naturally radioactive substances. How could a radioactive nucleus eject an electron with so much energy that the electron could not possibly have existed within the tiny nucleus in the first place? (Later, it was learned that the electron is actually created at the moment of the decay.) The problem was so difficult that, a friend of Meitner's claimed, it was like new taxes—better not discussed. Another drew a skull and crossbones next to every reference to the problem in his physics book.

Some of Meitner's work laid the experimental foundation for Wolfgang Pauli's prediction of the neutrino in 1931. The Austrian theorist disclosed his theory in a letter to Meitner and several other colleagues, addressed as "Dear Radioactive Ladies and Gentlemen." According to his hypothesis, a disintegrating nucleus emits two particles simultaneously—an electron and a light, scarcely detectable neutral particle later named the neutrino. Then he challenged Meitner to find it, a feat that was also impossible until the 1950s.

At the peak of her career in 1934, Meitner began the greatest experiment of the century. For four years, she competed against Enrico Fermi, Ernest Rutherford, and especially Irène Joliet-Curie. Unknown to Meitner, she was also racing for time against Adolf Hitler. When Hitler came to power in 1933, he began firing all "non-Aryans" from their university jobs. At first, Meitner was safe. As an Austrian, she was not covered by Germany's anti-Semitic laws, and the Kaiser Wilhelm Institute was controlled by powerful industrialists, who modified Nazi excesses within the lab. So she kept working.

In Rome, Fermi was bombarding heavy elements with neutrons, hoping that their nucleus would absorb a neutron and change into an even heavier element. He particularly hoped that the nucleus of uranium, the heaviest naturally occurring element, would absorb a neutron and turn into an even heavier, manmade element. His collisions produced so many different kinds of radioactivity that identifying the elements was extremely difficult. Nevertheless, Fermi assumed that he had discovered new "transuranic" elements heavier than uranium. His "discovery" sent nuclear physicists and chemists everywhere hunting for more elements heavier than uranium.

Meitner was fascinated and asked Hahn to team up with her again. During their twelve years apart, Hahn had lost touch with modern physics. But Meitner needed an expert radiochemist who could identify new and extremely heavy elements—the most difficult of all to identify, especially when only a few atoms at a time would be available for analysis. In 1934, after weeks of persuasion, Hahn agreed.

They began by bombarding uranium with neutrons; then they studied the particles emitted by the uranium. By the end of 1937, Hahn and Meitner thought they had identified at least nine different radioactive substances. From Paris, Irène Joliot-Curie reported others.

To help identify their minute samples, Hahn brought in a young analytical chemist, Fritz Strassmann. Extremely shy and gifted, Strassmann was an artist in the laboratory. When he worked, people gathered around to watch. He was virtually unemployable in Hitlerian Germany, though, because he refused to join any Nazi organizations. During the war, he and his wife, with their small child, sheltered a Jew in their apartment. Strassmann was honored posthumously by the Holocaust Memorial in Jerusalem.

Two other women completed the team. Clara Lieber, an American from St. Louis, Missouri, had received her bachelor's degree from Smith College. Chaperoned by her mother, she earned a Ph.D. from the University of London. Lieber was a guest chemist at the institute; she later married and quit science. A German, Irmgard Bohne, was the team's technician. Together, the five formed the most experienced group in the world studying the problem.

Strassmann and Hahn regarded Meitner as the intellectual leader of the team. "On all very difficult questions and calculations in physics, Hahn consulted Meitner," according to Günter Herrmann, who was Strassmann's student and friend. "The guidance of Lise Meitner over the four years is obvious.... She decided they should enter the field and what they should do next, including some mistakes." She not only defined the problem and laid out a logical course of action, she also kept the team working until it got a believable answer.

The most remarkable aspect of the race is that the competing teams in Italy, France, and Germany split uranium atoms in two—making fission—for four years without realizing that they were doing it. Fermi, Irène Joliot-Curie, and Meitner, Hahn, and Strassmann—they were all making uranium fission. Yet not one of them realized it. They all thought that the uranium atoms were absorbing neutrons and changing into heavier, manmade elements—what they called "transuranics."

How could some of the world's leading minds make such a mistake? First, they were not looking for fission. They were looking for something else entirely: heavy, transuranium elements. They assumed that when uranium was bombarded with neutrons and became radioactive, it changed only slightly. They knew that an atom could change gradually, step by step, from one element into another by losing one or at most two protons at a time. But no one dreamed that one blow could divide an atom into two roughly equal, middle-sized atoms. So no one checked whether the particles from the collision were middle-sized elements; expecting big atoms, they searched for big atoms. Fission is a classic example of a discovery's cropping up where least expected.

As a chemist, Hahn was delighted to have lots of new "transuranium" elements to study. "For Hahn it was like the old days when new elements fell like apples when you shook the tree," Meitner's nephew Frisch recalled. Physicists like Meitner, however, were bewildered and frustrated. Experiments were increasingly hard to explain. Thus, Hahn and Strassmann reported their results in chemistry journals using words like "unquestionable...leave no doubt...we are certain...no longer in doubt...needs no further discussion." Meitner, on the other hand, published in physics journals that the transuranic elements were "very difficult to reconcile with current ideas of nuclear structure.... Perhaps one must look elsewhere for an explanation."

Several near-misses almost solved the puzzle. Strassmann played around in the lab late one night in 1936 and found what looked like mid-sized atoms of barium. "Are you sure of your chemistry?" Meitner asked. Strassmann admitted that he was not. In a firm but friendly manner, Meitner rejected his results. Had they redone the experiment, the riddle might have been solved much sooner and Meitner might have gotten more credit. In Paris, Irène Joliot-Curie found particles that looked like middle-sized lanthanum, but Hahn pooh-poohed her results.

A Berlin chemist, Ida Noddack, suggested that, logically, the uranium nucleus could split into two mid-sized pieces. No one paid attention. Was Noddack's suggestion rejected because she was a woman? Hardly. Among those who ignored it were Lise Meitner, Irène Joliot-Curie, and Maria Goeppert Mayer. Neither Ida Noddack nor her husband Walter took her suggestion seriously either. If they had, they would have done the simple experiment needed to prove her point. It seemed too farfetched. In addition, the Noddacks' scientific reputation was already shaky; earlier, they had "discovered" a nonexistent element.

In the meantime, Meitner was losing her race against Hitler. According to the Nazis' motherhood cult, scientific training for women was both scandalous and laughable. At the same time, "German physics" was considered superior to "Jewish physics," which included Einstein's relativity theories and quantum theory. As a Jew, Meitner was banned from attending her beloved Wednesday colloquium and from teaching, attending conferences, publishing articles, and lecturing. When Hahn gave a talk outside the institute on his work with Meitner, he could not mention her name. Her reputation was being obliterated. In addition, at a time when it was safer for Jews to remain inconspicuous, Carl Friedrich von Weizäcker, an able physicist whose father was secretary of state under Hitler, became one of her assistants.

The ban on "Jewish physics" made official seminars so dull that another of Meitner's assistants, the future Nobel Prize–winning molecular biologist Max Delbrück, organized private meetings at his mother's home to talk about physics and biology. German scientists protested publicly only once during the Nazi years, at a 1934 memorial service for the Jewish chemist and Nobel Prize–winner Fritz Haber. As government employees, university professors were forbidden to attend. Hahn, who had resigned from the university in

Berlin in order to maintain his independence, read the memorial address. Meitner, Strassmann, and Delbrück attended. Only two professors came, although many sent wives in their stead.

Like most Jews, Meitner wanted to stay in Germany. She figured she was "too valuable to annoy." Regarding institute employees, she felt "a very strong feeling of solidarity between us, built on mutual trust, which made it possible for the work to continue quite undisturbed even after 1933, although the staff was not entirely united in its political views.…It was something quite exceptional in the political conditions of that day." In addition, her old friend Max Planck wanted her to remain at the institute as long as possible; in 1935, Niels Bohr arranged a Rockefeller Foundation grant for Meitner to leave Germany and spend a year in Copenhagen, but at Planck's request she turned it down.

Meitner and Hahn often argued about politics because she thought he should take a public stand against Nazism. "As long as it's only us [the Jews] who have the sleepless nights and not you, it will not be any better in Germany," she often told him.

The longer Meitner stayed in Germany, the harder it was to find a job abroad. At the same time, pressure was mounting to fire her from the institute. Hahn was caught in the middle. With Meitner the Jew and Hahn the anti-Nazi in charge, the institute was considered politically unreliable. Hahn worried that he might be deposed as director. When the German army marched into Austria on March 12, 1938, Meitner's position changed overnight. Austria no longer existed, and she no longer had a passport. Legally, she had become a German citizen of Jewish descent.

"The Jewess endangers the institute.…She must go," institute Nazis warned Hahn. As he confessed later, "I rather lost my nerve," and discussed Meitner's resignation with an important politician. As a result, "Hahn says I must not come to the institute anymore," Meitner wrote angrily in her diary. Several days later, when the politician changed his mind, Hahn called her back. She was shocked that Hahn had put the institute and his position ahead of her welfare. In the back of his mind, Hahn explained, "I always remembered that Lise would have to give up her position [eventually] because she must see that she endangered the institute."

The president of the Kaiser Wilhelm Society, who greatly admired Meitner, asked the minister of education to let her go to "one of the neutral countries, e.g., Sweden, Denmark, or Switzerland." Meitner feared that bringing her case to the attention of the authorities would lead to her arrest, so she moved quietly to a hotel. The minister's response was chilling: "It is considered undesirable for well-known Jews to travel abroad where they appear to be representatives of German science or where their names and their corresponding experience may even demonstrate their inner attitude against Germany." Then her friend Max von Laue received a tip from the office of Secret Police Chief Heinrich Himmler: No university graduates, Jewish or otherwise, would be allowed to leave Germany. Meitner would have to stay in Germany—without work.

It was friends abroad, not Hahn, who arranged her escape at the last moment. Invitations were pouring in from Dutch and Swiss colleagues and from Bohr in Denmark for her to give seminars and attend conferences—anything to give her an official excuse to leave Germany. Dutch friends succeeded in arranging with the Dutch government to admit her without a visa or a passport. Professor Dirk Coster sent a telegram saying that he would soon be in Berlin, and Meitner understood that she would leave with him. That night, to avoid suspicion, she worked late at her office correcting a young associate's paper. Then, in the presence of Hahn and a friend, she quickly packed two small suitcases with a few summer dresses and ten marks, as if she were going on a holiday. She left behind all her belongings and scientific papers. Hahn gave her his mother's diamond ring for emergencies. On July 13, 1938, Meitner left Berlin. She was fifty-nine years old and had lived in Germany for thirty-one years.

Coster met her at the Berlin railroad station and together they boarded the train. At each stop along the way, German police checked identity papers. Several passengers were arrested and removed from the train. A Nazi military patrol studied her invalid Austrian passport for ten minutes while Meitner sat frozen with fear. Then one of the guards returned and handed the passport back to her without a word. Two minutes later, she crossed into a quiet country town in the Netherlands. After a short stay, she moved on to Denmark where Niels Bohr had offered her a position at his Copenhagen institute. Perhaps concerned about overshadowing her nephew Otto Robert Frisch, who had a job there, she stayed only a short time with Bohr.

Instead, she officially retired from the Kaiser Wilhelm Institute and accepted an offer from a Nobel Prize–winning Swedish physicist, Manne Siegbahn, to work with him at the Research Institute of Physics, now the Manne Siegbahn Institute in Stockholm. There she would be close to Berlin in case the situation improved. Siegbahn had just built the first European cyclotron, and she thought she could perform experiments with it. She arrived in Sweden, a friend said, looking like a "worried, tired woman with the tense expression all refugees had in common." Had Meitner known what awaited her in Sweden, she might have decided to stay in Copenhagen. Because Sweden was exile. And exile destroyed Meitner's career.

Later, Meitner regretted having stayed in Berlin so long. As she wrote Hahn, "by staying, I have supported Hitlerism." Nevertheless, she missed her old life desperately. Sweden was dark, the language was new, she was living in a hotel, her family was still in Austria, and Hahn was having trouble sending her papers and belongings. In Berlin, she had been the director of her own department; in Stockholm, Siegbahn controlled everything, and he enjoyed building machines more than using them for experiments.

Her self-confidence, built up after decades of struggle, was crumbling. As she wrote Hahn that fall of 1938, "I have none of my scientific equipment. For me that is much harder than everything else. But I am really not embittered—it is just that I see no real purpose in my life at the moment and I am very lonely.... Work can hardly be thought of. There is [no equipment] for doing experiments, and in the entire building just four young physicists and very bureaucratic working rules." Siegbahn, she explained, is "not at all interested in nuclear physics, and I rather doubt whether he likes to have an independent person beside him.... I often see myself as a windup doll, who does something automatically with a friendly smile, but has no real life in her."

Meitner always refused to discuss what happened next. "It is a rule of mine," she stated emphatically. Between letters and lab notes, however, it is possible to reconstruct what occurred. Although modern research partners often work at different institutions and collaborate by phone, fax, computer, and overnight mail, long-distance cooperation was rare in 1938. Nevertheless, Meitner was still a member of the Berlin team. Hahn was accustomed to consulting with her on all physics questions, so they wrote each other every other day; Berlin–Stockholm mail was delivered overnight. As Strassmann later said, "What does it matter that Lise Meitner did not take direct part in the 'discovery'...? [She] was the intellectual leader of our team and therefore she was one of us, even if she was not actually present for the 'discovery of fission.'"

Hahn and Strassmann, having read Irène Joliot-Curie's latest paper, decided to search among the collision debris for radium, an element almost as heavy as uranium. When they found some, Meitner wrote urgently for more details, searching for irrefutable evidence. If the project produced something other than transuranic elements, four years of work were in vain.

In Berlin, Hahn was under pressure too. His wife was showing symptoms of mental illness, and he could not arrange for Meitner's pension or furniture. To keep the experiment and his collaboration with Meitner secret, he isolated his team from the rest of the institute. No one was told about the experiment. On November 9, 1938, stormtroopers all over Germany burned synagogues, imprisoned twenty-seven thousand Jewish men, and murdered others as police officers watched. A few days later, Hahn traveled secretly to Copenhagen to confer with Meitner and Bohr. Meitner urged Hahn to double-check his radium data. "Fortunately, L. Meitner's opinion and judgment carried so much weight with us in Berlin that we immediately undertook the necessary control experiments," Strassmann reported. Thanks to her, they made a well-nigh incredible discovery.

The uranium-neutron collisions had produced middle-sized atoms that resembled barium. The night of December 19, 1938, Hahn stayed late in the laboratory, writing to Meitner. The uranium did not lose a few particles, he reported. Instead, it apparently changed into something like—or identical to—a middle-sized atom of barium. But Hahn could not understand what had happened and begged Meitner for an explanation. He wrote,

> Dear Lise,
>
> It is now practically eleven o'clock at night. Strassmann will be coming back at a quarter to twelve, so that I can get off home at long last.
>
> ...There is something about the radium isotopes that's so curious that we wanted to first tell it only to you. The half-lives of the three isotopes are now firmly established. They can be distinguished from *all* elements except barium. All the reactions agree....Our "radium isotope" behaves like barium....Perhaps you can propose some fantastic explanation. We know ourselves that it can't decay into barium.... Write as soon as you can.

Two days later, Meitner replied, "Dear Otto, your radium results are very puzzling—a process that goes with slow neutrons and leads to barium....However, we have experienced many surprises in nuclear physics, so that one cannot say without further consideration that it is impossible."

Reassured by Meitner's reaction but worried that Irène Joliot-Curie might beat them, Hahn and Strassmann sent their results to a German journal on December 22. Hahn mailed Meitner a carbon copy. The article would be published on January 6, 1939.

Their report was curiously roundabout. As chemists, Hahn and Strassmann concluded that uranium had changed into barium. "As 'nuclear chemists,' who are somewhat related to physicists, we cannot yet decide to take this big step, which contradicts all previous experience of nuclear physics. It is still possible that we could have been misled by an unusual series of accidents."

Rarely has such an important result been announced so indirectly. Many readers wondered if Hahn and Strassmann even understood what they had found. The article was written in a great hurry, and it may have been grafted onto an old draft about radium.

That New Year's, Meitner and her nephew Frisch met in a small resort town in western Sweden to celebrate the holidays with friends. It proved to be the most momentous visit of Frisch's life. When he came down for breakfast on December 30, he found Meitner brooding over a letter from Hahn. She was agitated, astonished, and disturbed. Hahn's barium opened a completely new scientific road and showed that their previous work had been in error.

Frisch was eager to talk about his own work, so he suggested, "Perhaps it is all wrong."

Meitner shook her head. "Hahn is too good a chemist," she answered. "I am sure this result is correct. But what on earth does it mean? How can one get a nucleus of barium from one of uranium?"

Frisch also wanted to go cross-country skiing. Because Meitner had not brought her skis, she offered to walk beside him while he skied. She was used to walking miles every day for exercise, so she easily kept up with him. As she walked, she explained that Hahn had found the particles produced by the uranium collision to be inseparable from barium.

A theoretician might not have believed Hahn, but Meitner was an experimentalist. Furthermore, she understood nuclear theory and was an expert in nuclear structure. So she knew by heart the masses, energies, atomic numbers, and so on, and could make rough calculations outside in the snow.

Uranium has 92 protons and barium has 56, she mused. How could uranium lose 36 protons all at once? A neutron could chip off one or two protons, but 36? Could the uranium nucleus be chiseled or sliced in two? No, the neutron could not act like a chisel, and the nucleus was not a solid object to be sliced in two. It was more like a liquid drop.

At that, Meitner and Frisch stopped walking and skiing and stared at each other. Perhaps the nucleus could get pulled into an oval shape with a waist in the middle. When the waist was thin enough, the drop might tear in two.

At first they thought that surface tension would hold the drop together. But 92 positively charged protons and approximately as many negatively charged electrons make a uranium nucleus electrically unstable. Maybe its surface tension would be weak. Meitner and Frisch sat down on a fallen tree, took out pencils and scraps of paper, and began to calculate.

The numbers worked. The uranium nucleus behaves like a large thin-walled balloon filled with water or like a large wobbly piece of jelly. Struck by a neutron, the nucleus splits in two. The drop could split several different ways, into barium with 56 protons and krypton with 36. Or it might become rubidium with 37 protons and cesium with 55 protons. Or other pairs of mid-sized atoms whose protons would total 92. Here at last was the reason why Meitner, Hahn, and Irène Joliot-Curie had discovered so many different nuclei. None of them was a transuranic element; they were all ordinary elements roughly half the size of uranium.

Figuring fast, Meitner discovered that when a uranium nucleus splits, it releases roughly two hundred million electron volts—twenty million times more than an equivalent amount of TNT. She was amazed. For the first time, more energy had come out of an experiment than had gone into it! Sixty years old and officially retired, Meitner had explained one of the greatest discoveries of the century.

Frisch returned to Copenhagen the next day and told the news to Bohr, who was on his way to the United States. Striking himself on the head, Bohr cried out, "Oh, what idiots we have all been!" Promising to say nothing until Meitner and Frisch had sent their results to a journal, Bohr rushed off. He was so excited, though, that he told a fellow scientist on board the ship—and forgot to say that the information was still a secret. Soon after the boat docked, the news spread all over the United States.

Bohr tried to make sure that Meitner and Frisch got credit for the explanation, but their articles were published weeks after Hahn and Strassmann's paper. Meitner and Frisch's vital contribution was lost in the excitement, although Frisch was also the first to do the experiment that actually confirmed what Hahn had discovered only indirectly. Frisch's name for the process stuck, though: nuclear fission, named for the biological process of cell division.

Meitner realized that the value of the work she had done over the past four years might be questioned. She wrote Hahn and Strassmann in the early hours of January 1, 1939, to congratulate them on their experiment. "You are in a much better position than I, since you and Strassmann have discovered it yourselves, while I only have years of work to refute—not a very good recommendation for my new beginning."

Two days later, she reiterated, "Believe me, although I stand here with very empty hands, I am nevertheless happy for the wonder of these findings....People will say that the three did nonsense and, now that one is gone, the other two made it right....I am gradually losing all my courage," Meitner confided. "Forgive this unhappy letter. I never wrote before how bad it really is. Sometimes I do not know what to do with my life. Most probably there are many people who have emigrated who feel as I do, but it is still very hard."

Meitner had reason to worry. Within a month Hahn was claiming that physics had impeded the discovery and that chemistry alone had solved it. As he confessed, "For me the uranium work [the discovery of fission] is a gift from heaven. Namely, I was fearful sometimes that...[I would lose] part of the institute."

During World War II, Meitner stayed in Sweden, unable to do any follow-up experiments on fission. She refused an Allied invitation to work on the atomic bomb. Hahn remained in Germany and, making it clear he did not like the Nazis, served on nuclear reactor committees. Hahn, Strassmann, and a few pupils continued to study fission products and publish their results throughout the war. At the institute, Strassmann played the fool so that he would be considered too unreliable for weapons research.

Even before the war was over, the Nobel committee voted secretly to give the 1944 Nobel Prize for chemistry to Otto Hahn—and only Otto Hahn. Few chemists or physicists object to Hahn's receiving a Nobel Prize. But physicists are virtually unanimous that Meitner should have received a Nobel Prize, too, whether in physics or chemistry. She initiated the experiment and, with Frisch, explained the process. Because of her, Frisch confirmed experimentally Hahn's indirect evidence. Others like Günter Herrmann believe that Strassmann also deserved to share Hahn's chemistry prize. And feminists like the German writer Renate Feyl question whether Hahn should have received a Nobel at all; she complained in 1983, "Lise Meitner's lifelong scientific accomplishments were crowned by the Nobel Prize for Otto Hahn."

Meitner lost for a number of reasons, but many scientists believe that the main reason was Manne Siegbahn. Swedish experimental physicists, including Siegbahn, controlled the physics Nobel Prize in its early years. And they believe that Siegbahn vetoed any prize for Meitner.

Timing was a factor, too. Sweden was a scientific backwater in 1939, and when the Nobel Committee voted, its members had been cut off from worldwide developments for five years. The atomic bomb had not been dropped on Japan yet. "In 1944, it was not officially known that nuclear fission was so important, and chemists were primarily interested in Hahn's work because he had disproved Fermi's Nobel Prize [for transuranic elements]," Wolfgang Paul emphasized. "If fission had been known to be so important, if the prize had been given after the war, it would have been clear that Meitner had to be included. It always takes at least several years for a great experiment to be understood. The experiment happened in 1939; a lot of the work on fission done in France, England, and the United States was immediately classified, and so its importance wasn't generally understood. She should have had the prize in physics the same year that Hahn got one in chemistry because she changed the theory immediately within a week of getting hard data that she trusted."

While vacationing with friends in central Sweden on August 6, 1945, Meitner heard a radio announcer broadcast the news that an atomic bomb—using the fission of uranium atoms—had been dropped on Hiroshima. Meitner was an instant celebrity. In a joint transatlantic broadcast with Eleanor Roosevelt a few days later, she noted, "It is an unfortunate accident that this discovery came about in time of war." In an interview with the *Saturday Evening Post*, she stressed that "I myself have not in any way worked on the smashing of the atom with the idea of producing death-dealing weapons. You must not blame us scientists for the use to which war technicians have put our discoveries."

Hahn, upset with the press attention Meitner was getting, issued a press release a few days after the bomb was dropped denying her contribution before and after she left Berlin. He complained that, "as long as Prof. Meitner was in Germany, there was no discussion of the fission of uranium. It was considered impossible. Based on extensive chemical investigations of the chemical elements caused by irradiating uranium with neutrons, Hahn and Strassmann were forced to assume at the end of 1938 that uranium splits into two parts...." Nor did Hahn change his mind with the years.

Meitner never complained publicly or privately about her failure to win the Nobel Prize. She was, however, distraught that German friends—Hahn especially—refused to apologize publicly for wartime atrocities. "I do not think they comprehend just what fate has befallen Germany through their passivity. And they understand even less that they share responsibility for the horrible crimes Germany has committed.... How shall the world trust a new Germany when its best and intellectually most prominent people do not have the insight to understand this and do not have a burning desire to make whatever amends are possible?"

When Meitner heard the first radio accounts from the concentration camps, she wailed and could not sleep. Writing Hahn, she complained, "You have all worked for Nazi Germany and you never once even attempted a passive resistance. Certainly to buy off your conscience, you have helped a single person here and there, but you let millions of innocent people be murdered and never a protest is voiced. I and many others with me believe that the right way for you would be to make a public declaration that you accept that through your passivity you bear a partial responsibility for the happenings and that you have the responsibility to help make right the wrongs...."

"Finally," she charged, "you betrayed Germany itself. When the war was totally hopeless, you didn't even oppose the senseless destruction of Germany.... You really can't expect the rest of the world to pity Germany."

She fought with Hahn the entire year before he came to Stockholm, in December 1946, to collect his Nobel Prize. When he arrived, they argued again. In public interviews, he did not once mention her name. In his Nobel speech, he emphasized that chemistry alone had solved the problem. Stressing that Germany needed help, he declared that the Nazis had victimized Germany before and during the war and that the Allies were victimizing it after the war.

Meitner was pained that he said nothing about her or their thirty years together. "He is convinced that Germans are being treated unjustly, the more so in that he simply suppresses the past," she wrote a friend. "As for me, I am part of that suppressed past."

When Hahn denied any German responsibility for the atomic bomb, Meitner reminded him that Germany had done other terrible things. He did not respond. German scientists supported Hitler's philosophy, she charged; the Academy of Science and the German Physical Society had expelled Einstein because he was Jewish. When Franck resigned in protest against Hitler's anti-Semitism, scientists signed a petition claiming that Franck had sabotaged the Third Reich. Scientific conferences condemned "Jewish mathematics." Hahn "suppresses the past with all his might, even though he always truly hated and despised the Nazis.... Since he does not have a very strong character and is not a very thoughtful person, he deceives himself," she protested.

In 1947, Meitner refused Strassmann's invitation to return to Germany to direct her old institute, which had been renamed the Max Planck Institute for Chemistry and moved to Mainz. "The Germans still do not comprehend what occurred, and they have forgotten all atrocities that did not personally happen to them. I think I could not breathe in such an atmosphere," she wrote.

After the war, a veritable personality cult grew up around Hahn. One of the few German scientists trusted by the Allies, he became president of the Max Planck Society. He appeared on German medals, buildings, coins, and stamps. Meitner and Strassmann faded from view. When the Max Planck Society identified her as Hahn's "junior colleague," she demanded of Hahn, "Will my scientific past also be taken away from me?" The Deutsches Museum in Munich exhibited her experimental apparatus—the equipment she had designed for the fission experiment—and identified it as Otto Hahn's worktable. The museum's sign did not include Meitner as a team member until the American chemist Ruth Sime complained publicly at a historians' convention held in the museum itself in 1989.

Despite the eradication of her reputation, Meitner remained friends with Hahn. They were like brothers and sisters who maintain family ties despite deep philosophical disagreements. She even returned to Germany several times, for conferences and awards.

Meitner lived in Sweden for twenty-two years, visiting the United States several times to work and see relatives. When she was seventy-five, she fell down a flight of slippery stairs at a physics conference, picked herself up, and continued her conversation. She did research and climbed mountains until she was eighty-one. In 1960, she retired and moved to Cambridge, England, to be near her nephew. By then she had published approximately 150 scholarly articles. The United States Atomic Energy Commission awarded its Enrico Fermi Award in 1966 to the entire fission team: Hahn, Meitner, and Strassmann. It was the first time that non-Americans had received the prize and the first time that a woman had won it.

A few days before her ninetieth birthday, Lise Meitner slipped almost imperceptibly away. Although she had never talked about the Nobel Prize, she left her letters and papers to Cambridge University. She made sure that her side of the story would be told.

EPILOGUE

Years after Meitner's death, physicists in Darmstadt, Germany, fused two isotopes of bismuth and iron to make Element 109, the heaviest known element in the universe. In 1992, the physicists named their new element meitnerium—in honor of Lise Meitner.

"Lise Meitner should be honored for her fundamental work on the physical understanding of fission," declared Peter Armbruster, the physicist leading the Darmstadt team. "She should be honored as the most significant woman scientist of this century."

* * *

4

Emmy Noether

March 23, 1882–April 14, 1935

MATHEMATICIAN

STORM TROOPER ERNST WITT, resplendent in the Brownshirt uniform of Hitler's paramilitary, knocked on a Jew's apartment door in 1934. A short, rotund woman opened the door. Emmy Noether smiled, welcomed the young Nazi into her home, and started her underground math class. The Brownshirt was one of her favorite pupils.

While Marie Curie and Albert Einstein worked in a glare of publicity, Emmy Noether lived a life of academic subterfuge. She attended German universities before it was legal for women to get degrees. She lectured under another mathematician's name after the Prussian government refused her permission to teach. She submitted her articles for publication through a friend; and she was an anonymous editor of a leading German mathematics journal. After she was fired by the Nazis, she held secret and illegal seminars in her apartment. When she fled Germany, her school—the world's best in abstract algebra—disappeared. Throughout World War II, German mathematicians used her "Jewish" mathematics illegally, without mentioning her name. It was almost as though she had never lived.

Luckily, Noether herself was so broad and solid, so loud and full of life, that no one who met her could ever have denied her existence. She was the female version of Albert Einstein. Utterly uninterested in appearances, she ignored all the feminine conventions of the day. She was overweight, enthusiastic, and opinionated. She was messy, unfashionable, and comfortable. She was also loving, utterly unselfish, and friendly. A famous contemporary thought she was "warm like a loaf of bread.... There radiated from her a broad, comforting, vital warmth." The influence of her mathematics is still strong. She was a founder of abstract algebra, one of the largest and most active fields of mathematics today. As a progenitor of the New Math, she has touched the lives of almost every schoolchild in the United States.

Emmy Noether before going to Göttingen.

Fritz and Emmy Noether vacationing at the Baltic seashore in the summer of 1933—their last meeting before exile. Left to right: Mrs. H. Heisig, Fritz Noether, Emmy Noether, Dr. H. Heisig, Mrs. Fritz Noether.

The only picture of Emmy Noether and her entire family, on a summer walk in the country shortly before World War I. In the foreground, Emmy and her three brothers (her mathematician brother, Fritz, is on the right) with her mother and father, the mathematician Max Noether, in the background.

Amalie Emmy Noether was born March 23, 1882, the first of four children in a well-to-do Jewish family in the small Bavarian university town of Erlangen. Her mother, Ida Amalia Kaufmann, came from a wealthy family of wholesale merchants. Emmy's father, Max Noether, came from a prosperous family of iron importers.

Max, who limped as a result of childhood polio, left trade to become the first of three generations of Noether mathematicians. An important algebraic geometrist, he was one of approximately two hundred Jewish professors in Germany. Germany, which had expanded its universities and technical schools throughout the nineteenth century, permitted Jews to attend them during the second half of the century. Thanks to their traditional respect for learning and abstract reasoning, the Jews benefited enormously. They accounted for less than one percent of the German population, but roughly 7 or 8 percent of its university students. Anti-Semitism was widespread throughout Europe, however, and relatively few Jews became professors. The faculty of the University of Zurich complained in 1901, for example, that Albert "Einstein is an Israelite and...to the Israelites among scholars are ascribed (in numerous cases not without cause) all kinds of unpleasant particularities of character, such as intrusiveness, impudence, and a shopkeeper's mentality."

Protected by Max's high position, Emmy's upbringing was warm, companionable, and—above all—mathematical. Max was a professor at Erlangen for forty-six years. His son Fritz became a prominent applied mathematician, and another son studied chemistry. Max Noether's close friend was Max Gordan, an eminent Erlangen mathematician famed for his idiosyncracies. Gordan liked to walk and talk mathematics at the same time, muttering long calculations and gesturing violently as he went.

Max Noether indulged his own mild eccentricities, too. Something of a miser, he lectured a relative who bought her son a postage stamp for his collection, and he canceled a family outing when he discovered that one of the children was too tall to ride the bus free.

As a girl, Emmy was considered ordinary, myopic, and plain. Playing number games once at a children's party, Emmy was the first to answer: "Oh, but of course, the answer is—" No one in the family took this, however, as a sign that Emmy needed more education. Instead, family stories about Emmy's childhood concentrated on her appearance and personality. She was remembered as "clever, friendly, and rather endearing." As a child she had a slight lisp. As a young woman, she was said to have loved parties and dancing.

While Emmy absorbed her family's preoccupation with mathematics and education, German law prevented her from pursuing either. German schools for middle-class girls were no more than finishing schools. Until they were fourteen or fifteen, girls studied some arithmetic and conversational French and English while concentrating on household management, elementary pedagogy and child care, religious instruction, and "aesthetic subjects." Germany had virtually no college-preparatory high schools for girls until well into the 1920s. Although laws varied from German state to state, the majority of German universities were not open to women until 1910, when Noether was twenty-eight years old. By then American women had been attending colleges for half a century.

In other respects, German women were also far behind their sisters in the United States, Great Britain, and France. German women could not assemble, attend public meetings where politics was discussed, make public speeches, or join political parties until 1908, when Noether was twenty-six years old. Luckily for her, she reached adulthood just as Germany was beginning to liberalize—ever so slightly—its laws regarding women's education. Much of her life was spent waiting for German law to catch up with her. Unfortunately, it caught up with her with a vengeance in 1933 with Hitler's anti-Semitic legislation.

Taking advantage of the few educational opportunities available to women, Emmy studied to become a language instructor in a girls' school. For three years she attended a teachers-training program that was still below the level of instruction in a boys' high school. At age eighteen, however, she passed the five-day examination certifying her to teach girls French and English. Her grades were "very good" except in classroom teaching—a skill some thought she never mastered.

Then, belying a friend's later claim that "there was nothing rebellious in her nature," Noether did the unexpected. At age eighteen, she took charge of her life in a totally unconventional manner. She did not become a schoolteacher. Instead, she spent two years auditing classes at the University of Erlangen.

When she started, German women could not get credit or degrees from a university, but they could ask a professor's permission to sit in on his lectures. This represented considerable progress because, in the past, women had had to get permission directly from the minister of education. Still, professors could—and did—refuse to admit women to their lectures. In the midst of Germany's rapid urbanization and industrialization, they regarded themselves as custodians of Germany's traditional culture. As a history professor explained, "Surrendering our universities to the invasion of women...is a shameful display of moral weakness." In an 1895 survey, a majority of German professors agreed that universities were "unwomanly" and beyond a woman's mental abilities. Two years before Noether began auditing classes at Erlangen, its faculty senate had declared that the presence of women students "would overthrow all academic order." Luckily, Noether was asking permission from family friends, and they agreed.

Noether was one of only two women auditors among nearly one thousand male students at Erlangen between 1900 and 1903. Still intending to teach in a girls' school, she signed up for more foreign-language classes. But for the first time, Emmy Noether also studied mathematics. As a result of those classes, she totally revised her plans. In July 1903, after more than two years auditing classes, she took and passed an entrance examination to Bavarian universities.

Then Emmy Noether took another, even more radical step: she left home. Instead of continuing at her hometown university, she enrolled in Göttingen—again as an auditor—at the age of twenty-one. This time, though, she intended to study mathematics. Göttingen had Germany's leading mathematics department. Furthermore, it was administered by the celebrated mathematician Felix Klein, an ardent supporter of higher education for women and a close friend of her father's. She remained in Göttingen only one semester. Apparently she became ill, because she moved back to Erlangen, which by then permitted women to earn degrees. There, in October 1904, she enrolled as Erlangen student number 486. There were only five other women at Erlangen the entire time Noether was a student there—three studying for degrees and two auditing classes. She was one of eighty full-time women students in all of Germany.

At Erlangen, Noether studied primarily with her father and his friend Max Gordan. Gordan, who supervised her thesis, shunned any suggestion of broad principles and avoided defining even basic concepts in his lectures. He wrote articles consisting of twenty pages of solid formulas; friends inserted the words. Nevertheless, Gordan was "the king of invariant theory," invariants being those quantities that remain the same, even when the variables have been changed.

Inspired by Gordan, Noether produced a virtuoso's thesis filled with computations and formal manipulations. It ended with a flashy cadenza of 331 ternary quartic covariant forms. With earthy frankness and good humor, she later called it "crap," "shit," and "a jungle of formulas." It impressed Erlangen's examiners, though. On December 13, 1907, when Noether was twenty-five years old, they awarded her highest honors. Although Noether eventually developed a different and totally noncomputational approach to mathematics—what Gordan would have called *"Theologie, nicht Mathematik"*—she revered him and kept his portrait on her study wall.

Noether spent the next eight years at home, working at the university without pay, position, or title. As her father's health deteriorated and he took to a wheelchair, she gradually took over his duties. She published half a dozen Gordan-style papers that are now considered classics, supervised doctoral students, joined several international mathematics organizations, gave talks abroad, and developed a modest international reputation. Noether loved the informal socializing and mathematical talk at conventions and workshops. By 1913, when she was thirty-one years old, she was lecturing at Erlangen as her father's substitute. She also picked up some of his domestic habits. When relatives came to visit, she checked the kitchen; discovering a few bits of leftover goose, she declared them enough to feed the crowd. The guests knew better and dined out. Max Noether's tightfistedness and Max Gordan's disregard for social conventions would come in handy when Emmy became self-supporting.

Noether's mathematics, like her career, ripened slowly. Her early work was brilliant, but it depended on the ideas of others; she was still known as Max Noether's daughter. Later, their relationship reversed itself. Today Max is known as "Emmy Noether's father."

Early twentieth-century intellectuals—including mathematicians, artists, architects, musicians, dancers, writers, and physicists—were fascinated by the concept of abstraction. Eager to strip reality of its special, individual peculiarities, they sought general principles that always hold true. David Hilbert, considered the greatest mathematician since Carl Friedrich Gauss, was using highly abstract methods at Göttingen. In Erlangen, Noether began applying his approach to algebra. During 1913–1914, Emmy and her father paid a long visit to Hilbert and Klein in Göttingen to write the official obituary of Max Gordan. Recognizing Emmy's talents, Hilbert and Klein invited her to join their team. She moved to Göttingen in 1916 and remained eighteen years.

Hilbert and Klein were working with Albert Einstein to develop a mathematical formulation of his general theory of relativity. Einstein had underestimated the difficulty of the mathematics involved, and Hilbert was intensely interested in atoms and relativity. He believed that "physics is much too hard for physicists." Thanks to Gordan's training, Noether was an authority on invariants. Invariants are an important aspect of relativity, which concerns what changes and what does not change when a phenomenon is seen by observers who are in different places or who are traveling at different speeds in different directions.

By this time, Noether dressed to match the unconventionality of her position in German life. Her mother's brothers had established a trust fund for their unmarried niece, but Noether did not spend its income on clothes. According to a small Viennese boy who saw

Noether about the time of her move to Göttingen, she looked like a country parish priest. In her mid-thirties, she wore a black coat that hung almost to her ankles and a shoulder bag slung crosswise like a railroad conductor's. Her hair was cropped short, a decade before it was fashionable. The little boy thought her hat was a man's. In an era when elaborate codes dictated every detail of a woman's clothing, deportment, and speech, Noether shocked the child. Happily absorbed in mathematics, she valued comfort, convenience, and price above transitory details like fashion.

From Göttingen, Noether wrote home good-humoredly that Hilbert's team was working on extremely difficult calculations for Einstein and "none of us understands what they are for." Eventually, she provided an elegant formulation for several concepts of Einstein's general theory of relativity. Einstein wrote Hilbert, "You know that Frl. Noether is continually advising me in my projects and that it is really through her that I have become competent in the subject." Hilbert replied, "Emmy Noether, whom I called upon to help me with such questions as my theory on the conservation of energy..." As for her delighted father, Max Noether wrote, "I see every day how her abilities mount, and I expect to get much pleasure from it." Hilbert and Klein were soon convinced that Emmy Noether deserved faculty status at Göttingen.

It took three attempts, four years, a German revolution, and the intercession of Albert Einstein to get Emmy Noether the most junior faculty position possible—without pay. The obstacle was a 1908 Prussian law prohibiting women university lecturers. In order to lecture at a German university, scholars undergo a process called the "Habilitation," which involves a formal lecture and faculty approval. Noether was the first woman to make the attempt at Göttingen.

The ensuing controversy stretched over many months and faculty debates but boiled down to one issue: most professors agreed in principle with the government that women should not be university lecturers. As one professor put it, "Only exceptionally can a woman's brain be mathematically creative." On the other hand, they wanted an exception made for Noether. At one extreme, an astronomer argued against anything that kept women from their patriotic duty of producing strong sons to fight Germany's wars. At the other extreme was Hilbert, the only professor who argued that merit, not politics or social issues, should decide faculty appointments. (His advocacy of women's educational rights was considered such a joke that on his fiftieth birthday, he was awarded the lifetime presidency of a fictitious organization, the "Union of Women Students.")

When Noether gave her lecture on November 9, 1915, every academic in town turned up. "Even our geographer came to hear it and found it rather too abstract," Noether wrote in a letter home. "The faculty wants to make sure the mathematicians aren't selling them a pig in the poke," or, as she put it in German, "a cat in the sack."

When the faculty met, the mathematicians announced that they wanted to hire Noether as a *Privatdozent,* the lowest faculty rank. Other departments objected. "How can a woman become a *Privatdozent?*" someone asked. "Having become a *Privatdozent,* she can then become a professor and a member of the university senate. Is it permitted that a woman enter the senate?" The senate was the policy-making body of the university.

Referring to Germany's soldiers then fighting in World War I, another complained, "What will our soldiers think when they return to the university and find that they are expected to learn at the feet of a woman?"

With biting sarcasm, Hilbert retorted, "*Meine Herren,* I do not see that the sex of the candidate is an argument against her admission. After all, the senate is not a public bathhouse." Göttingen's swimming establishment was segregated by sex.

Eventually, the faculty appealed to the Prussian minister of religion and education for a special exception. The ministry refused to grant one, but permitted a compromise. Noether could lecture—as Hilbert's assistant—although she could not be a faculty member. For the next three years, Hilbert listed Noether's classes under his name. Thus, the 1916–1917 university catalog read, "Mathematical-physical seminar. Theory of invariants: Prof. Hilbert, with the assistance of Frl. Dr. E. Noether, Mondays 4–6 p.m., free of charge." She could not even earn the usual student lecture fees.

Two years later, a new university at Frankfurt tried to lure Noether away from Göttingen with the offer of a faculty position. Frankfurt was not a state institution and expected to be exempt from the government ban against women instructors. Noether was inclined to go. "Greatly disturbed by the thought that our treasured colleague might be lost," Göttingen's mathematicians applied to the ministry again in 1917. Six days later, the ministry responded, "Your fear that she'll go to Frankfurt is unwarranted, because she won't be allowed to become a lecturer in Göttingen, Frankfurt, or anywhere else."

Thanks to Albert Einstein and Germany's defeat in World War I, Noether's chances improved dramatically in 1919. Einstein wrote Felix Klein in December 1918, "On receiving the new work from Fräulein Noether, I again find it a great injustice that she cannot lecture officially. I would be very much in favor of taking energetic steps in the ministry. If you don't think this is possible, I'll do it by myself. Unfortunately, I have to leave for a month, but I ask that you give me a short note when I get back. If something has to be done beforehand, you can do it over my signature."

By then Germany had lost the war, a revolution had exiled the Kaiser, the Socialists in power had given women the vote, and women sat in Parliament. Socialists had been in the vanguard of women's rights in Europe for decades, and Noether belonged to two socialist political parties between 1919 and 1922; she may have belonged to some pacifist organizations as well. Times had changed.

The next month, Klein asked the Prussian ministry whether "changed political circumstances" might not make it possible to appoint Noether to the faculty. The bureaucrats stalled until May, when they finally withdrew their objections. Within a few weeks, the university completed all necessary formalities and Noether was a *Privatdozent.* She was delighted. Thirteen years after getting her doctorate, at age thirty-nine, she could lecture legally under her own name. Although still unpaid, she began teaching that fall.

During the rush through Göttingen's red tape, Noether gave her second Habilitation lecture. It included Noether's Theorem, which is both her most famous and her least famous accomplishment. Physicists who know nothing else about Noether rely extensively on Noether's Theorem. Most mathematicians, on the other hand, have never heard of it. This powerful principle is a foundation stone of quantum physics. One of the great intellectual achievements of the twentieth century, quantum physics describes the behavior of atoms, nuclei, and subnuclear particles. Noether's Theorem proves that fundamental laws about the conservation of energy, momentum, angular momentum, and so on, are identical with the laws of symmetries. As a result, the laws of physics are independent of time or place; they hold true next year or in a thousand years, and a problem has the same outcome no matter where it is done. As Einstein wrote Hilbert, "Yesterday I received from Miss Noether a very interesting paper on invariant forms. It amazed me that one could view these things from such a general standpoint. It would not have done the Old Guard at Göttingen any harm, had they picked up a thing or two from her. She certainly knows what she is doing."

Quantum theory has had an enormous influence on twentieth-century physics and science. Its effect is all-pervasive because it concerns all matter in the universe. In comparison, Einstein's general relativity theory has had rather little effect on either physics or science in general. Relativity is seen only in rare and extreme situations. Nevertheless, it was Einstein who won a Nobel Prize in 1921. Nobels are given for physics but not for mathematics, so Hilbert, Klein, and Noether did not win a Nobel for their mathematical formulation of Einstein's theories. Nor did Noether receive a Nobel Prize for her contribution to the development of quantum theory. Physicists who did win include Werner Heisenberg, P. A. M. Dirac, and Erwin Schroedinger.

Noether could not submit her lecture about Noether's Theorem for publication herself. Göttingen's mathematicians had no mathematics department, building, offices, or budget until a few years later, and the Royal Göttingen Academy of Science functioned as the meeting place and cheap publishing house for local scientists. The Academy refused to admit Noether, so Klein submitted her articles for publication for her. At the Academy one day, Hilbert suggested, "It is time that we begin to elect some people of real stature to this society." Then he added thoughtfully, "*Ja,* now, how many people of stature have we indeed elected in the past few years?" As he looked slowly around the room, he concluded, "Only—zero. Only zero." As a woman, a Jew, a pacifist, and a Social Democrat, Noether was much too extreme for local tastes.

The glorious title of *"nichtbeamteter ausserordentlicher professor"*—translatable as "unofficial, extraordinary professor"—was bestowed on Noether in 1922. This mouthful of a promotion made Noether a volunteer professor without pay, pension, or privilege. Göttingen wits poked fun at its absurdity, observing that "an extraordinary professor knows nothing ordinary, and an ordinary professor knows nothing extraordinary." The government finally made a special exception for Noether. Normally, a *Privatdozent* had to wait six years before becoming a professor, but the Prussian ministry agreed to give her credit starting from 1915, when she had first applied for a lectureship.

In case Noether developed delusions of grandeur, the Prussian minister of science and art spelled out her lowly status. He wrote her, "This title does not signify any change in your present legal position. In particular, your position as *Privatdozent* and your relationship to your faculty remains unchanged; neither are you to receive the salary due an official position." She was one of two female and 235 male faculty members at Göttingen. Hertha Sponer, a physicist friend of Maria Goeppert Mayer, was the other female *"nichtbeamteter ausserordentlicher Professor."*

Göttingen never made Noether a regular professor. She was paid a tiny salary, only two hundred and fifty marks a month, hardly enough to live on—"alms," in the words of a German biographer. She was Göttingen's lowest paid faculty member during the 1920s. She was never covered by the state civil service system and never received benefits or a pension. Moreover, her salary came up for governmental review yearly. After postwar hyperinflation destroyed the value of her inheritance and her father died in 1921, she was in dire financial straits. Nevertheless, Noether considered her new stipend worth celebrating, and when a friend advised her to get a new wardrobe in honor of it, she did.

Noether helped turn algebra in a totally new direction and forged a radically different, conceptual approach to mathematics during the 1920s. Both developments are still, to this day, enormously productive. She was one of the leading founders of abstract algebra—now a major area in mathematics—and also worked in group theory, ring theory, group representations, and number theory. Group representations has proved extremely useful for both physicists and crystallographers.

Noether's single most important paper, "Theory of Ideals in Rings," was published in 1921. In it she thought only in concepts, comparing and contrasting them in grand and abstract schemes. Formulas, numbers, physical examples, and computations fade away. It is as if she were describing and comparing the characteristics of buildings—tallness, solidity, usefulness, size—without ever mentioning buildings themselves. Numbers and formulas actually seemed to hinder her understanding of mathematical laws and proofs.

"She saw the connections between things that people hadn't realized were connected before. She was able to describe in a unified manner many ideas that people had thought were different. She saw their underlying similarity," explains University of Texas algebraist Martha K. Smith.

"Her group was incredibly productive. It was the first time that anyone had taken a large number of phenomena and found the governing body of principles.... They stripped out the individual features of lots of questions in algebra, geometry, linear algebra, and topology [the study of surfaces]," according to Richard E. Phillips, a Michigan State University algebraist.

Today, Noether's creativity is universally recognized among mathematicians, but at the time her approach was highly controversial. When Noether visited her brother Fritz in Breslau, the two argued for hours about whether mathematics should be applied or abstract. Fritz, an applied mathematics professor, thought that mathematics should describe the physical universe. Emmy contended that math should be developed "for the fun and intellectual interest of it. If she knew how useful her mathematics had become today, she'd probably turn over in her grave," observed Fritz's older son, Herman Noether.

Emmy spent holidays, Christmas, and Easter with Fritz's family at the Baltic seashore or in Czech ski resorts. At teatime, the adults played elaborate mathematical guessing games with scientific friends. At Easter, Emmy brought the children matzos, which the boys ate with butter and jam. Fritz's wife was Catholic, and their two sons knew nothing of their Jewish heritage. "We didn't know what the matzos were for or why; my education had nothing to do with having a Jewish background. I discovered that in 1933," Herman Noether said.

Although her brother's circle in Breslau included several women scientists, in Göttingen Emmy Noether was the only woman in a crowd of men. Sturdy and frank, she was "one of the boys" there. Flocking noisily around her "like ducklings around a motherly hen," her students were dubbed "the Noether boys"—even though one of "the boys" was Fräulein Grete Hermann. Noether defended them fiercely against the slightest injustice. "The entire reservoir of her maternal feelings went to them," the Russian topologist P. S. Alexandrov recalled.

Stories about Noether's appearance, dress, weight, and so on are legion—testimony to the curiosity that a woman university teacher attracted in the early twentieth century. The male establishment jokingly called her *"Weichbild von Göttingen"*—comparing her weight to Göttingen's surroundings as they spread out around the town. Even good friends like Alexandrov and the famous mathematician Hermann Weyl called her *"der Noether"*—*"der"* being the article for a masculine noun. Weyl said they used the term jokingly, with "a respectful recognition of her power as a creative thinker who seemed to have broken through the barrier of sex."

Although today the nickname has lesbian overtones, it did not at the time; it was meant as a compliment, to stress that she was businesslike and straight-to-the-point without any of a woman's frailties. "With all the animosity to her in the 1920s, if there'd been even a hint of any lesbianism, she would have been accused of it. Germany of the 1920s was very free-speaking; there was none of the prudery of the 1930s. Her enemies would have brought it up without any question, and there was never any hint of it," observed Emiliana Pasca Noether, a twentieth-century European historian who married Emmy Noether's nephew Gottfried, Fritz Noether's younger son.

That is not to say, however, that no one discussed her sex life. Not even Weyl could resist speculating about it in his eulogy at her memorial service shortly after her death in 1935. Weyl considered her asexual: "She was a one-sided being who was thrown out of balance by the overweight of her mathematical talent. Essential aspects of human life remained undeveloped in her, among them, I suppose, the erotic."

Mothering her "boys," she shared her few belongings with them. She lent her tiny apartment to left-wing student groups for meetings and let Jacob Levitzki, a poor Israeli, use it when she visited Russia. Trying to find Levitzki a job in Göttingen or India, she recommended him as "unusually capable and likable...with nothing unpleasantly Jewish about him."

Free of vanity, she shared her ideas with her students as generously as she did her possessions. And then she promoted "their" work. Although she eventually published forty-four papers, many of her ideas took form only in the publications of others. Often a casual remark of Noether's wound up as a basis for a major article by one of her associates. "A large part of what is contained in the second volume of Bartel L. van der Waerden's *Modern Algebra* must be considered her property," Weyl said. When she solved a problem before van der Waerden, she let him publish it as his own. He acknowledged, "She always wrote the introductions to our papers, formulating for us the principal ideas which we, as beginners, could never have grasped and pronounced with her clarity."

Noether approached outings with as much vitality and enthusiasm as she showed when she talked. Weyl remembered long post-seminar walks "through the cold, dirty, rain-wet streets of Göttingen." Striding quickly through town with Noether, Emil Artin asked her to repeat her explanations over and over again until she tired and slowed down enough for him to understand them. His wife, Natasha Artin, accompanied Noether and friends on the Hamburg subway in 1934, after Hitler had come to power. Noether talked loudly and excitedly, oblivious to the fact that in Nazi Germany her words had acquired sinister double meanings. With the subway crowd agog and Natasha Artin terrified, Noether boomed on about *Idealtheorie, Ideal, Führer, Gruppe, and Untergruppe.*

On her limited income, Noether managed a simple and orderly life. "She didn't have very much money, but also she didn't care," Herman Noether explained. She contributed financially to caring for her youngest brother until his death in a sanitarium in 1928 and then saved for her nephews' education. She lived in a boarding house until she was thrown out because the student boarders objected to dining with a "pro-Marxist Jewess." Later, she rented an attic apartment. She and Nina Courant, wife of mathematician Richard Courant, ignored the "men-only" sign and swam daily in all kinds of weather in the public swimming pool by the Leine River; professors biked there at lunchtime to swim and eat frankfurters and rolls at its snack bar. Six days a week, Noether ate the same dinner at the same time at the same table in the same cheap restaurant. On Sundays, she cooked dinner for her students, took a long walk with them, and for dessert made *mathematische Papp*—mathematical mush pudding. Then she scandalized the men by leaving the dirty dishes for her house cleaner. The fact that she was entertaining a crowd of hungry young men on a low budget with limited spare time and no interest in housekeeping does not appear to have struck the students. They were accustomed to servant-filled homes orchestrated by the fulltime wives of Göttingen's well-paid professors. Her informal lifestyle became the butt of many jokes. A favorite story explained why she did not repair her umbrella like a proper hausfrau: if the sun was shining, she forgot about the umbrella, and if it rained, she needed it.

Göttingen's social life revolved around parties nominally hosted by the professors but actually organized by their wives. Noether contributed her share, specializing in children's parties and mathematical teas or evenings when she served lots of sweets, tea, or Rhine wine. They were relaxed, comfortable affairs where students mixed with eminent mathematicians like Hilbert, Edmund Landau, Richard Brauer, and Weyl. She must have created a cozy atmosphere because her living room was selected as the site of a carefully planned reconciliation between two giants: Hilbert the formalist and L. E. J. Brouwer the intuitionalist. A group watched nervously as Alexandrov skillfully engineered the conversation around to a man whom both loathed. United in antipathy, Hilbert and Brouwer became bosom friends—at least while they remained at Noether's.

Noether pared her life of inessentials and conventions, just as she stripped mathematics of its inessentials. Alexandrov admired her "extraordinary kindness of heart, alien to any affectation or insincerity; her cheerfulness and simplicity; her ability to ignore everything that was unimportant in life." But not all his colleagues agreed. They complained because her voice was not soft and refined; it was "loud and disagreeable." Another thought she looked "like an energetic and very nearsighted washerwoman." Still others chided that "her clothes were always baggy." As the historians Robert P. Crease and Charles C. Mann remarked, "Had Noether been a man, her appearance, demeanor, and classroom behavior would have been readily recognized as one of the forms that absent-minded brilliance frequently assumes in the males of the species."

Weyl recalled her unconventional behavior as fond foibles in his eulogy following her death. He dwelt long—and some may feel patronizingly—on her appearance and mannerisms. "It was only too easy for those who met her for the first time, or had no feeling for her creative power, to consider her queer and to make fun at her expense," Weyl said. "She was heavy of build and loud of voice, and it was often not easy for one to get the floor in competition with her.... No one could contend that the Graces had stood by her cradle.... She was not clay, pressed by the artistic hands of God into a harmonious form, but rather a chunk of human primary rock into which he had blown his creative breath of life."

"Oh, she was very much overweight," her nephew Herman

Noether agreed. But Emmy Noether did not care. "If I don't eat, I can't do mathematics," she said. "Her mathematics was the most important thing there was," her nephew recalled.

Her enthusiasm extended to teaching. "This winter," she wrote a friend, "I'm giving a course on hypercomplex numbers, which is as much fun for me as it is for my students." She did not lecture about completed theories. She talked about her current research, thinking along with the students as she developed her ideas, wiping the blackboard clean almost as rapidly as she wrote. She spoke loudly and fast, condensing many syllables into one or two. "As one listens, one must also think fast—and that is always excellent training," wrote home Saunders Mac Lane, an American student who became a distinguished algebraist.

As the mathematical excitement mounted, Noether reached into her bosom for a handkerchief to wave around enthusiastically. Her blouse "easily became untidied by the animation with which she spoke," a scandalized observer reported. Hairpins came loose. "Soon enough little wisps would begin to stick out here and there." Two horrified women students tried to tidy her up during a break after two hours of lecturing, but Noether was so busy talking to students that they had to give up. She resumed her lecture "in a state of disarray."

Her speech could not keep up with her thinking. Working on a new proof for a theorem, she got disappointing results and flew into a rage, hurling the chalk to the floor and stamping on it. "There," she protested, "I'm forced to do it the way I don't want to." And then she waded through the traditional method—flawlessly.

Generally only a small audience of loyal followers—perhaps five or ten—attended. Her work was too abstruse for crowds; only the best survived. Regulars sat in front, visitors in back. When newcomers departed, the front bench announced triumphantly, "The enemy has been defeated; he's cleared out." Once she found more than one hundred students waiting for her. She told them frankly, "You must have the wrong class." When they shuffled their feet in the traditional signal to begin, she gave her lecture anyway. Afterward, a Noether boy passed her a note, "The visitors have understood the lecture just as well as any of the regular students." Not even state holidays stopped her; when the university was closed, she simply walked her students through the woods to a coffeehouse, talking mathematics all the way.

The best young German mathematicians studied with her and, during the 1920s and 1930s, foreigners flocked to her lectures as well. She appealed to an international élite far more than to the average hometowner. When a Dutchman joined her class, she exclaimed, "Ah, another foreigner! I get only foreigners!" The Russians wore proletarian shirtsleeves dubbed "the Noether Guard uniform." They shocked Göttingen's dark-suited burghers but set the dress code for generations of mathematicians and physicists.

Although Noether had only seven official and thirteen unofficial doctoral students, most of them became eminent mathematicians. Her students were Grete Hermann, Heinrich Grell, Werner Weber, Jakob Levitzki, Max Deuring, Hans Fitting, and Otto Schilling. Deuring became what she could never be—a professor at Göttingen. Levitzki was one of the founders of modern ring theory, while Helmut Hasse's book is still the standard reference on number theory. Fitting and Witt laid the foundations for much of group theory and geometry. Van der Waerden's book expounding Noether's ideas to other mathematicians was especially important. "The freshness and enthusiasm of his exposition electrified the mathematical world—especially mathematicians under thirty like myself," recalled Garrett Birkhoff, a leading American mathematician of the 1930s. Van der Waerden's book became a foundation piece of abstract algebra and made Noether famous.

In the 1950s Birkhoff and Saunders Mac Lane popularized van der Waerden's book in a textbook for college undergraduates. When the Russians shook up American education by launching the Sputnik satellite, the Birkhoff-Mac Lane text was adopted nationwide; its abstract approach changed the face of American math education from elementary to graduate school. "The degree of its influence would be hard to overstate, and there's been only a small retreat since then," according to Richard Phillips. Noether's approach, greatly watered down and popularized twice over into the "New Math," has affected almost every schoolchild in the United States.

Despite Noether's absorption in mathematics, she was also interested in politics, especially Soviet life. The Russian revolution, with its antiaristocratic proletariat and its opposition to Fascism, appealed to many liberal westerners during the 1920s and 1930s. Noether spent the winter of 1924–1925 in Moscow, living in a modest dormitory and teaching at the university and a local institute. Alexandrov, the president of the Moscow Mathematical Society, acknowledged his great debt to her; at her suggestion, he and Heinz Hopf introduced group theory into combinatorial topology, starting its transformation into algebraic topology.

As the fame of her school spread, Weyl mounted another campaign to get Noether a real professorship. "I earnestly tried to obtain from the Ministerium a better position for her, because I was ashamed to occupy such a preferred position beside her, whom I knew to be my superior as a mathematician in many respects," Weyl said. "In my Göttingen years, 1930–1933, she was without doubt the strongest center of mathematical activity there, considering both the fertility of her scientific research program and her influence upon a large circle of pupils." By then, there was a host of Noether eponyms: Noetherian rings, the Noetherian theorem, Noetherian problems, Noetherian modules, Noetherian scheme, Noetherian space, Noetherian factor systems, and so on.

Olga Taussky, a young Czech mathematician who in 1971 became the first woman professor at the California Institute of Technology, met Noether at a mathematics conference in 1930. "As soon as I had finished [my talk], Emmy jumped up and made a quite lengthy comment which, unfortunately, I was unable to understand because of insufficient training. However, Helmut Hasse understood it and replied to it at some length, and there developed between these two mathematicians some sort of duet which they enjoyed thoroughly. Clearly, Emmy was pleased and I even overheard some nice remarks she made about my lecture. She spoke to me frequently later.... She was very friendly; so was Hasse."

"When lunch came," Taussky said, "I sat down next to Emmy, to her left.... Emmy was very busy discussing mathematics with the man on her right and several people across the table. She was having a very good time. She ate her lunch, but gesticulated violently when eating. This kept her left hand busy too, for she spilled her food constantly and wiped it off from her dress, completely unperturbed.... Emmy was having a great time." Table manners were never Noether's strong point; she sometimes drooled.

The next time the two women met, Noether made a disparaging remark about Taussky's thesis advisor, and Taussky came "violently" to her professor's defense. German students were not supposed to contradict a professor, and Taussky was shocked at her own daring. But "Emmy was completely calm, and I feel certain was not in the least angry with me. I recognized for the first time that she was *a person who did not mind criticism.*"

The year 1932, was a happy, banner year marking Noether's international acceptance as a mathematician. She won a five hundred-mark mathematics prize and was the first and only woman invited to address a general session of the International Congress of Mathematics in Zurich, Switzerland. Taking Taussky's suggestion, she began her talk with some simple examples that non-algebraists could understand. The speech was widely praised. Zurich represented a "full recognition of her ideas," Alexandrov said. "Her accomplishments were lauded on all sides."

Back home, however, many in Göttingen's establishment still did not accept her or her work. Noether was kind but often naive and thoughtlessly frank. Her bragging also irritated people, and Taussky heard a senior professor speaking roughly to her. Although a local journal published the birthdays of other mathematicians, it ignored Noether's fiftieth in 1933. Joking about the slight, Noether remarked, "I suppose it is a sign that fifty does not mean old."

Meanwhile, mathematics—like everything else in Germany—was becoming highly politicized. Few German academics opposed Hitler's rise to power. "The great majority of scholars viewed the Weimar government with icy reserve: they were willing to serve the German state, but not the Social Democrats. They regarded parliamentary politics as sordid and factional," the historian Alan D. Beyerchen observed.

One of Noether's graduate students, Werner Weber, organized a Nazi boycott of Professor Edmund Landau because Landau was Jewish. Weber, who was Landau's assistant, stood outside the professor's lecture hall and not one of his seventy students entered. "Aryan students want Aryan mathematics and not Jewish mathematics," another student explained to Landau. Nazism was popular among Germany's university students, and many attended class wearing storm troopers' brown shirts and swastikas. Hitler had given the SA, his militia, the "freedom of the streets," carte blanche to enforce his political agenda.

An ardent pacifist, Emmy Noether was uninvolved in politics. The Nazification of science was becoming hard to ignore, however. The famous Brauer-Hasse-Noether theorem involved two friends—one a Nazi and the other a Jew—who clashed over a publication. Helmut Hasse, who applied for Nazi party membership a few years later, had commissioned Richard Brauer to write articles for a mathematical encyclopedia. When the encyclopedia's publisher refused to pay Brauer because he was Jewish, Hasse refused to intervene on behalf of his collegue.

When Noether's brother Fritz complained about a painting portraying Germany's president in military uniform in front of burning Russian villages, Fritz's students criticized him. Then Nazis murdered one of Fritz's best students because the youth was wearing a gray jacket, the color of the jackets worn by the liberal Social Democrats. When Fritz spoke at the boy's funeral, Nazis in Breslau decided that Emmy Noether's brother was dangerous.

A few months after Noether's triumphant speech at the International Congress in Zurich, Hitler became chancellor of Germany. Jewry, he charged, "grasped in its hands all key positions of scientific and intellectual as well as political and economic life." Moving against the "satanic power" of the Jews, he began firing Jewish professors.

In early May, the Prussian Ministry for Science published a list of professors with Jewish ancestry, including six from Göttingen. Noether was on the list. Within a few days, they were fired and prohibited from lecturing in any German university. The expulsion of Jewish scholars devastated the University of Göttingen. Most of the leaders of its mathematics and physics institutes were Jews: James Franck, Max Born, and Richard Courant. Among others who eventually fled Göttingen were Hertha Sponer, Edward Teller, Landau, Weyl (whose wife was Jewish), and Noether.

Fourteen of Noether's students appealed to the ministry for her reinstatement on the grounds that all her pupils were Aryan and that her mathematics represented "Aryan thinking." Others wrote too. Recalled Weyl, "I suppose there could hardly have been in any other case such a pile of enthusiastic testimonials filed with the Ministerium as was sent in on her behalf. At that time we really fought; there was still hope left that the worst could be warded off. It was in vain."

Why was Noether one of the first six professors fired? Certainly, the Nazis were strongly anti-feminist. In *Mein Kampf*, Hitler wrote, "The message of women's emancipation is a message discovered solely by the Jewish intellect, and its content is stamped by the same spirit." As Hitler's minister of propaganda Josef Goebbels wrote, "The mission of women is to be beautiful and to bring children into the world."

University documents show, however, that Noether was dismissed because she was Jewish and politically liberal. The university's chief administrator, Kurator J. T. Valentiner, declared that Noether was too much of a socialist and a left-winger to ever make a good Nazi. Her official notification of dismissal cited her Jewish background.

Noether's response, as always, was simple and direct. She and Weyl formed the German Mathematicians Relief Fund, and she moved her seminars to her apartment. "She harbored no grudge against Göttingen and her fatherland for what they had done to her. She broke no friendship on account of political dissension," Weyl said. Ernst Witt, a student she had particularly liked, showed up regularly at her apartment seminar in his SA uniform. If Noether was disturbed, she did not show it—and laughed about it later. When Hasse wrote her in May, she answered, "Many thanks for your dear, compassionate letter! I must say, though, that this thing is much less terrible for me than it is for many others. At least I have a small inheritance (I never was entitled to a pension, anyway) which allows me to sit back for a while and watch."

Throughout the troubled, chaotic summer of 1933, Weyl said, "Her courage, her frankness, her unconcern about her own fate, her conciliatory spirit were, in the midst of all the hatred and meanness, despair, and sorrow surrounding us, a moral solace.... Her heart knew no malice; she did not believe in evil—indeed it never entered her mind that it could play a role among men."

The Nazis "in a matter of weeks scattered to the winds everything that had been painstakingly created over the years," Alexandrov said. As usual, Hilbert had the last word. At a banquet, the Nazis' newly appointed minister of education unctuously inquired, "And how is mathematics in Göttingen now that it has been freed of the Jewish influence?" The old man thought a moment. "Mathematics in Göttingen?" he asked. "There is really none any more."

Noether's friends began a frantic search to find a job abroad for a liberal, Jewish, pacifist, woman mathematician. Jews of either sex were difficult to place in the 1930s. Her first choice was the University of Moscow, but its bureaucrats were too slow. Weyl tried to secure Rockefeller Foundation funding for her at the Institute for Advanced Study in Princeton, New Jersey, where he and Einstein had gone. A prominent mathematician informed the Rockefeller Foundation that, with the exception of the seventy-one-year-old Professor Hilbert, "Miss Noether is unquestionably the most important teacher in Germany."

Somerville, a women's college at Oxford University, and Bryn Mawr College, a small women's college in suburban Philadelphia, competed for Rockefeller Foundation money to pay Noether in 1933–1934. Bryn Mawr offered a modest four thousand dollar salary—two thousand dollars of it from Edward R. Murrow's Emergency Committee to Aid Displaced German Scholars. Oxford offered only twenty-four pounds but had the firm conviction that Bryn Mawr would not be a "suitable place" for Noether. Noether put the college off until the last minute, hoping for Russia or Oxford. Finally, she gave up and accepted a temporary, one-year post at Bryn Mawr.

"Bryn Mawr is a very fancy school, where you must wear a hat," friends assured her. Noether preferred berets, but she bought a conventional hat and wore it when she arrived in the fall of 1933. Getting into a car with Anna Pell Wheeler, who had invited her to Bryn Mawr, Noether took one look and exclaimed, "You don't have a hat on." At that, she took hers off and tossed it into the back of the car. It never reappeared.

Anna Pell Wheeler and Emmy Noether became close friends. Wheeler was an elegant, twice-widowed mathematician whose first husband had been a Russian counterspy. Wheeler had studied mathematics at Göttingen, spoke German, and knew the personalities and atmosphere of German mathematics. She was also the best-known woman mathematician in the United States, a linear algebraist with degrees from Radcliffe College and the universities of Nebraska, Iowa, and Chicago.

"Mrs. Wheeler and Emmy had a wonderful relationship," recalled Ruth Stauffer McKee, Noether's only American graduate student. "Both had been a woman alone in mathematics. Emmy just thought Mrs. Wheeler was wonderful, she went to her with all her problems, and she wanted to do everything the way Mrs. Wheeler did it." Whenever friends arrived from abroad, Noether took them to visit "her good friend Mrs. Wheeler and they had a very good time."

Encountering Einstein at tea in Princeton, Noether amused him with funny stories in German about their mutual acquaintances. As she carefully explained to German friends, she traveled the short distance between Bryn Mawr and Princeton each week to give seminars at the Institute for Advanced Study, not at Princeton University. Noether strongly opposed single-sex schools. Of Princeton, she wrote judiciously, it is "a men's university where nothing female is admitted. In part their standards in mathematics are already very good. Princeton will receive its first algebraic treatment this winter, and a thorough one at that." She would go slowly, however, because her approach had already driven away a few locals.

Back at Bryn Mawr, Wheeler arranged tiny scholarships for three postgraduate students to work with Noether. Wheeler told them "to work as we had never worked in all our lives.... She also told us never to say anything unkind about [Noether]. Wheeler said she was worried that people would say the wrong thing about her, and I think we were all very careful to try to protect her," McKee said.

In any event, her Bryn Mawr students regarded her appearance as utilitarian, not mannish. Her hat and "man's" shoes turned out to be a beret and "nice comfortable oxfords," McKee said. Noether had a large ovarian tumor that made her look pregnant so she wore neat, dark dresses that hung loose from the shoulders and were tied with a belt. She had three pieces of jewelry—a watch, a pin, and a necklace—and wore very thick glasses.

Her Bryn Mawr students were inspired, not confused, by her lectures. They had never been exposed to abstract algebra, so Noether assigned a short section in the first volume of van der Waerden's *Modern Algebra*. A day or two later she stopped by the seminar room and asked the students how it was going.

"Well," responded McKee, "I'm having trouble knowing how to translate all these technical terms such as '*Durchschnitt*.'"

"Ah-hah," Noether replied. "Don't bother to translate. Just read the German."

"So we simply accepted the German terms and thought about the concepts behind the terminology; we never did learn the English words, and we conversed about math in a strange mix of English and German terms," McKee said. "The strange phenomenon, as I look back on it, was that from our point of view, she was one of us, almost as if she too were thinking about the theorems for the first time. Miss Noether was a great teacher! I just have never gotten over how she made me feel that I was on the same par with her. And she was so concerned. Mrs. Wheeler had told her how I was trying to earn my way. She slipped a ten-dollar bill into my pocket. Mrs. Wheeler said, 'I think she does that.'" Noether's salary was quite low by American standards, but Noether thought she was rich and saved half her earnings for her nephews.

Her "methods of thinking and working, they were simply a reflection of her way of life: that is, to recognize the unessentials, brush them aside, and enter wholeheartedly into the present," McKee concluded. "Bitterness and jealousy were rejected by her as unessential. There was never any indication of bitterness toward Germany.... Nor was there any sign of jealousy because of her treatment as a woman, even in the end when her colleagues from Göttingen went to the Institute for Advanced Study at Princeton with possibilities of many promising advanced young students."

Emmy Noether, with what a relative called "her usual buoyancy," adjusted to Bryn Mawr. Eager to fit into its social life, she asked the owner of her boardinghouse to arrange a tea for her students and Wheeler. "I must tell you that her living arrangements were modest but comfortable," McKee related. "Mrs. Hicks was a kind, thoughtful person, very interested in caring for Emmy, her clothes, and belongings. She planned a lovely tea party, with Mrs. Wheeler as the honored guest pouring." Noether was so disturbed to see her friend working that she marched into the kitchen and brought Mrs. Hicks out to relieve Mrs. Wheeler. On nice days Noether sat in the yard, with her eyes closed, thinking. Her landlady thought she was sleeping. During 1934–1935, her second year in Bryn Mawr, Noether took the women on long hikes, through Bryn Mawr and the surrounding countryside. She often stopped in the middle of the street loudly explaining a point in an English-German mixture while her students tried to nudge her like a mother hen to the curb.

The summer of 1934, Noether returned to Göttingen to close up her apartment, ship her desk to Bryn Mawr, and say farewell to her brother Fritz and his family. A Swiss committee formed to help political refugees had found him a job in Tomsk, Siberia. Most of her old colleagues avoided her, and she realized that she would not be back home for a long time.

That fall, Noether invited Olga Taussky to Bryn Mawr. Emmy was not in the best mood, Taussky recognized immediately. German politics were deteriorating rapidly. She had left her homeland in her fifties—then considered a rather advanced age for a woman—and had no job prospects for the following year. She did not want to teach Bryn Mawr's undergraduates, and her friends seemed to be moving very slowly on her behalf. Sometimes she seemed frightened that Taussky would find a job before she did. She also had medical problems.

Noether was not depressed that winter, but her moods were changeable. Taussky could never predict how she would react. She did not like Taussky's scarves knotted in front, for example. Bryn Mawr women tied them in back like gypsies. Noether informed Taussky that she looked like a Berlin *Droschkenkutscher,* a horse-cart driver.

Finding Noether a permanent position was proving to be extremely difficult, however. There were too many Jewish refugees, too many Jews still in Germany, and too few places willing to hire them. With Noether safe in the United States, others had to be considered first.

Her mathematician friends persisted. S. Lefschetz of Princeton University noted that "were it not for her race, her sex, and her liberal political opinions (they are mild), she would have held a first-rate professorship in Germany.... She is the outstanding refugee German mathematician brought to these shores, and if nothing is done for her, it will be a true scandal." Norbert Wiener of the Massachusetts Institute of Technology wrote, "She is one of the ten or twelve leading mathematicians of the present generation in the entire world.... Of all the cases of German refugees, whether in this country or elsewhere, that of Miss Noether is without doubt the first to be considered."

It was clear that Bryn Mawr, with its emphasis on undergraduate instruction, was not the place for Noether. She needed more advanced students. Birkhoff wrote, "As far as undergraduate work is concerned, she will be probably of no use at Bryn Mawr." Mrs. Wheeler confided later to McKee, "We didn't know what we were going to do about her."

Warren Weaver of the Rockefeller Foundation wrote Bryn Mawr's president, "It has become clear that Noether cannot possibly assume ordinary academic duties in this country. She has no interest in undergraduate teaching, has not made very much progress with the language, and is entirely devoted to her research interests.... There is no hope whatsoever of absorption at Bryn Mawr, but there appears to be a fair chance for absorption at the Princeton Institute, this being an ideal disposition of the case."

By April 1935, enough funds from various sources had been patched together to support Noether at three-quarters pay for two more years at Bryn Mawr. Still not a permanent job, but progress. There were also vague plans to have her move to the Institute for Advanced Study after that. No sooner was the patchwork plan approved than, on April 10, Noether went to the hospital for surgery.

Noether had been operated on for benign fibroid tumors once before in Germany. She considered herself a "tough guy," insisted that she was not worried about the operation, and was looking forward to being slimmer. She consulted her landlady's physician, who advised immediate surgery. Before she left for the hospital, she was persuaded to make a list of belongings to be distributed among friends—just in case. She wanted her books given to mathematical friends; money and furniture to Mrs. Hicks and her German landlady; and money to a mathematician's widow "during the next few years until her children earn enough so that there will be no further difficulties." She told a Princeton friend that "the last year and a half had been the very happiest in her whole life, for she was appreciated in Bryn Mawr and Princeton as she had never been appreciated in her own country."

Noether had high blood pressure, and the doctors did not consider her a good operative risk. Removing an ovarian cyst the size of a large cantaloupe, they left two small tumors to avoid prolonging the operation. At Noether's request, they removed her appendix.

For three days, she had a normal convalescence. Then, on Sunday, the fourth day, she suddenly lost consciousness and her temperature spiked to 109 degrees. Philadelphia consultants diagnosed a stroke, but a postoperative infection is more likely. There was a wide disparity in the quality of treatment available in American hospitals during the 1930s. Perhaps if she had been treated instead at the University of Pennsylvania, then one of the finest hospitals in the country, she would have survived. Instead, Emmy Noether died on April 14, 1935, at the height of her creative powers, exiled from her family, her homeland, and the school she had created.

The German mathematical journal *Mathematische Annalen*—which had never included Noether's name on its masthead although she had been one of its editors for years—courageously published her obituary. Einstein wrote a letter to the *New York Times*. The letter, published May 4, 1935, noted that "Fräulein Noether was the most significant creative mathematical genius thus far produced since the higher education of women began." (There is a story that Weyl wrote the letter but that a *Times* editor objected, "Who's this Weyl? Get this signed by Einstein and we'll run it.")

In 1937, two years after Emmy Noether's death, her brother Fritz was arrested in Siberia, taken to a slave labor camp near Moscow, and, a month before Germans captured the camp, executed as a spy. Weyl and Einstein had tried to secure his release but failed. They did succeed in bringing his sons, Gottfried and Herman, to the United States, where they attended college and graduate school. In 1989, Soviet President Mikhail Gorbachev declared that Fritz had been unjustly executed.

By the 1970s, interest in Emmy Noether had revived, thanks in large part to an Austrian biography by a mathematics teacher, Auguste Dick. Although Bryn Mawr had lost Noether's personal papers, it had stored her ashes. When the Association of Women in Mathematics had a symposium in 1982 on the one-hundredth anniversary of Noether's birth, her ashes were buried under a brick walk in the library cloisters. The following year, the collected works of Emmy Noether were published, edited by Nathan Jacobson of Yale University. In an introduction, the French mathematician Jean Dieudonné declared, "The publication of Emmy Noether's collected papers has long been overdue, since she was by far the best woman mathematician of all time, and one of the greatest mathematicians (male or female) of the twentieth century."

That same year, the city of Erlangen dedicated a new school and named it the Emmy Noether Gymnasium. At the dedication, her nephew Gottfried remembered how difficult it had been for his aunt to get an education. As he said of the new coeducational, college-preparatory school, "I think that Tante Emmy would have approved."

* * *

SECOND GENERATION

5

Gerty Radnitz Cori

August 15, 1896–October 26, 1957

BIOCHEMIST

Nobel Prize 1947

GERTY CORI'S HUSBAND was offered the job of his dreams at an American university—provided he stop working with his wife. When he refused, his would-be employers were appalled. Taking Gerty aside, they gravely informed her that she was ruining her husband's career. Sternly, they lectured the Czech-born woman, "It is un-American for a man to work with his wife."

In private later, Gerty burst into tears. Her husband Carl reassured her. Collaborating with one's wife in the 1920s was not un-American, he said. "It is merely unusual."

The University of Rochester passed up a good opportunity when it turned Gerty Cori away: she became the first American woman to win a Nobel Prize in science. Only Marie Curie and her daughter Irène Joliot-Curie had won science Nobels before her.

For thirty-five years, Gerty and Carl Cori formed such a close scientific partnership that it was hard to tell who had contributed what. The Coris conferred with each other constantly. When one started a sentence, the other finished it. Listening to them talk, friends had the impression that two voices were expressing their ideas from one brain.

Together, the Coris laid the foundation for our understanding of how cells use food and convert it to energy. The Cori cycle has become such a basic part of high school science that it is easy to forget how revolutionary it was during the 1920s. For the first time, it was possible to show how muscles use sugar for quick energy and how the muscles and liver store excess energy until it is needed.

Their other discoveries were even more important. At the time, little was known about enzymes, the proteins that enable cells to function, grow, and reproduce. Nevertheless, the Coris were able to identify and isolate the enzymes in charge of converting sugar from the form that muscles use for energy to another form, which the body uses for storing energy. They were pioneers in the study of both enzymes and hormones, and their work had major implications for the understanding of diabetes. Gerty Cori began the study of inherited disorders with her studies of diseases caused by enzyme deficiencies. Their work ultimately proved to scientists and physicians alike that an understanding of basic chemical processes is important to biology and medicine.

Their influence continues to this day: They trained a generation of leading biological scientists; their lab produced eight Nobel Prize winners, including Carl and Gerty Cori.

Gerty Cori rarely discussed her early life. She was born Gerty Theresa Radnitz to a moderately wealthy Jewish family on August 15, 1896, in Prague, Czechoslovakia, then part of the Austro-Hungarian empire. Her father, Otto Radnitz, was a chemist and businessman who managed several beet-sugar refineries. The eldest of three daughters, Gerty was privately tutored at home until she was ten years old. Then she was sent to a girls' finishing school to learn social graces and a smattering of culture, enough to converse pleasantly with a husband and his friends. It was Gerty's uncle, a pediatrics professor, who encouraged her to attend medical school. Officially, women were allowed to attend Carl Ferdinand University in Prague, but in practice, most could not. Girls' schools at the time did not teach Latin, mathematics, physics, or chemistry, all of which were required subjects for entry to the university. Before Gerty could enter medical school, she would have to learn eight years of Latin and five years of mathematics, physics, and chemistry.

Vacationing with her family in the Tyrol the summer of her sixteenth birthday, she met a high school teacher who offered to get her started in Latin. By the end of the summer, she had mastered three years of Latin. By the end of the following year, she was ready to take the test: "the hardest examination I was ever called upon to take," she said. European students who wanted to do biomedical research attended medical school. So when Gerty passed the entrance exam, she enrolled in the medical school at the German branch of the university in Prague. The university also had a Czech-language branch. The year was 1914, and she was eighteen years old.

Gerty met the two loves of her life—biochemistry and Carl Cori—during her first year in medical school. From the moment she caught a glimpse of biochemistry, she was fascinated. Science was one human endeavor that promised to help mankind; and biochemistry was an exciting new science that applied the principles of chemistry to biological problems. She met Carl Cori in anatomy class. Carl was tall

Gerty Cori.

Gerty and Carl Cori, 1947.

Nobel Prize winners at Washington University (seated, from left) Carl Cori; Joseph Erlanger; Gerty Cori; and Arthur H. Compton, chancellor of Washington University, 1952.

Gerty and Carl Cori in her lab, 1948.

and handsome, fair-haired and blue-eyed. Gerty was pretty with reddish-brown hair and brown eyes. While Carl was shy, she was vivacious, exceedingly quick, and brilliantly intelligent. In public, she seemed to overshadow Carl because he was slower at digesting material and she was more aggressive. Gerty's Jewish background did not bother Carl. Raised in the polyglot Adriatric port city of Trieste, where his father directed a marine biological station, Carl was immune to the anti-Semitism so prevalent in Eastern Europe.

Carl was smitten. He thought Gerty was "a young woman who had charm, vitality, intelligence, a sense of humor, and love of the outdoors." Gerty, who derived a deep sense of satisfaction from imposing order on a chaos of data, dreamed of becoming a research scientist. Together, they studied their way through medical school, spending their vacations mountain climbing and skiing, until Carl was drafted into the sanitation corps of the Austrian army during World War I.

When Carl returned from the war, he resumed medical school and courtship Cori-style. He and Gerty collaborated on a research project involving a component of blood and published their first joint paper. After their graduation from medical school, Gerty converted to Catholicism so that she and Carl could be married in a Roman Catholic church; the Cori family opposed their marriage because they thought that Gerty's Jewish ancestry would cripple Carl's career.

By the time of their marriage in 1920, Eastern Europe was in shambles. Starvation and semistarvation were rampant. The victorious allies had dismantled Austria-Hungary, and borders and populations were shifting back and forth. One night Carl and his friends dressed as laborers and secretly dismantled a research institute to move it from Czechoslovakia to Hungary, because its founder was Hungarian. In the newly established nation of Czechoslovakia, basic research was a low priority and physicians were needed more than researchers. Even in Vienna itself, research opportunities were slim. Thanks to a case of frogs sent by his father, Carl was the only doctor able to conduct any research at the university.

For most of 1921, Gerty and Carl had to live in different cities. Gerty had a job at the Karolinen Children's Hospital in Vienna, where she studied and published several papers on cretinism, now called congenital thyroid deficiency.

As pay for her work, she was given her dinners. The hospital's physicians had voted against accepting a free dietary supplement for the children, however, because it was provided by American relief organizations. As a result of the poor food, Gerty developed xerophthalmia, now known to be caused by vitamin A deficiency. Fortunately, a visit to her parents' home in Prague, where the diet was more nutritious, cured her.

At the University of Graz, Carl had to prove that he was not Jewish in order to get a job. Aware that anti-Semitism was on the upswing, Carl became convinced that Europe would soon be fighting another war.

The Coris sought comfort in long weekend walks together in the countryside and in visits to Vienna's magnificent art museums. "Art and science are the glories of the human mind," Gerty declared late in life. "I see no conflict between them. In the past, they have flourished together during the great and happy periods of history.... Contemplation of the great human achievements through the ages is helpful to me in moments of despair and doubt. Human meanness and folly then seem less important."

Eventually, the Coris decided that they had to leave Europe at any cost. Desperate, they even applied to the Dutch government to work for five years among the native population of Java. While they waited to hear about the job, the director of a cancer center in Buffalo, New York, interviewed Carl for a position. The New York State Institute for the Study of Malignant Diseases operated a hospital where cancer patients were treated with X rays and radium radiation. The institution later became a distinguished research center, now known as Roswell Park Memorial Institute. At the time, the director wanted a German-trained biochemist to run the laboratory because German chemistry was the best in the world. But American opinion after World War I was so vehemently anti-German that the director decided to compromise by hiring an Austrian.

Carl's professor recommended him highly, and Carl got the job for three thousand dollars a year. He sailed steerage for Buffalo in 1922, leaving Gerty behind at her Viennese job for six more months. Only after Carl had secured a position for her as an assistant pathologist at the institute did she join him. The Coris were twenty-five years old when they arrived in Buffalo. During their nine years there, they established their scientific reputations and became American citizens.

"The high state of development of biochemical methods in the United States came as a revelation," she recalled. Since their laboratory duties were minimal and supervision was slight, both Coris were free to pursue and publish their own research. During her first two years in Buffalo, Gerty studied the effects of X rays on the skin and on the metabolism of body organs.

Research in Buffalo involved a certain amount of skullduggery. When a new director of the institute put his name on their manuscripts without even reading or trying to understand them, the Coris secretly removed his name and submitted the articles to journals under their own. The new director believed that parasites caused cancer, and each month he assembled his staff to admonish them, "Gentlemen, it behooves us to find the cause or cure of cancer—and it's got to be intravenous." Gerty refused to humor him. She protested indignantly that she could find no parasites in the stools of the hospital's patients. Angrily, the director warned her that she would be fired if she did not stay in her own laboratory and quit working with Carl. She obeyed for a while. But soon, when no one was watching, she brought out her microscope and began studying her own research slides again. After waiting out the storm, the Coris resumed work together. They were determined to collaborate.

Both Gerty and Carl had become interested in how the body sends energy from one place to another. It is difficult today to realize how little was known at the time about the body's ability to maintain a constant supply of energy between meals and bouts of exercise. A French physiologist had discovered in the nineteenth century that both the liver and the muscles contain a starchlike substance that he called glycogen: the "sugar maker." But he did not know that each molecule of glycogen consists of hundreds of glucose sugar molecules chemically bonded together. When the body needs energy, it breaks apart the glycogen molecule to make the sugar molecules available for immediate energy.

Throughout the 1920s, the Coris carefully measured minute amounts of sugar, glycogen, and two controlling hormones in laboratory animals. The precision and accuracy of their measurements became the hallmark of their work. And it was Gerty who was "undoubtedly primarily responsible for the development of the quantitative analytical methodology," wrote Gerty Cori's biographer, Joseph Larner of the University of Virginia.

By 1929, after six years of intensive work, the Coris could explain in general terms how mammals get their energy for heavy muscular exercise. They spent the rest of their careers filling in the details. According to their theory, energy moves in a cycle from muscle to the liver and back again to muscle. When a runner starts to sprint, for example, glycogen in the muscles is converted to sugar, specifically glucose. The muscles extract most of the energy from the sugar, but leave some in the form of lactic acid. To conserve its resources, the body recycles the lactic acid back into glycogen in a series of elaborate steps. First, it is sent from the muscle to the liver. Next, the athlete pants heavily to supply the body with oxygen so that the liver can convert the lactic acid back into sugar. The sugar then returns to the muscle, where it is converted back into glycogen for storage. The Coris called their theory "the cycle of carbohydrates." Everyone else calls it "The Cori cycle."

The Cori cycle had a profound effect on the treatment of diabetes. Insulin had been discovered in 1921, but little was known about how the human body uses it or sugar. Thanks to the Cori cycle, physicians had some sense of how healthy bodies maintain a balance between exercise, food, and blood sugar. The Coris' popular fame rests on their cycle. Among biochemists, however, they are also revered for their later studies of the specific enzymes that operate the

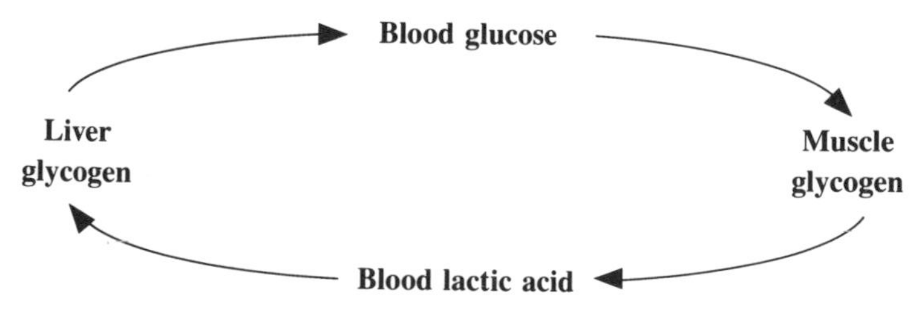

Figure 5.1. The Cori cycle.
The Coris explained how the energy of a heavily exercising mammal moves from the muscle to the liver and back to muscle.

cycle, the hormones that affect it, and the hereditary diseases that can result from the absence of or damage to the enzymes.

The Buffalo years were enormously prolific. Gerty and Carl Cori wrote fifty papers together there. Carl's name was listed first on some, Gerty's name was first on others, depending on who had done most of the research for that particular article. In addition, Gerty published eleven articles on her own, and Carl was the author of another thirty.

The Coris became a smooth-working team in Buffalo. As Carl discovered, working with an independent woman like Gerty was "a delicate operation which requires much give and take on both sides and occasionally leads to friction, if both are equal partners and not willing to yield on a given point." After the first year, they adjusted to each other. They became an affectionate team. Neither competed with the other, and each trusted the other's work. When a colleague said enthusiastically, "Wasn't that a wonderful seminar Gerty gave," Carl shrugged and replied, "Well, of course."

The talents of one complemented the skills of the other. "Carl was not the lab genius," observed William Daughaday of Washington University School of Medicine, who worked in the Cori lab. "Carl was the visionary. Gerty was the lab genius, omnivorous in her interests, gobbling up new issues. They discussed everything together. Gerty read enormously widely and deeply. She was his initial processor, and he got many of his ideas from her outreach." The role she played is a critical one in biology, where no fact exists in complete isolation and seemingly unrelated clues from other fields may be important. Gerty's reading also suggested new methods and new approaches to problems. As for Carl, Daughaday said, "he set the information into concepts."

As a team, the Coris were stronger than either would have been singly. Both were ambitious, but Gerty had motivation enough for both. She was tense, while Carl was more relaxed; without her he would not have worked quite as hard—or accomplished as much. With Carl's contemplative talents and Gerty's instinct for ferreting out information in the laboratory, they made one discovery after another.

At home, for entertainment, they read voraciously. In the evening, they enjoyed reading American literature aloud. Carl preferred archaeology, poetry, and art. Gerty read biographies and history. They both loved the out-of-doors. They swam in Lake Erie and toured the Adirondacks and the White Mountains. They enjoyed their Buffalo years, but, as time went on, it became difficult to justify the study of carbohydrate metabolism in an institute devoted to cancer research. Together, they decided they had to leave Buffalo.

Despite their teamwork, it was Carl—not Gerty—who began getting job offers from universities. Among those who tried to hire Carl were Cornell University and the University of Toronto, where Frederick Banting and Charles Best had discovered insulin in 1921. But Carl turned them all down because they refused to offer a position to Gerty too. She wanted recognition as well as a laboratory. The University of Rochester medical school offered Carl a job with three conditions. He had to take speech lessons, give up work on insulin, and stop collaborating with Gerty. Carl flatly refused, declaring that "the last two requirements are not acceptable." It was during their final visit to Rochester that Gerty was taken aside and told that it was un-American for a man to work with his wife.

Actually, husband-and-wife research teams were as American as apple pie. The academic countryside was littered with scientific couples studying botany, genetics, chemistry, and other sciences. Professor husbands and their low-ranking, low-paid wives often worked together for decades. Other men formed lifelong collaborations with unmarried women scientists who dedicated their lives to their protectors in return for the privilege of working. In either case, the women were generally low-level instructors, lecturers, or research assistants while their male partners were professors with tenure. A woman had a permanent position only as long as her personal relationship with the man continued. In case of divorce or disaffection, the woman could be fired.

Many universities simply refused outright to hire wives at all. Nepotism rules banned two family members from working for the same department or university. The rules—in some areas, they were state laws—were often enforced more strictly against women than against male relatives. Of course, the rules did not prevent wives from working as unpaid volunteers in their husband's labs. In effect, the regulations merely prevented women from gaining recognition and status from their work. Nepotism rules enjoyed widespread political support from both men and women in the United States. Thus, despite Carl's reassurances to a tearful Gerty, it was by no means clear that they could find jobs together.

In 1931, Carl Cori was offered a professorship by a university with relatively liberal policies regarding working wives. The Washington University School of Medicine in St. Louis, Missouri, was a private institution and could bend its nepotism rules. Washington University had already hired several women scientists when it offered Carl the chairmanship of its pharmacology department. Even Gerty could have a position. Regarding her as little more than a technician, the medical school offered her a lowly job as a research associate. Her token fifteen hundred dollars salary was twenty percent of Carl's pay.

The offer was far better than they were likely to get elsewhere. Gerty desperately wanted some kind of recognition and was anxious to accept the job, but the dean had second thoughts. He asked Carl to come to St. Louis to give a seminar so the faculty could judge his caliber.

The Coris arrived in St. Louis during a scorching heat wave. Carl stretched out on the hotel bed and tried to ignore the heat with an electric fan and a Jane Austen novel. Gerty was frantic; she wanted him to polish his all-important talk. Even without any last-minute rehearsal, however, his lecture was well received. Only one faculty member, an anatomist, still opposed Carl's appointment. When Carl visited the anatomy lab, he casually fingered one of the bones littering a desk. Not realizing that Carl was the son of a marine biologist, the professor cagily inquired whether Carl could identify the bone. "Yes, it's the inner ear of a whale," Carl answered without hesitation. "Having thus passed my anatomy examination, I heard of no further opposition."

At age thirty-five, Carl became a full professor and Gerty a "research associate." She remained one for thirteen years. Carl had tenure, meaning that he could be fired only for extremely serious reasons. Gerty's only tenure was her marriage to Carl.

Even by the standards of the 1930s, the pharmacology laboratory at Washington University School of Medicine was primitive. Equipment was practically nonexistent, workers shared what little there was, and there were no lab technicians or glass washers. Today, chemists simply order purified chemicals from supply houses. The Coris made their own biochemicals, doing precise quantitative syntheses themselves. Nothing was automated or air-conditioned, and they often designed their experiments to conserve both material and time.

To insure the quality and consistency of the chemicals they used, Gerty maintained tight control over how they were made and stored. Newcomers to the lab spent several months under her tutelage, relearning basic procedures and her own delicate techniques. That way, everyone in the laboratory knew exactly what the chemicals would do and what results meant. Once Gerty had the routine organized, she got started.

Smoking incessantly, scattering ashes over the lab tables, Gerty had an infectious exuberance. When she read or learned something exciting, she raced down the hall, clack-clacking in her pumps to tell Carl. With her insatiable curiosity, investigative zeal, and tremendous drive, she found research exhilarating. The highest compliment she could pay was, "So-and-so is a good worker!" Once, when some samples had been mislabeled, a colleague told Gerty that an experiment had to be redone. "A whole day wasted!" she moaned. Every day was important; a day without results was a day lost. At the time, most biologists and physicians considered biochemistry irrelevant to their work. The Coris, on the other hand, understood that precise measurements of the chemical processes occurring inside the human body would revolutionize both biology and medicine. Nothing short of perfection would do.

One afternoon, when an important breakthrough appeared in a German journal, graduate student Jane H. Park was dispatched to the library across town to bring back the article at once. She could have waited until the next morning, when she had a class near the library, but the Coris wanted to read the article immediately. It could not be checked out, of course, and photocopying machines had not been invented yet. So Park dutifully translated the article into English, copied it out by hand, and brought it back to the Coris. But no, they needed it in the original language in order to extract every nuance and morsel of information. So Park was sent back across town to recopy the article *in German* and was told to come *right back* and not to *dilly-dally.* As Park emphasized years later when she was a professor at Vanderbilt University, "That happened every day in every way in every experiment and topic that was discussed: Nothing less than perfection, and every experiment was a burning and exciting event." Carl Cori had the same sort of commitment, but it was expected in a man. In a woman, some thought it was unseemly.

Before their arrival in St. Louis, the Coris had studied carbohydrate metabolism in laboratory animals. Dealing with such complex organisms had introduced many variables and uncertainties into their studies. As Carl noted, "One could only make guesses as to what was going on." To understand what exactly was happening, the Coris decided to study the muscle tissue itself. Soon, they were seeking even simpler systems, with even fewer variables and complications. As Gerty explained, "In the intact cell there are taking place so many simultaneous reactions that it is impossible to trace the individual steps." Using scissors, the Coris gently minced the frog muscle, soaked it in cold distilled water, and extracted the soluble components. Then they studied the reactions in the solution, free from any cells or living organisms.

During one of the hottest summers on record in St. Louis, Gerty was forty years old, pregnant, and working on a major breakthrough. The lab had no air-conditioning, and its temperature hovered around 100 degrees Fahrenheit all through August 1936. But Gerty and Carl hoped to explain how the body could break down glycogen into sugar on its own, without using a lot of extra energy to do so. Analyzing the extract from the minced frog muscle, they discovered a totally new glucose compound, glucose-1-phosphate, also called the Cori ester. Eventually, they discovered that glycogen breaks down into sugar in three steps, each one of which requires only a little energy. The Cori ester was one of those steps.

Gerty worked on the ester until the last moment before she went to the maternity hospital. And three days after Tom Carl Cori was born, she returned to work. Somewhat to the Coris' surprise, Tom did not grow up to be a scholarly and cultured European. Instead, he turned out to be a typical American boy with a passion for baseball. Carl and Gerty applied tremendous pressure on him to become a research scientist, but he refused. Their relationship with him deteriorated during his adolescence. He earned a Ph.D. in chemistry and later became the president of Sigma-Aldrich, a large and prestigious specialty chemical company that supplies laboratory chemicals like those his parents had to make themselves.

Gerty and Carl worked all day during the week and half days on Saturdays, which were considered normal work days. But they spent their evenings and the rest of their weekends at home, where they tried to avoid shoptalk. They went skating and swimming or played tennis and attended concerts. They entertained almost every weekend in a modern house on a wooded lot in suburban Glendale eight miles from the lab. They gave breakfast lawn parties where everyone helped in the garden. They mixed artists, musicians, novelists, and business friends with their science colleagues. Young unmarried scientists, male and female, were always included. And every summer the Coris went mountain climbing in Colorado or the Alps or they visited Europe, especially Italy, where one of Gerty's sisters was a painter.

In 1938 and 1939 Gerty changed the direction of their research: toward enzymology. Today enzymes are known as large protein molecules with knobs and claws that grab, hold, stretch, and bend the molecules they act on. At the time, however, few had been identified. Little was known about their behavior either, except that they direct and control most of the chemical reactions that characterize life. But if enzymes were triggering the conversion of glycogen to sugar, Gerty wanted to know about them. According to her biographer, Joseph Larner, she was the decisive influence in the Coris' seminal move into enzymology. Of the ten papers published by the Coris during 1938 and 1939, she was the major contributor to seven, Carl was the major author of two, and another colleague wrote one.

Soon after, the Coris discovered phosphorylase, the enzyme that breaks glycogen down into the Cori ester. Phosphorylase tears apart the bonds that hold glycogen's hundreds of sugar molecules together. The Coris' discovery of phosphorylase marked the first time that carbohydrate metabolism was studied at the molecular level. In 1942, an expert protein chemist, Arda Green, crystallized phosphorylase in the Coris' laboratory so they had ample to study.

In 1939, the Coris thrilled the biology world when they made glycogen in a test tube. During a lecture at an international conference in Toronto, Carl Cori picked up a test tube, mixed phosphorylase with the Cori ester and some other compounds, incubated the mixture at room temperature for ten minutes, and then tested it to prove that it had turned to a starchlike material. Then he passed the test tube around the audience for everyone to see. Forty years later, at an anniversary conference held in the same Toronto hotel, speakers were still talking about Carl Cori's drama. Physiologists had been told for years that large molecules could only be made within living cells. Yet the Coris had executed the first bioengineering of a large biological molecule in a test tube. As Carl Cori noted with understatement, "A most exciting period of biochemistry began to unfold.... Nothing comparable to this period happened until [the 1960s] when it became possible to explore the genetic apparatus of the cell."

Twentieth-century medical science was "a succession of hunters," as the Nobel Prize–winning enzymologist Arthur Kornberg observed. Early in the twentieth century, the hunters searched for microbes and vitamins. During the middle of the century, enzyme hunters laid the foundation for today's studies of genes and nerve cells. The Coris were leaders in the enzyme hunt.

Gerty and Carl discovered one enzyme after another, revealing the conversion of glycogen to sugar and back as a multi-step process involving a variety of complicated enzymes. By now, they were studying purified enzymes, biochemicals totally isolated from living cells and animals. They were exploring the relationship between the structure and the function of molecules and how enzymes turn chemical reactions on and off. Their work had wide implications for the treatment of diabetes and other diseases. They were among a few, select pioneers who first proved that biochemistry can explain many biological and medical phenomena. Yet, at the same time, they were pushing biochemistry closer and closer into modern molecular biology.

Gerty may have been a world leader in science, but she was still a lowly research associate at Washington University. Unfortunately, it took a world war to improve her status and that of other women scientists in America's labs. While Carl worked on defense projects, she kept their research going, but manpower shortages made it difficult to staff the lab. For the first time, women scientists were in demand. To keep Arda Green, for example, the university was forced to give her a full-time faculty position. Green's promotion may have forced the university to give Gerty Cori equal treatment, too. At any rate, in 1944, Gerty was promoted to associate professor and given tenure at Washington University.

As the war was ending, Harvard University and the Rockefeller Institute in New York City offered Carl *and* Gerty professorships. At one point, a Harvard official who did not know that the Rockefeller Institute was trying to lure the Coris away, confidently informed Rockefeller that the Coris were moving to Massachusetts. The Rockefeller proposal was especially tempting. In response, Washington University apparently made Carl Cori a deal: he could chair a new and enlarged biochemistry department in larger quarters, and Gerty Cori could be a professor. They accepted.

By 1947, the Cori lab was the world's liveliest center for the study of enzymes. It was more a research center or institute of scholars than a school. The Coris accepted few graduate students. Instead, advanced researchers flocked from around the world to work for a year or more beside the Coris. Eight alumni of the Cori lab won Nobel Prizes: besides the Coris, the list includes Arthur Kornberg, Earl W. Sutherland, and Edwin G. Krebs of the United States, Severo Ochoa of Spain, Christian R. de Duve of Belgium, and Luis F. Leloir of Argentina.

"This was classic biochemistry, when enzymes were isolated and their properties studied—enzymes that by now are so old-hat to biochemistry students that they probably forget that they ever had to be discovered. But a lot of them were discovered and crystallized in Gerty Cori's lab," recalled a Cori lab veteran, David Brown, of Washington University School of Medicine.

The lab made so many discoveries so quickly during the late 1940s and early 1950s that Carl worried a bit. "It makes me a little uneasy, but you have to accept them," he conceded. Gerty herself worked like an artist who brushes in the broad outlines of a painting before focusing on the details; she liked to make an exciting discovery first and only later go back to do precise controlled experiments. Sometimes she became so excited, she jumped up and down.

By then, Gerty was running the laboratories. Carl had quit working in the lab and was busy writing and supervising junior researchers. Gerty, on the other hand, spent every working day there—except for a few days when the university finally convinced her that, after twenty-five years, the rooms needed new paint. When she was not working at the lab bench, she was reading articles. Somehow, she had convinced Washington University's librarians to send her journals as they arrived, before they were put on the shelves. During lulls in experiments, she raced through their articles and rushed the journals on to the library.

Much of the life of the department was conducted—or at least overheard—in the two-hundred-foot-long hallway between Gerty's laboratory and Carl's office. The two conferred incessantly, puffing on cigarettes, pacing through the smoke. If she had really hot news, she started calling him while she was still several doors away from his office, "Carlie! Carlie!" When a colleague made a particularly exciting find, she ran down the hall to Carl shouting, "It's free glucose! It's free glucose!" The department had only two telephone lines. One was Carl's private line. The other was out in the hall itself, where everyone could overhear Gerty ordering groceries, planning menus with the housekeeper, or discussing after-school activities with her son.

Gerty also arranged for the Mercantile Library, a private lending library in St. Louis, to deliver between five and seven books each week to her office. By Friday, she had finished the books and prepared her order for the next week's reading. She did this week after week. Gerty could talk knowledgeably about anything from political theory and sociology to art, literature, and grocery shopping. Colleagues were amazed at the breadth of her knowledge and the depth of her understanding. As Gerty said, "I believe that the benefits of two civilizations, a European education followed by the freedom and opportunities of this country, have been essential to whatever contributions I have been able to make in science."

Both Coris read deeply as well as widely outside science. On their way to Stockholm to pick up their Nobel Prize, they stopped off in Buffalo and talked with a young sociologist who had written a scholarly article on the theory of revolutions. Gerty still screened the scientific literature for the team, too. In 1947, Gerty Cori read Oswald T. Avery's paper showing that DNA was the chemical basis of heredity. She went clicking down the hall to tell Arthur Kornberg, "You must read this. It is *very* important." She recognized the significance of DNA five years before the famous molecular biology group at the California Institute of Technology, according to Kornberg.

Brown-bag lunches with the Coris were legendary, not for the cuisine, but for the conversation. Lunch was a one-hour tutorial every day. Talk ranged from research reports by eminent visitors to the latest book the Coris had read. Carl spoke five languages, wrote poetry, played the cello, and could taste a bottle of wine and name its vineyard. Gerty Cori was lively and vivacious, fun, and humorous. "I really got my education there," recalled Neil Madsen of the University of Alberta.

Gerty Cori may have been good-humored, but she did not like to be scooped. When the Coris' friend Luis Leloir discovered a new enzyme in Argentina, Gerty marched into the office of the young man who had been working on a similar project. He had been mildly distracted by his marriage and honeymoon during the month when Leloir made his discovery, and Gerty disapproved. "You missed it! You missed it!" she complained. Another day Kornberg was sitting in the lab when Gerty burst in waving a journal. "We've been attacked," she cried. Sheepishly, Kornberg looked over the article, trying to find the offending section; the point at issue was trivial.

"Her behavior was not abnormal, although it would be regarded so for somebody who was not supposed to have that kind of personality, like a homemaker or a companion," Kornberg stressed. Furthermore, as tough as Gerty Cori was, Carl Cori was tougher. "In so many ways, she was very understanding, humane, and generous. It seems to me that that kind of dedication and determination to do something creative and important is part of the job. She'd have achieved less and been less recognizable as a creative scientist had she done something different."

"Some people didn't get along with her," agreed her friend Mildred Cohn, now professor emeritus of the University of Pennsylvania and a member of the National Academy of Sciences. "That's an understatement. One of the reasons was that she wasn't all sweetness and light. Though she was a kind and caring person, she had a very sharp tongue and was a very intelligent woman, and she didn't suffer fools gladly.... But when she was critical, it was on an intellectual level. She had very high standards, and it always amazed me that she was so emotionally involved."

Gerty Cori had always been a passionate woman, kind and gentle as well as intense and volatile. But when a young researcher forgot to turn off her water bath and it boiled dry, she raged, "It's ruined, ruined, absolutely ruined! Haven't you any responsibility?" Edwin Krebs, now professor at the University of Washington in Seattle, thought his career was ruined as well, but the eminent biologist Viktor Hamburger (chapter 9) was standing nearby, and he was clearly amused. Afterward, Hamburger consoled Krebs. "You mark my words, she'll think more about this, and she'll realize that she was unduly harsh. But she won't apologize," Hamburger warned. "Instead, when she comes in in the morning, she'll stick her head in the door and say 'How's the work going?' And you should accept that as an apology."

Krebs protested, "But she's never stuck her head in the door before on her way in." The next morning, though, Krebs heard her heels click-clacking down the hallway. Then she poked her head through the door, and she asked, "How's work going?"

Years later, Jane Park had a similar experience. Park had stored some vials in Gerty Cori's refrigerator. When she removed them, Gerty came storming down the hall, convinced that some of her own solutions had been taken. Park told her, "Mrs. Cori, I can't talk to you about this right now. My mother is in the hospital, and she's dying. This isn't that important, and we have to leave it at that." So Gerty clacked on back up the hall in her heels, and the next day, after Park's mother had died, Gerty poked her head in Park's lab to say, "I'm very sorry to hear about your mother." Park replied, "She was quite ill. She had emphysema." Gerty answered kindly, "Yes, I know what that means." That was the way Gerty said she was sorry. But the point, Park emphasized, is that Gerty had to know how the solutions had been treated or else she could not trust the results of anyone's experiments. The integrity of the lab's data was at stake.

Gerty refused to compromise her scientific standards. "Gerty Cori could cut you down to shreds if your mind wasn't working right, and her standards were pretty high. But I thought it was wonderful," noted Luis Glaser, now provost of the University of Miami. As a graduate student, Glaser once spoke at the Coris' Friday afternoon journal club, when recent articles were summarized and discussed. The tradition was that no one left the room before Gerty Cori. Glaser was so nervous that he raced through his fifty-minute talk in a half hour. Arthur Kornberg tried to keep the comments going but gave up after ten minutes. With ten minutes remaining, Gerty turned to her neighbor and started to talk in order to teach Glaser a lesson: Professionals know how to fill a lecture hour. "Everybody else sat there, sort of staring into space," Glaser remembered.

Another day, Glaser was scheduled to give the lunch seminar. As he walked to the seminar room with Gerty, she asked what he planned to talk about. When he told her, Gerty replied flatly, "That bores me."

"I went on and gave the talk. What else could I do?" Glaser laughed. "But she was right. In that sense, she could be quite ruthless. She didn't ever hold back on telling you. But she told everybody, whether it was a Nobel Prize winner or a graduate student. I thought it was wonderful, and I still do. If she was just beating on a poor, meek graduate student, that would have been unconscionable, but to beat on everybody was wonderful."

Given Gerty's insistence on the purity of her experimental data, one can only imagine her reaction to one of the laboratory's insulin articles during the mid-1940s. The experimental evidence cited in the article proved to be false—some thought fraudulent. And Gerty spent almost a year holed up in her lab with Carl trying to straighten out the fiasco—despite the fact that she was not a coauthor. Carl had been, even though he had not participated in the experiments. Those had been done by others.

The Coris' insistence on intellectual integrity made their lab an island of toleration in the midst of anti-Semitism and sexism. They wanted discoveries that were accurate, but they did not care who made them. St. Louis, like most American universities during the 1950s, was anti-Semitic. The community was only beginning to lift its bars against Jews. Even as the Coris were hiring both women and Jews, another Washington University department rejected a job candidate because his wife was Jewish.

Gerty Cori was particularly sympathetic to women researchers. When Mildred Cohn joined the lab in 1946, Gerty made it clear that she approved of working mothers. Her first words were, "I understand you're more fortunate than I. You have both a daughter and a son, and I have only a son." Said Cohn, "She endeared herself to me the moment she said it."

When Barbara Illingworth Brown's babysitter quit, Gerty sent her own housekeeper over to pinch-hit. "She was a very passionate woman who could blow up, but she was very caring and kind," Cohn decided. Gerty found proper medical care for a sick young secretary from New Zealand, slipped money to young people in need, and helped visitors find housing. When she learned that Cohn was not a member of the prestigious American Society of Biological Chemists, she nominated Cohn and got her in that year. And when Cohn was invited to give her first talk at another university, Gerty was as excited as Cohn's mother. "She was very aware of women's problems," Cohn concluded. Gerty once made a slighting remark about another woman to Cohn and came in the next day conscience-stricken. "Forget what I said yesterday," she begged Cohn. "It really wasn't fair."

As far as Gerty Cori was concerned, women competed with men in science by being equally serious, knowledgeable, and committed. She wore severe business suits to work and looked askance at women who dressed fashionably. The women in the lab, whose clothes were as dowdy as Gerty's, staged a contest to choose the worst-dressed woman among them. Gerty Cori won easily. "She was not a vain woman," Park decided.

For the times, Washington University was good for women. When its nepotism rules were recodified during the 1950s, husbands and wives could work at the university if they were in different departments. Many universities at the time banned wives from employment anywhere in the institution. When the new rules were published, the chancellor wrote Gerty a personal letter saying that the regulations were not aimed at her. Carl Cori's attitude about hiring had always been that rules were meant to be broken.

Gerty Cori had had firsthand experience with discrimination. When she first came to Washington University, the administration had regarded her as Carl's assistant. Carl Cori, on the other hand, was a towering figure in biochemistry. Every year he got a sizable grant from Eli Lilly & Company, simply by writing a one-sentence letter: "I would like to continue my studies in carbohydrate metabolism." He won award after award, most of them alone. For example, he was elected to the National Academy of Science and the Royal Society of London without Gerty; he received the prestigious Albert Lasker Basic Medical Research Award in 1946 without her, too. Even after Gerty won the Nobel Prize, organizations like the American Chemical Society continued to honor Carl but not Gerty.

In 1944, Washington University's dean asked Gerty to prepare a report "on Cori's work" to accompany a grant application to the Rockefeller Foundation. Gerty's response referred to "the Coris" and how "*they* discovered... *they* isolated," and so on. To his credit, the dean passed the report on unedited. At the Rockefeller Foundation, however, someone crossed out "the Coris" and "they" and penciled in "Dr. Cori" and "he." If Gerty had known, she would have been furious.

In 1946, Carl became chair of the new biochemistry department. He promoted Gerty to full professor as of July 1, 1947. After sixteen years, she finally had tenure, status, and pay more commensurate with her abilities.

The year 1947 brought Gerty both the worst of news and the best of news. That summer, the Coris left for their annual month's vacation in the Rocky Mountains. While they were hiking on Snowmass Mountain, Gerty fainted. Snowmass is a fourteen-thousand-foot-high peak near Aspen, Colorado. As physicians, the Coris immediately guessed that she must have a problem with the hemoglobin in her blood and that it had been exacerbated by the high altitude. Back in St. Louis, Gerty visited a physician at the medical school, but he was unable to explain the incident.

A few weeks later, on October 24, the Coris learned that they had both won the Nobel Prize. They received the prize for discovering the enzymes that convert glycogen into sugar and back again to glycogen. "Your synthesis of glycogen in the test tube is beyond doubt one of the most brilliant achievements in modern biochemistry," the Nobel Committee declared. The Coris shared their award with their friend

Bernardo A. Houssay of Argentina, who had shown that the pituitary gland plays a key hormonal role in sugar metabolism.

The morning of the announcement, reporters hovered around the department in disbelief as the Coris went about their business as usual. They had planned to attend a laboratory lecture from eight to nine A.M., and nothing could persuade them to skip it. Gerty was a realist about the prize. As she noted, "Salaries and prestige [for basic research] are not very high in our society. Of course, where the excellence of the work is certified by some medal, prestige follows, but it is then not so much attached to admiration for 'useless' work but to the medal."

A few weeks before the Coris were to leave for Stockholm in triumph, Gerty's physician told them his devastating diagnosis. Gerty was suffering from an unknown, fatal type of anemia. Her body was not producing red blood cells, and her bone marrow was being replaced gradually by fibrous tissue. Why, no one could say. The official diagnosis was agnogenic myeloid dysplasia, meaning only that the bone marrow was malfunctioning for unknown reasons. She would be increasingly dependent on blood transfusions for the rest of her life. Just when the Nobel Prize signaled the end of Gerty Cori's long struggle for recognition, she began her long struggle to survive and keep working.

Today, Larner thinks that her illness may have been triggered by the X rays that she studied in Buffalo during the 1920s. Excessive amounts of radiation are known to cause fibrosis of the bone marrow.

Despite the tragic news, Gerty and Carl continued with their plans to attend the Nobel ceremonies in Stockholm as if nothing had happened. They stopped off in Buffalo and visited parts of Europe on the way. As with all their work, they split up their Nobel Lecture. Carl gave the first and third parts, and Gerty presented the second part. Only Carl spoke at the banquet. Looking at Gerty, he thanked the Nobel Committee for including her in the prize. "That the award should have included my wife as well has been a source of deep satisfaction to me. Our collaboration began thirty years ago when we were still medical students at the University of Prague and has continued ever since. Our efforts have been largely complementary, and one without the other would not have gone as far as in combination."

Back home, they shared part of their $24,460.50 prize with several coworkers on the phosphorylase project, including Sidney Colowick, Arda Green, and Gerhard Schmidt. Colowick had published articles with the Coris and, when his name appeared between theirs, quipped that he was "the meat in the Cori sandwich." Green, who had crystallized phosphorylase for the Coris, used her money to buy a Chinese rug. Schmidt bought a car and called it his Nobel Prize.

For the next ten years Gerty Cori survived on blood transfusions, Carl's care, and work. Carl monitored her hemoglobin incessantly and often administered her transfusions. Gerty kept a small, metal-framed army cot with a thin mattress in her office to rest on, although she usually spent the time reading scientific journals. She and Carl traveled all over the world trying different treatments, but each one had damaging side effects. Transfusions during the 1950s were made of whole blood, and over time her body produced more and more antibodies to the cellular elements in the blood. After each transfusion, she felt sick and feverish. Carl devised a treatment that helped for about a year. Later, when her spleen was removed in an experimental procedure, her liver took over and grew so big she looked almost pregnant. But the operation kept her alive and working for several more years.

Publicly, she ignored her illness. Only once in ten years did she say anything downbeat to her friend Cohn. "You know, Mildred," she said then, "if something like this happens to you, it would be better if a ton of bricks fell on you." Later, with gallows humor, she joked that she had gone to a party "to squelch the rumors that I'm dead."

Above all, she kept working. After she received the Nobel Prize, she was elected to the National Academy of Science, eight years after Carl. President Harry S Truman appointed her to the newly formed National Science Foundation board. She served two terms, getting massive blood transfusions before each long, turboprop flight to Washington, D.C. She attended meetings when other patients would have stayed in bed. The sessions were important, though, because the federal government had decided to support scientific research, and policy guidelines were being established.

Gerty Cori was instrumental in defeating a proposal made by the president of Harvard University, James B. Conant. Conant did not want the NSF to fund researchers with M.D. degrees for experiments involving animals. He wanted only Ph.D.-trained researchers to do animal research. Gerty argued back, citing M.D.-trained scientists like herself who had won Nobel Prizes. Eventually, Conant's proposal died for lack of support. At another NSF board meeting, she cornered its executive director; she was furious that one of her postdoctoral fellows had failed to win an NSF fellowship. "We seldom make mistakes on choosing our people," the director protested. "Well, you made a mistake this time," she argued. The young man got his fellowship.

With iron determination, Gerty Cori did some of her most

important work during her illness. She started with a bet with Joseph Larner. Larner had come up with an idea about a group of children's diseases called glycogen storage diseases. At the time, they were thought to be one disease because the children accumulate abnormal amounts of glycogen in their tissues, especially in the liver. Larner suggested that the condition might be caused by the lack of a particular enzyme. Gerty bet that another enzyme was missing. It turned out they were both right. Larner's missing enzyme caused one form of the disease while Gerty's caused another. Gerty was thrilled. She jumped up shouting, "This is a molecular disease!" The only other disease known to be caused by an altered molecule was sickle cell anemia, which Linus Pauling had just discovered.

Gerty devoted her remaining years to sorting out the glycogen storage diseases. Eventually, she proved that four different diseases are caused by a defect in or absence of a particular enzyme. Today, ten or more glycogen storage diseases are known, and some of them are still fatal. Her work opened the entire field of genetic diseases for study. As Glaser observed, "Now that inherited diseases are a great growth industry, it is hard to imagine how revolutionary her accomplishment was—that someone could take a piece of liver in a biopsy from a patient and determine what was wrong with them and why they had particular symptoms. It was a major finding, really a milestone in biochemistry and our understanding of a whole category of diseases."

It was also a spectacular piece of molecular biology at a time when the field had hardly begun. It remains "an unmatched scientific achievement," declared their colleague Herman Kalckar. Gerty's work had come full circle, back to her early interest in children's medicine in Vienna.

As Gerty weakened, friends from the old days sensed her anger and bitterness. Her old cheerfulness had gone, and she was more volatile and more inclined to explode with frustration. When Carl hired nurses to look after her, she fired them. "I think she wanted him to take care of her," Cohn said. "They were so close.... It was the most amazing closeness." On a visit to London for a scientific congress in 1955, Gerty almost collapsed while sightseeing. In Oxford, she fell on the street and was carried to a friend's home to recuperate.

The summer of 1957, Gerty Cori published her last article, a review of children's glycogen storage diseases. By fall, Carl was carrying her from room to room at the lab. The month before she died, she was bedridden at home. Carl cared for the household, looked after her, and went to work late. Gerty, the woman who had gobbled up five weighty books a week for decades, was reading mysteries. "I can't read anything serious anymore," she complained. On October 26, 1957, Gerty Cori died at age sixty-one, at home, alone with Carl.

Scientists from around the world arrived for her memorial service. Houssay flew up from Argentina, an English friend came from London, and Ochoa arrived from Spain. In his eulogy, Ochoa remembers Gerty as "a human being of great spiritual depth…modest, kind, generous and affectionate to a superlative degree and a lover of nature and art." Just before a string quartet played Beethoven, a record was broadcast. Gerty had made it for Edward R. Murrow's *This I Believe* series. Through the church came Gerty's voice in a ringing affirmation of her love for science, truth, and humanity. She said:

> Honesty, which stands mostly for intellectual integrity, courage, and kindness are still the virtues I admire, though with advancing years the emphasis has been slightly shifted and kindness now seems more important to me than in my youth. The love for and dedication to one's work seems to me to be the basis for happiness. For a research worker, the unforgotten moments of his life are those rare ones, which come after years of plodding work, when the veil over nature's secret seems suddenly to lift and when what was dark and chaotic appears in a clear and beautiful light and pattern.

EPILOGUE

Carl Cori retired from Washington University School of Medicine in 1966 and took a position with Harvard University and the Massachusetts General Hospital. His friends were happy when he remarried. Although his scientific achievements after Gerty's death were never as great as they had been during the years of their partnership, he continued research until shortly before his death at age eighty-eight in 1984. During his last illness he told a visitor, "You know, Gerty was heroic."

* * *

6

Irène Joliot-Curie

September 12, 1897–March 17, 1956

RADIOCHEMIST

Nobel Prize in Chemistry 1935

THE BELGIAN ARMY SURGEON felt his way through the mangled flesh of a young soldier's leg. Searching in vain for shrapnel, the doctor met the cool and steady gaze of an eighteen-year-old girl, Irène Curie. Pointing to an X ray of the leg, she calmly observed that, according to the logic of three-dimensional geometry, the doctor should enter the patient from another angle. When the surgeon finally took her advice, he immediately located the shrapnel.

Irène Curie had spent weeks working alone in an Anglo-Belgian field hospital a few kilometers from the front during World War I. She was in charge of teaching the hospital staff how to X ray the wounded for shrapnel and fractures. One of her hardest jobs was convincing military surgeons that X rays and geometry could save patients.

In the midst of carnage, a teenage girl was teaching veteran military officers what they should have learned in peacetime but had rejected as too high-tech for battlefields. The Belgian surgeon was one of her worst pupils. He was ignorant of even the most elementary geometry. When she visited another battlefront hospital for a few days to repair a broken X-ray unit, she told her mother that he would probably destroy her equipment while she was gone.

Irène's mother, physicist Marie Curie, had not hesitated to leave her teenage daughter alone at the front. "My mother had no more doubts about me than she doubted herself," Irène recalled later. And Marie Curie was right.

A short time later, Irène traveled by train to Amiens to install another X-ray unit. When she arrived, military authorities announced that her equipment could not be unpacked for at least fifteen days; the city had been bombarded by aeroplanes and was in chaos. Coolly requisitioning a sergeant and a medical student, she unpacked the railroad car and, in less than an hour, installed the unit. For protection from the X rays, she donned cotton gloves and stepped behind a wooden screen. Then she began X-raying the wounded. When necessary, doctors operated directly under her X-ray beam. As Irène put it, she "surmounted the little difficulties of the moment." She celebrated her eighteenth birthday alone, within earshot of artillery fire.

When not at the front, Irène taught 150 women X-ray technicians how to adapt their equipment to a wide variety of French electrical systems and how to calculate mathematically the location of each wound. In her free time, she studied for her examinations in mathematics and physics and began her doctoral thesis.

Irène's wartime work gave her massive doses of radiation and contributed to her early death from leukemia. It left her with a lifelong hatred of war. It also cemented an already close relationship with her mother. A series of traumatic crises during Irène's childhood had turned her into her mother's closest friend; the war made them professional colleagues as well. The extraordinary calm and self-possession that enabled her to handle the military during World War I stood her in good stead later in life. As an adult, she battled a life-threatening case of tuberculosis for twenty years—at the same time that she was a mother, a research chemist, and an influential public leader.

Irène Curie was born a month prematurely on September 12, 1897, after her parents, the Nobel Prize–winning scientists Marie and Pierre Curie, had taken a long bicycle ride. Irène was a difficult baby; Marie called her "my little queen" and "my wild one." She saw to the baby's care, recording Irène's progress as carefully as her own experiments. She wrote that at eleven months, "Irène said 'Thank you' with her hand; she's walking well on all fours; she says, 'gogli, gogli, go.'" Vacationing in the country at twelve months, "Irène cut her seventh tooth, bottom left.... For three days now, we have been taking her to bathe in the river." Their lives settled into a quiet routine. "In the evenings we are busy with the baby.... During the whole year, we have not been out to the theatre, to a concert, or even to pay a visit."

A few weeks after Irène's birth, Pierre's mother died and his father, Eugène, moved in to help care for the baby while Marie was at work. She discovered polonium and radium while Irène was still an infant. Her famous backbreaking work isolating radium salts in a ramshackle shed was conducted when Irène was still a preschooler. As a result of Marie's long hours at work, Irène's grandfather became her second father and mother. He raised her as he had raised Pierre

Irène Joliot-Curie.

Irène Joliot-Curie in a military hospital at Amiens, 1916.

Irène Joliot-Curie as government minister, 1936.

Frédéric and Irène Joliot-Curie, 1955, a few months before her death.

Curie: to love nature, poetry, and radical politics. He and Marie Curie both disapproved of tight social and academic restraints on children. Neither corrected Irène when she failed to greet visitors. Instead, they encouraged her to play outside with her friends. Athletic and tall for her age, she was happiest running out-of-doors.

As Irène explained years later, "My spirit had been formed in great part by my grandfather Eugène, and my reactions to political or religious questions came from him more than from my mother." Politically, he was an anticlerical republican opposed to the power of the conservative French aristocracy and the Catholic church. He was more anticlerical than Marie Curie, who told her daughter, "I raise you without religion but you will see later if you want to take one when you are older." Irène adopted her grandfather's intransigent attitude about religion. "I think like him," she explained later. As long as she lived, she never once entered a church, not even to see a work of art. "Although I respect very much sincere belief, I could not truly feel myself *en famille* with one of my children having different ideas from mine on such a profound question," she wrote.

Like her father, Pierre Curie, Irène talked little and thought slowly and deeply. Puzzling over dinosaurs and mammoths, she decided to ask her grandfather about them: "He's old. He will have seen one." Scrutinizing a Rembrandt etching of a worn old hag, she burst into tears, "Oh, the poor old woman!"

She was a childish mix of timidity and boldness. When Marie brought her students home for tea, Irène hid in her mother's skirts and announced defiantly, from time to time, "You must take notice of me." At the beach, she cautiously told a friend of her mother's, "I do not know you very well."

When her parents won the Nobel Prize in 1903 and journalists descended on the Curie house, they tangled with Irène. Finding no one else at home, a reporter interviewed six-year-old Irène and her black-and-white cat. "Where are your parents?" he inquired. "At the laboratory," she replied soberly. Thus began Irène's long and distrustful relationship with the gentlemen of the press.

When Irène was seven years old, her sister, Ève Denise Curie, was born. Sixteen months later, their father was killed, run over by a horse-drawn wagon. Irène was playing next door when Marie entered the room and quietly told Irène that her father was dead. The child listened a moment and continued playing.

"She's too small; she doesn't understand," Marie Curie murmured. At that, the child burst into tears and raced into her mother's arms. Years passed before Marie Curie could even mention Pierre to her children.

As a single parent and the sole support of two daughters, Marie Curie made sure that the girls learned to cook, knit, and sew. She taught them unconventional skills, too: mathematics and how to swim, row, bicycle, skate, ski, ride horseback, climb mountains, and practice gymnastics. Fresh air helped control her incipient tuberculosis and her radiation sickness, and she was obsessed with her children's health. She also hired Polish governesses to teach the girls her native language.

The two sisters had diametrically opposite talents, interests, and personalities. At home, Irène was shy and socially awkward. Marie realized that Irène "resembled her father in her intelligence. She was not so quick as her sister, but one could already see that she had a gift of reasoning power and that she would like science." At ten, Irène was ash blond, gray-eyed, and tall, and when she smiled—which was not often—she was quite pretty. She idolized her mother. Her French biographer, Noelle Loriot, concluded that a modern family would send such a solemn child to a psychiatrist.

Ève, on the other hand, was charming and flirtatious. A gifted piano player and writer, she was determined to resist the lure of science. Yet Marie raised both daughters to find useful work. The Curies were not a kissing family, but Marie Curie was tender and affectionate with her children, and the three of them remained close all their lives.

Long after Irène's death, her children found a packet of letters written by Irène and Marie Curie to each other over a thirty-year period. Reading them, it is possible to trace their developing relationship. Practical women, they were interested in substance, information, facts, and details. Their letters are organized and to the point; neither rambles or gossips. Yet they are also tender, affectionate, and respectful of one another. And over the years, their mother-daughter relationship blossomed into deep friendship.

Irène wrote her mother while vacationing with a governess or with relatives or while her mother was on business. After she asked for a correspondence course in mathematics, her letters often contained algebra problems and solutions. She wrote with dignity and good humor, without pleases, whines, or wheedling. In August 1909, at the age of eleven, Irène asked her mother ten questions: "I would like to know the exact date of your arrival here at Puyravaud. 2) Will my uncle Jacques be at Sceaux soon? 3) Write me if my palm tree is doing well; also my monkey-puzzle plant and if the palm has new leaves.... I have made you ten questions. Reply to all when you write me. I love you. Irène."

Marie Curie strongly disapproved of the rigid French educational system. Influenced by Pierre's home schooling, she complained about the French lyceum and its long hours, poor lighting and heating, rote instruction, cold lunches, and the lack of physical exercise, art, and laboratory science. Children should think and play more and memorize less, she argued. Judging a nation by its education budget, she found France lacking.

To minimize the time Irène spent in public school, her mother organized a private cooperative school. Ten children from six professors' families joined, and each family agreed to contribute a class each week. Physicists Jean Perrin and Paul Langevin taught physical chemistry and mathematics, Marie Curie taught experimental physics, and so on through literature, art, natural sciences, English, and German. Traveling between classes, Irène entertained her schoolmates by translating the bestseller *Quo Vadis* from its original Polish into French. The cooperative lasted only two and a half years, but Irène remained close friends with her classmates for the rest of her life. With them she felt comfortable and happy as she never did among strangers.

Marie tutored Irène in mathematics long after the cooperative ended. Once, when Irène daydreamed, her mother became uncharacteristically impatient and hurled Irène's notebook out the window into the garden below. Irène calmly rose, walked quietly downstairs, picked up the notebook, and without changing her pace, climbed the stairs, and answered the question.

During her last two years of high school, Irène attended a private girls' school that taught modern languages instead of Greek and Latin. Marie and Pierre Curie found classical languages, which dominated the French public schools, a great loss of time. By the end of a year, Irène was reading seven books at a time in French, English, and German. "When I have a book," she admitted, "I devour it." To compensate for the school's lack of laboratory science, Irène lectured her classmates on the biological details of conception and childbirth.

In 1910, when Irène was thirteen years old, Grandfather Eugène Curie died after a difficult, yearlong illness. His death left her to face adolescence and three traumatic family crises without his support. Her letters changed tone, as if she were afraid that her mother too might disappear. "Dear Mé," as she called her mother, "I shall be so happy when you come because I badly need to caress somebody." Visiting Polish relatives during the summer of 1911, she wrote, "I love you very much, you see, and I would like you to come. Come quickly or at least write me when you are coming. Your BIGGGG Irène is impatient to see you again." Like a young lover, she stared longingly at her mother's portrait and sighed that her vacation would be pleasanter "if a sweet Mé were here, near me, to look at." Marie Curie was Irène's idol, soul mate, and safe haven.

Marie Curie's nomination for a seat in the prestigious French Academy of Sciences erupted early in 1911 into a national debate between the antifeminist, xenophobic Catholic right-wing and the republican, anticlerical left, which supported Curie. Later that same year, another even more catastrophic crisis occurred. Irène was attending an after-school gym class when a friend handed her a newspaper story: "A Love Story. Mme. Curie and Professor Langevin." Marie Curie and Irène's former mathematics teacher were accused of adultery. Turning white, Irène fainted. Taken to her mother, she clung to her skirts dry-eyed but terrified until she was dragged away sobbing, "I don't want to leave Mé. I have to stay with Mé." No one listened.

At the peak of the scandal, on November 8, 1911, Marie Curie learned that she had received a second Nobel Prize. Close to collapse, she pulled herself together long enough to take Irène to Sweden and deliver her Nobel lecture. Irène was dazzled. For the first time, she sensed her mother's fame and importance and her standing in the scientific community.

Back in France, Marie collapsed, gravely ill with a kidney

infection. For a time she considered suicide. For almost a year, she lived in seclusion, and not even her daughters were permitted to see her. Polish governesses and an aunt cared for them. Frantic that she had sullied Pierre Curie's name and reputation, Marie Curie insisted that her daughters write to her under another name in care of a friend. Irène was stricken and appalled. When she was finally allowed to write her mother as "Madame Curie," she was greatly relieved.

The disasters solidified Irène's distrust of outsiders and cemented her already close friendship with her mother. Irène became a woman with two distinct personalities. In public, she was aloof and brusque. With family and close friends, she was calm, relaxed, and even smiling. She could cry over a good story or a sad memory, but otherwise she possessed an extraordinary calm. She simply ignored what did not interest or please her. She had eyes only for what was important.

Irène spent her fifteenth birthday at L'Arcouest, a tiny fishing village on a promontory in northern Britanny. "The country is wild and superb," she wrote her mother. "On the edge of the sea there are big rocks we call castles. Each of us has her own." Irène returned to L'Arcouest every year thereafter. Marie and Ève came, too, and their friends. The village, discovered by Sorbonne professors, soon attracted so many scientists that the press dubbed it "Fort Science" or "Sorbonne Beach." L'Arcouest became Irène's haven, one of the few places outside her laboratory where she let down her guard. Sometimes she took the long and difficult train trip just to spend forty-eight hours there; even a taste revived her spirits.

During the summer of 1914, as Europe edged closer to World War I, Marie Curie sent Irène and Ève to L'Arcouest with two Polish maids. She promised to join them in August. Working on mathematics, Irène wrote her mother,"Great event!! I start to understand something about series.... The weather is magnificent. Irène."

The next day she confided, "I hate with all my heart Taylor's formula; it's the ugliest thing I know." Two days later, "The derivations are going well; the inverse functions are adorable."

As war became inevitable, Marie Curie wrote Irène, "Things seem to be turning bad.... You and I, Irène, we must look for a way to make ourselves useful." In September, Marie was still in Paris. The German army had attacked neutral Belgium and was marching on the French capital. "Paris is calm and gives evidence of holding firm," she reported to Irène.

In Britanny, a local malcontent overheard Irène and her sister speaking Polish with the maids and accused Irène of being a German spy. The mayor intervened, but the incident reminded Irène of the terrible year when the right-wing press had attacked her mother as a Pole.

"It means more because you yourself were accused of being a foreigner," Irène reminded her mother, "and we don't have anyone in the Army....They say I'm a German spy....I'm not very frightened about all this, but I'm very upset. It makes me sad to think people take me for a foreigner when I'm so profoundly French and I love France more than anything else. I can't help crying every time I think about it....I would a hundred times over prefer to be alone with Ève right now than with Walcia and Jozia [the maids]." She begged permission to return to Paris.

Marie urged patience. "My sweet dear...I sense how you have already become a companion and friend to me," she wrote September 6, 1914, less than a month after war started. "If you cannot work for France today, work for its future. Many people will be missing after this war; it will be necessary to replace them. Do your physics and mathematics the best that you can."

When Irène returned to Paris a few weeks later, she requisitioned a horse-drawn cab to help her mother move into her new Radium Institute. Then she went to work installing X-ray units in military hospitals and training personnel to operate them.

After the war, Irène became her mother's assistant at the Radium Institute. She was thrilled with research. Whenever she saw a brightly colored precipitate form during a chemical procedure, she confessed to an almost childish joy. Watching a radioactive substance glow in the shadows, she said she felt like an adventurous explorer. She worked for her own enjoyment, regardless of the problem's importance. Competition and success did not matter. She never wondered whether she could achieve at her parents' level. Like her father, Pierre Curie, she simply studied science for the personal pleasure of understanding nature's beauty.

Working with her mother, Irène became more than just a daughter, friend, and collaborator. She functioned as a substitute husband, protecting and caring for a wife. Pondering their relationship, Irène concluded, "I was very different from her, more like my father, and that's perhaps one of the reasons that we got along so well." Ève was their housewife; she lived with her mother but entertained literary and artistic friends in another apartment of her own. Each morning, Irène woke early, warmed breakfast, and took it to her mother. They enjoyed the quiet moments talking about poetry, the theater, books, or work at the laboratory.

"I think often of the year of work opening before us," Marie Curie wrote her daughters. "I hope it turns out well. I think also of each of you and how you both give me sweetness, joys, and solicitude. You are really, for me, a great richness, and I hope that life spares me a few more good years to live with you both together." To Irène alone, she wrote, "You know, my child, that you are an excellent friend for me and that you make my life easier and sweeter. I look forward to work with more courage thinking of your smile and your ever joyous face."

As far as Ève was concerned, her sister was "calm and constant good humor. I never heard her say a bad thing, and, to my knowledge, she has never lied in her life. She is exactly what she shows us, with all her merits and demerits, without embellishing anything to please us." Irène knew her scientific worth but was not conceited. For the Curies, Irène explained, "fame was something from the outside. It really had no connection with us." Her education had been unconventional, and it did not occur to her to justify herself to anyone else. Although Ève could never make her sister angry, nothing could make Irène do anything that bored her either.

Irène was mad about athletics. She danced at an École Normale ball until eight A.M., "a record," she told her mother. With a Corsican friend, Angèle Pompéi, she took fifteen-day backpacking trips through the mountains each summer, snapping photos all the way. Alone, she swam the Australian crawl in the River Seine. Mountaineering, skiing, and the crawl—all were men's sports in the twenties.

She prized convenience and loathed wasting time. She liked high-tech camping equipment—oiled-silk raincoats and lightweight purses—and wanted everyday clothes as utilitarian as hiking garb. In fashion-mad Paris, she wore baggy dresses, cut loose in the sleeves for ease of movement and full in the waist so that she could eat "like a pig" when she wanted. Her slips had handkerchief pockets, and she appalled visitors by fishing under her lab coat and skirt for a handkerchief to blow her nose. To save time, she cut her naturally curly hair herself.

To her coworkers in the lab, Irène Curie seemed intimidating and robust, both intellectually and physically. Her imperturbability and her knowledge of physics and math seemed well-nigh incredible for someone age twenty-five. Never verbal or socially conventional, she was often direct and brusque, qualities hardly unusual in a man, but shocking in a young woman. "Irène not only called a spade a spade but felt free to analyze the spade's defects," a wit observed. Although she smiled with her family, she seldom did elsewhere. Jealous of her privileged position in her mother's laboratory, coworkers called her the "Crown Princess."

Marie Curie's 1921 fund-raising trip to the United States to collect a gram of radium was a dazzling break in routine. Ève bought a new wardrobe and wowed the reporters, who called her "Miss Radium Eyes." Wearing her usual baggy clothes and cotton stockings, Irène sometimes substituted for her mother and occasionally spoke and received honorary degrees on her own account. When bored, she simply yawned and found a quiet place to read. Their pony ride to the bottom of the Grand Canyon made it all worthwhile, and Irène brought home a raccoon to sleep in a basket in her room.

In 1925, Irène Curie put on a baggy black dress and marched with businesslike stride to the Sorbonne to defend her dissertation. A thousand people crowded into the auditorium to see Marie Curie's daughter. To avoid upstaging Irène, Marie Curie stayed home.

Irène Curie had analyzed the alpha particles emitted by radioactive polonium as it disintegrates. She had focused particularly on the controversial question of how the alpha particles are slowed down as they move through matter.

Polonium, an element discovered by her mother in 1898, was extraordinarily useful for scientific research up to the mid-1930s. Until then, the only way to learn about the atomic nucleus was to bombard it with radiation from a naturally radioactive element. Physicists set a sample of radioactive radium or polonium next to a nonradioactive material and watched for interesting particles to measure. If all went well, a particle ejected by the radioactive element would knock another interesting particle out of its target. Polonium was especially useful for these studies because it emits only one kind of radiation, alpha particles. These are the nuclei of helium atoms, and they move so fast that they can smash into other nuclei. Irène Curie was one of only a handful of experts in radioactivity during the 1920s, and her mother's Radium Institute was one of only a few world centers in the field. As Irène marched off to the Sorbonne to defend her thesis, she had every reason to be confident that her thesis would be acclaimed as a valuable contribution to science. She had dedicated it "To Madame Curie by her daughter and pupil."

Congratulated by her admiring examiners, Irène returned to the Radium Institute for a garden reception and champagne served in laboratory beakers. Her degree was world news. The *New York Times* reported it, and a French woman journalist appeared to ask whether a scientific career might not be too taxing for a woman.

"Not at all," Irène Curie replied matter-of-factly. "I believe that men's and women's scientific aptitudes are exactly the same.... A woman of science should renounce worldly obligations...."

"And what about family obligations?" asked the reporter.

"These are possible on condition that they are accepted as

additional burdens....For my part, I consider science to be the paramount interest of my life."

And radium's dangers? Irène, who had already suffered several radium burns, said that research offered fewer risks than industry. "We know better how to protect ourselves."

Then Marie Curie whisked her daughter away incognito for a tour of Algeria to escape further publicity.

Soon thereafter, Marie interviewed a young army officer for a job at the institute. Frédéric Joliot had had no research training, but he was handsome, outgoing, and debonair. More important, Marie Curie's old flame, Paul Langevin, strongly recommended him. Fred Joliot started the next day.

Joliot's introduction to the lab was hardly auspicious. Marie Curie's colleague André Debierne gave him a tour and informed the young man that nothing more could be learned about radioactivity. "Everything has already been discovered," the older man announced loftily. Irène Curie was cool, too. "Good morning," she said, and then told him at top speed precisely what to do.

Fred and Irène were "opposites in *everything*," emphasized their daughter, Hélène Langevin-Joliot. Fred came from a large family of small business owners. He had wide and varied interests, from hunting and fishing to painting and piano playing. He was charm personified: a chain-smoking ladies' man, an extrovert who loved public adulation.

"Relationships between people were important to him; he would guess quickly if someone had a problem, and he wanted others to understand him," his daughter recalled. "Mother could walk right through something and see nothing. If somebody didn't shake hands with him, he'd worry; my mother wouldn't have noticed. Connections with people were tighter and more complicated with him; they were a nuisance for my mother. My mother was very like Pierre Curie, more stable. She needed time to think, and she'd do only what she wanted."

But, as Irène Curie soon realized, she and Fred Joliot agreed on important issues. They adored outdoor sports. Politically, they were antiwar leftists. As a youth, he had worshiped Marie and Pierre Curie; clipping their portraits from a magazine, he hung the pictures in his homemade laboratory. Fred's sociability softened and humanized Irène. Most important, they loved science and each deeply respected the other's abilities. They were scientific partners for the rest of her life and collaborated during their most productive years.

Working together, Fred said, "I discovered in this girl, whom other people regarded somewhat as a block of ice, an extraordinary person, sensitive and poetic, who in many things gave the impression of being a living replica of what her father had been. I had read much about Pierre Curie. I had heard teachers who had known him talking about him and I rediscovered in his daughter the same purity, his good sense, his humility."

One morning, Irène entered her mother's bedroom with breakfast and news: she was engaged—a fait accompli. She was twenty-eight years old.

When Fred and Irène married in October 1926, they lunched at Marie Curie's apartment and then returned to work. Fred slept at his mother's home that night, and Irène stayed at her mother's. The next day, Marie Curie left for a meeting in Copenhagen and Fred moved in. Later, the young couple occupied an apartment in a building partly owned by Marie Curie; so many scientists lived there, it was called the Sorbonne Beach Annex.

Marie Curie was a classic, difficult mother-in-law for the first two years of Irène's marriage. As Irène observed with understatement, "When I married, my mother was certainly pained to find us partly separated." Fred was two years younger than Irène, and Marie Curie did not know how long the marriage would last, so she insisted on a marriage agreement. (French law gave husbands control of their wives' property.) To secure Irène's research future, Marie Curie also double-checked that her daughter would inherit the use of the radium at the laboratory. When prominent guests came to the Institute, she introduced everyone but Fred. Or she introduced him as "the man who married Irène."

Marie Curie eventually adjusted. Fred and Irène ate their dinners with her several days a week. Soon, Irène discovered, "my mother and my husband talked often with such ardor, with such rapidity, that I could never say a word." Irène's first child, Hélène, helped too. However, it was Fred's skill that won over Marie Curie. "The boy is a skyrocket," she said proudly. She was right. Within twenty years he was one of the most powerful men in France.

In the beginning, Fred was eclipsed by his wife. Irène Curie had an established career before he even started his, and for years she was credited with their successes. Critics called him "The Prince Consort" and "Irène's gigolo." The perception, biographer Rosalynd Pflaum wrote, was that "at least in the beginning, this was a typical European marriage of convenience from his point of view." A coworker thought he was "the most ambitious man since Richard Wagner....[He] wanted to be Beethoven...Shakespeare, and Caesar all rolled into one." As for Fred, he admired great achievements and thought that people without ambition often relied on the efforts of others.

His signature reflected his insecurity. His wife signed her scientific publications as "Irène Curie," and he signed his "F. Joliot." They were Joliot socially. She did not care what she was called. Fred did. Close friends thought he should have remained plain "Joliot." For popular articles and political statements they used Joliot-Curie. Marie Curie's opinion of the addition can hardly have been favorable. Inevitably, English wits called the couple the "Jolly-Curios." Toward the end of his life, he was thinking seriously of changing the family name to Joliot-Curie so that it would continue after his death.

Fred never felt like one of the French scientific establishment. He felt more at home with the fishermen at L'Arcouest than with the scientific luminaries who vacationed there. After ten years of marriage, he asked a colleague, "Why are people so nasty? Why do they claim that I don't love my wife and that I have married her just for the sake of my career? But I do love my wife. I love her very much." Another coworker overheard him predict, "No, I have not yet been unfaithful to my wife, but if I am, it will not affect our marriage." Despite rumors of casual attachments, he and Irène apparently forged a happy and secure marriage. Fred wrote her teasingly, "I miss you in spite of your terrible... bossiness. Perhaps that is what I miss." And when she was away, she complained, "Every day I regret that I have too much room for my things.... I even regret that I don't have to make your bed, something I've never liked doing."

With the birth of Hélène and Pierre, Irène had everything she wanted: research and children. After Hélène was born, Irène told a friend, "If I had not brought children into the world, I would never have forgiven myself for having missed such an astonishing experience." When Fred called the Wilson cloud chamber "the most beautiful phenomenon in the world," Irène corrected him, "Yes, my dear, it would be the most beautiful phenomenon in the world—if it were not for childbirth." She was a feminist who defined her role as a woman in terms of both work and children. At home, she remained a traditional wife and mother. After Hélène's birth in 1927, doctors told Irène that she had contracted tuberculosis and should rest and have no more children. Instead, she returned to work almost immediately and gave birth to Pierre in 1932. For twenty years, she was chronically ill with tuberculosis.

The added expense of children and Irène's health precipitated a financial crisis. Fred's and Irène's combined salaries were so small that Fred almost left research to work in industry. When the rise of Fascism convinced the French government to begin investing in research, Fred got one of the government's new grants and joined Irène in full-time research. From then on, the Joliot-Curies campaigned unceasingly for more government support for scientific research.

They made a good team. Although Fred is considered a physicist, his doctoral thesis was pure chemistry; Irène is considered a chemist, but her thesis was pure physics. Fred thought fast, swinging back and forth between positions. Irène thought slowly, following an idea surely and resolutely in a straight line to its logical conclusion. The combination of approaches made for successes—although the notion of success never occurred to Irène.

Building on Marie Curie's original stock, the Joliot-Curies produced the world's largest and most powerful supply of polonium—almost enough to dust a square inch. Preparing polonium made them expert radiochemists, but the work was extremely dangerous. As scientists today know, polonium is highly toxic, and the body concentrates it in the lungs, spleen, and liver. The Joliot-Curies were too engrossed in their work to worry about possible dangers. Using Marie Curie's technique, they often mouth-pipetted radioactive materials, sucking them up into tubes to measure and transfer elsewhere.

By the early 1930s, foreign physicists were reading French scientific journals for the first time in years, just to see what the Joliot-Curies were doing. They were competing with a handful of other nuclear physicists, including Ernest Rutherford in England, Lise Meitner at the Kaiser Wilhelm Institute in Berlin, and Niels Bohr in Copenhagen. With every new discovery, more physicists entered the field and the Joliot-Curies had to quicken their pace, publishing short reports every few weeks.

At the time, physicists did not know that the nucleus of the atom is filled with protons and neutrons. Instead, they thought that it contained protons and *electrons*. The neutron had not been discovered, and a host of puzzles and contradictions remained unsolved. As the Austrian-born Nobel Prize–winner Wolfgang Pauli despaired, "Physics is once again very fouled up, and for me it is so difficult that I wish I were a film comedian or something like that and had never heard of physics in the first place."

Irène, by then pregnant with her son Pierre, took the first step toward solving the mess when she read an article by a German physicist, Walther Bothe. Bothe, who later won a Nobel Prize, had bombarded beryllium, a lightweight steely metal, with particles from radioactive polonium. To his surprise, rays emerged from the beryllium with so much power that they could penetrate lead two centimeters thick. He thought he had discovered a new type of gamma ray.

To study the new rays, the Joliot-Curies placed their polonium source of alpha particles next to a bit of beryllium. Just as the German had reported, the beryllium emitted powerful rays that could speed through lead. Then the Joliot-Curies tried placing different elements in the path of the new rays. When the rays bombarded paraffin wax, which is a rich source of protons, the wax ejected energetic protons at a tenth the speed of light. What would be powerful enough to make protons burst out of the paraffin so fast?

The Joliot-Curies concluded, incorrectly, that the mystery rays were gamma rays, the energetic cousins of X rays that move at the speed of light. They published an article saying so in January 1932. When Ernest Rutherford read that the Joliot-Curies had observed gamma rays pushing around protons with enormous amounts of energy, he exploded, "I don't believe it." Gamma rays have no mass and cannot make heavy particles move that fast. When James Chadwick repeated the Joliots' experiment in Rutherford's lab with some polonium from Lise Meitner and others, he discovered the neutron. Physicists finally knew the main constituents of the nucleus: protons and neutrons. The discovery of these massive, neutrally charged particles within the nucleus began the science of nuclear physics in earnest.

The Joliot-Curies had produced experimental proof of the existence of neutrons but had not understood their own data. Chadwick won the Nobel Prize for discovering the particle. Unfortunately, this was not the last time that the Joliot-Curies missed the prize by a hair. Irène would get three more chances.

The great Italian physicist Enrico Fermi quickly realized that neutrons would make fabulous research tools for investigating nuclei. Unlike protons and electrons, neutrons have no electrical charge and are not repelled by the negatively charged shield of electrons that surrounds each atom. And once a fast-moving neutron penetrates the nucleus, it is massive enough to eject a proton.

Following Fermi's lead early in 1932, the Joliot-Curies decided to study neutrons with Fred's favorite instrument. Using a Wilson cloud chamber to study subatomic particles is a bit like studying jet airplanes by analyzing their contrails. The apparatus enables scientists to follow indirectly the paths of charged atomic particles and atomic collisions by looking at the trails of foglike droplets that condense in their wake. Just as important, the apparatus photographs the events for later analysis.

Experimenting with neutrons, the Joliot-Curies saw tracks of electron-sized particles behaving in an odd and unusual way. The tracks looked as if they had been made by a positively charged particle the same size as a negatively charged electron. There were two possible explanations: a normal negatively charged electron could have entered the chamber through its wall and made a beeline for the neutrons. Or the mysterious particle could have been a positively charged electron—a positron, the first proof of the existence of antimatter. Unfortunately, the Joliot-Curies guessed wrong.

When C. D. Anderson in the United States did the same experiment a few months later, he guessed right and won a Nobel Prize. The particle had indeed been a positron. Once again, the Joliot-Curies had produced the experimental evidence of a major discovery but failed to understand what they had seen. For her contribution, Irène was nominated for a Nobel Prize with Anderson. (The romantic story that the Joliot-Curies missed the positron while working in a cliffside laboratory in the Alps is unfortunately incorrect, according to their daughter, Hélène Langevin-Joliot, who is a nuclear physicist. The mountain lab had no Wilson cloud chambers with large magnets.)

Later that year, the Joliot-Curies put their polonium next to a thin sheet of aluminum foil, expecting hydrogen nuclei to burst out. Instead, neutrons and positrons emerged. By now, the scientific community realized that Irène's consort was a talented physicist, and both he and Irène were invited to a prestigious week-long Solvay Conference in Belgium to report on their polonium experiment. Almost all the key nuclear physicists attended the conference in October 1933, including Marie Curie and Lise Meitner. Of forty participants, twenty were past or future Nobel Prize winners.

When the Joliot-Curies described neutrons and positrons streaming out of aluminum foil, Lise Meitner announced firmly that she had done the same experiment and had been unable to find "*a single neutron.*" Meitner was such a skillful experimentalist that few believed the young French couple. "After the session we were quite downhearted," Fred confessed. "But at that moment, Professor Niels Bohr took us aside, my wife and me, to tell us that he found our results very important. A little later Pauli gave us the same encouragement."

Returning to Paris in 1934, they redid their experiment and got the same results again. At first, they assumed that the aluminum nuclei, while under bombardment from the polonium's alpha particles, were emitting neutrons and positrons simultaneously. To check their assumption, Fred pulled the aluminum away from the polonium and tested it with a Geiger counter. As anticipated, neutrons had stopped coming out of the aluminum. He was greatly surprised to discover that—even though the aluminum was far away from the polonium—positrons were still streaming out of the foil by themselves. Something was happening inside the aluminum nuclei independent of the polonium and the bombardment by alpha particles.

Racing to Irène's lab, Fred brought her back to watch. When he again moved the aluminum away from the polonium and measured it with a Geiger counter for positrons, the instrument went click, click, click, dying down gradually over several minutes. They knew that sound: it meant radioactivity.

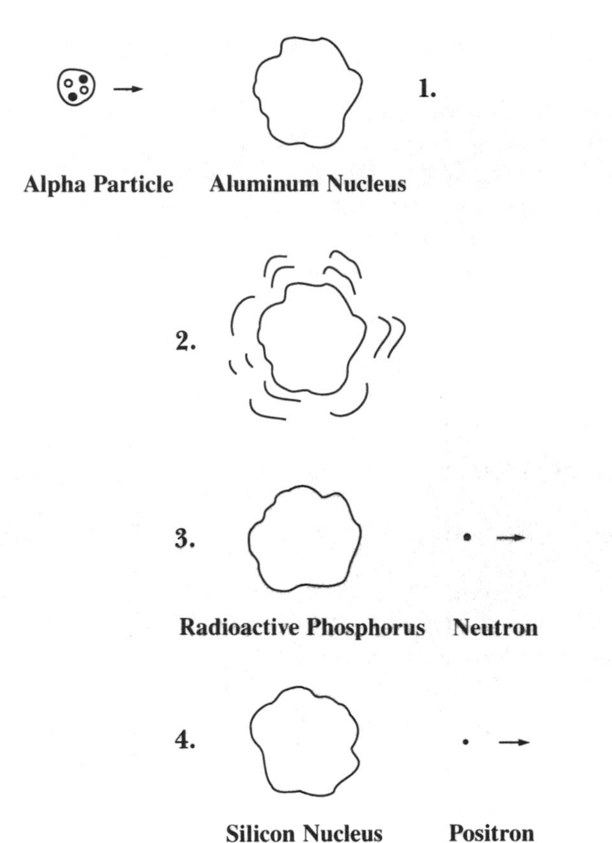

Figure 6.1. Artificial Radioactivity

1. An aluminum nucleus is bombarded by an alpha particle.
2. The alpha particle enters the aluminum nucleus.
3. A neutron is ejected immediately, and the nucleus is now radioactive phosphorus.
4. A few minutes later, a positron is ejected and the nucleus becomes silicon.

The aluminum nuclei had absorbed alpha particles from the polonium, ejected neutrons, and, in the process, for a few short moments, changed into a heavier element—an artificial type of phosphorus. The phosphorus nuclei were so unstable, however, that they soon ejected the positrons and changed again into a stable form of silicon. The Joliot-Curies had induced a naturally stable element to become artificially radioactive.

Immediately, they began searching for ways to prove chemically that the aluminum had indeed changed into phosphorus on its way to becoming silicon. They had only trace amounts of phosphorus, however, and half of it decayed within three and a half minutes. But Irène devised a three-minute chemical test—a piece of technical virtuosity—and it worked. Fred ran and jumped for joy around the institute basement like a child. With the neutron, they had misunderstood; with the positron they had misunderstood; with artificial radioactivity, they knew what they were doing. Their original report had been right; Lise Meitner had been wrong.

Irène brought her mother, and Fred invited Paul Langevin, his former professor and Marie's former lover. The two silently watched the demonstration, asked a few questions, and left. For the rest of his life, Fred remembered demonstrating artificial radioactivity to his idol Marie Curie and his adored teacher Langevin.

After the Joliot-Curies had isolated the artificially radioactive substance, Fred gave Marie Curie a tiny test tube of it. Taking it in her radium-scarred fingers, she held it up to a Geiger counter. When she heard the characteristic click, click, click, her face glowed with joy. She knew it meant a Nobel Prize for her daughter and son-in-law. It was surely the last great satisfaction of her life. Within months, she was dead of leukemia.

The Joliot-Curies were nominated for the physics Nobel Prize in 1934 for artificial radioactivity. Passed over that year, they won the Chemistry Prize in 1935. Irène was thirty-seven years old. When she heard the news, her childhood dread of reporters enveloped her, and she fled to the Bon Marché department store, where she bought oilcloth for the kitchen table. She was right to be worried about the press; the prize was popularly attributed to Fred's talent and her assistance.

Their Nobel Prize brought her family's total to three. When Ève's husband, diplomat Henry R. Labouisse, accepted a Nobel Peace Prize for the United Nations Children's Emergency Fund in 1965, the total for the entire Curie clan rose to four.

At an institute tea for the young couple, Irène remarked, "In our family, we are accustomed to glory." It was not a boast, but a simple and objective statement of fact. Nevertheless, when the trumpets sounded in Stockholm and the orchestra played Debussy in honor of France, neither she nor Fred could hide their delight. To emphasize how they divided the work equally, Irène gave the physics portion of their Nobel address and Fred gave the chemistry section. The rest of the ceremony was not as pleasant. A German winner cast a pall over the occasion by ending his speech with the Nazi salute. Irène soon became bored by the reception; when the king of Sweden asked where she was, Fred found her in a corner, reading a book. The prize money was welcome, though. It paid for a large new house, tennis court, and garden in a Parisian suburb. The year before her death, Marie Curie had bought the land intending to build two houses, one for herself and another for Irène and Fred.

After the Nobel Prize, their research team broke up, Irène was propelled into politics, and her health deteriorated. Thanks to the Nobel, Fred received a job offer from the Collège de France, the most prestigious research institution in France. He accepted the position partly to show what he could do without Irène and partly because he realized that nuclear physicists would need accelerators to split open the nucleus. The Collège de France had space for him to build a cyclotron. The move ended his scientific collaboration with Irène. From then on, he was involved in administration, accelerator design, his own research, and fund-raising for French science. Irène became a professor at the University of Paris and continued as research director at the Radium Institute. Among all their different positions, the Joliot-Curies controlled nuclear science in France. Despite her prestige, Irène maintained an informal atmosphere at the lab, chatting with coworkers in the front hall while she warmed her back on a radiator. To help the desperately poor family of a lab employee, for example, she rented a gymnasium and everyone at the institute took judo lessons from their black-belted colleague.

Using her new prestige, Irène Curie joined several organizations for women's rights. "I am not one of those...who think that a woman [scientist]...can disinterest herself from her role as a woman, either in private or public life," she remarked. While Fascists urged women to stay home, she pointed out that many women needed to earn money because they were unmarried or widowed or because their family incomes were inadequate.

As the Fascists gained power throughout Europe, she entered politics. The Popular Front was an anti-Fascist coalition of French centrist moderates, Socialists, and Communists elected in 1936. The Front asked Irène to become under secretary of state for scientific research. The post combined Irène's interest in women, funds for science, and anti-Fascism. She took the job, calling it "a sacrifice for the feminist cause in France." She wanted to advance "the most precious right of women...to exercise under the same conditions as men the professions for which they're qualified by education and experience." She was one of the first female cabinet ministers in France. Ironically, she could not vote. French women were not enfranchised until 1945. By prearrangement, she resigned after three months, turning the office over to a friend.

Irène Curie was a most undiplomatic politician. Answering invitations yes or no, she refused to use the customary flowery language employed in French business letters. She would not say that she was "desolated" to miss meetings that she did not wish to attend, and she declined to send correspondents her "deepest homage" or her "profound respects." She also hated wasting time. When meetings lost their focus, she stood up, gathered her belongings, and departed. When she arrived early at a ministerial meeting, she sat on the steps outside and worked. A patronizing janitor tried to shoo her away, "Out, little lady! We're waiting for the ministers."

"I am one," Irène Curie coldly replied.

The Joliot-Curies lost faith in the Popular Front after it refused to help the Spanish Republican government fight a Fascist revolt. At a 1938 meeting where the French Communist party urged more funds for scientific research, Irène and Fred sat center-front in the auditorium's most conspicuous seats.

As Fred became more active in politics during the late 1930s, however, Irène's tuberculosis worsened and she was increasingly confined to the sidelines. From 1934, she spent several weeks or months each year in rest cures, generally in the Alps. In Paris, she often worked lying down at home. Her exposure to X rays in World War I and to polonium in the laboratory may have lowered her resistance to the tuberculin bacillus. During World War II, when good food and home heating were unobtainable, her condition deteriorated. Neither she nor her family ever referred to her health.

Irène was happy working and still enjoyed outdoor sports and reading light literature, especially Colette, mysteries, Rudyard Kipling, and her all-time favorite, *Gone With the Wind.* The last two—romanticized views of imperialism and slavery, respectively—seemed peculiar choices for a committed leftist. With household help, she revived a childhood pleasure: her parents' Sunday open house. About one P.M. each Sunday, Irène served a simple dinner. She was not an expert cook, and her table seated twenty, so, to Fred's regret, she usually served stew.

Irène had a near miss at a second Nobel Prize in 1938. During the late 1930s, physicists were scrambling to unravel the puzzle of what happens when neutrons bombard uranium, the heaviest naturally occurring element. Enrico Fermi did an experiment in which, he thought, a uranium nucleus had absorbed a neutron and become an artificial element heavier than uranium. Soon after Fermi's experiment, Irène read about a follow-up project by her competitors Lise Meitner and Otto Hahn in Berlin.

Like Fermi, they bombarded uranium with neutrons and tried to identify the emerging radioactive nuclei. Irène disagreed with their results. With Fred at the Collège de France, she chose a young Yugoslavian physicist, Paul Savitch, to work with her. Focusing on one of the elements, they concluded that it resembled a known element called lanthanum. They assumed—like everyone else—that the element could not be lanthanum itself. Lanthanum atoms are lighter than uranium atoms, and everyone thought that uranium would absorb the neutron and become an even heavier element.

When Irène Curie published her conclusions, Hahn and Meitner were incredulous. Hahn and Joliot met at a conference in Rome, where Hahn took the Frenchman aside. "Between us, my dear," the smooth-talking German said, "it's because your wife is a woman that I haven't permitted myself to criticize her. But she is wrong, and we intend to prove it."

When Irène heard about Hahn's reaction, she redid the experiment and got the same result—and published it again! According to some French versions of the story, Hahn was furious and complained that Irène Curie was wasting his time with "her mother's old-fashioned methods." He dubbed her new particle "curiosum," a pun suggesting that the Curies were an odd curiosity.

But a young chemist, Fritz Strassmann, insisted that Hahn and Meitner take her evidence seriously. Only then did Hahn redo his experiment using Marie Curie's "old-fashioned methods." Then they, too, observed Irène Curie's results. They also found barium, another element lighter than uranium. Mystified, they wrote Lise Meitner for an explanation; she was Jewish and had fled from Nazi Germany to neutral Sweden. Talking it over with her nephew, Meitner realized that the uranium atom had broken into two smaller pieces. Her nephew called it fission, and Hahn won a Nobel Prize. Luckily for him, he had Lise Meitner to explain his work; Irène Curie did not.

When Irène read Hahn and Strassmann's article, she was furious that Joliot had not been working with her. She rarely used obscenities, but this time she exploded, "Oh, what dumb assholes we've been!" Once again, she had almost made a major discovery. She had come to the point where her evidence contradicted all the known rules of physics, but she had not been able to figure out which rule was wrong. The team of Hahn and Strassmann, also expert radiochemists, had reached the same point; but they had had Lise Meitner to explain what had happened. Irène had worked by herself, with no one like Meitner to consult. With a little help, she might well have beaten the Hahn-Strassmann-Meitner team. If she had, she would have won a second Nobel Prize, like her mother. Fred and Irène always regretted that they had not worked together on the experiment.

Fascinated by fission, Joliot rushed to study the process. He and his team worked out computations showing that a chain reaction would release enormous energy. He did not believe, however, that an atomic bomb could be produced in time to end the war with Germany. Instead, he saw fission as the solution to France's energy problems. In 1939, France imported nearly all its oil and a third of its coal. Today, thanks in part to Joliot, France gets eighty percent of its energy from nuclear power and exports energy to other European nations.

Worried that war would break out, Fred helped arrange in 1940 to ship large quantities of uranium and heavy water from the Continent to Great Britain and the United States. The heavy water, as deuterium-enriched water is called, became the foundation of the British-Canadian atomic bomb program. Unlike Ève Curie, who became a journalist abroad, the Joliot-Curies remained in France. As Irène announced, Marie Curie would never have left her laboratory. The Allies tried to persuade Fred Joliot to leave, but his elderly mother lived in Paris, Irène was in a nursing home at the time, and his English was poor. Moreover, his Alsatian upbringing had taught him that if the German occupation lasted decades, someone had to stay behind to keep French traditions alive.

Wartime food and fuel shortages aggravated Irène's tuberculosis. Her face permanently lined and wrinkled, she looked old in her mid-forties. While traveling, she became so exhausted one day that she simply lay down on the floor where she was—surrounded by people—and fell asleep. Yet she kept working. Like her mother, she would have felt lost without her laboratory. She continued her interest in women's affairs. After Fred—but not Irène—was admitted to the Academy of Sciences in 1943, she decided to apply every few years for membership. After each rejection, she publicized the academy's continued discrimination against women.

The Germans occupied Frédéric Joliot's laboratory in September 1940. When they asked for the heavy water, Fred said it had sailed for England on a ship that had sunk. The Germans left Joliot's friend and student, Wolfgang Gentner, in charge. That night, Gentner, who was fervently anti-Nazi, met Joliot in a student cafe. Together, they planned how to protect the lab from the Germans as much as possible.

In 1940, Joliot joined the Resistance and later became president of the largest, best-organized, and most belligerent of all the Resistance groups in France. When the Gestapo killed several of his close friends, including Paul Langevin's son-in-law, Joliot joined the Communist party. At the time, it was the most active anti-Fascist organization in France. "I became a Communist because I am a patriot," he said. It was a dangerous decision, because the Germans executed known Communists. Eventually, his Communist membership destroyed his career.

Toward the end of the war, Fred went underground and arranged for the Resistance to smuggle Irène and the children out of

France. Irène refused to leave the country until her daughter had completed her baccalaureate examinations. Hélène took them secretly in a small border village, and on June 6, 1944, Irène and her two children hiked over the Jura mountains into Switzerland. Irène was carrying a heavy physics volume in her knapsack. They were lucky: June 6 was D day. The day before and the day after the Normandy invasion, German border security was much tighter. On D day, the Germans had other things on their minds.

After the war, Fred was a French hero. He became head of the French government's science-funding agency and the French Atomic Energy Commission. Irène became a commissioner. Fred and the government were on a collision course, though. He insisted that nuclear power should be used only for peaceful purposes, while the French and United States governments were planning a hydrogen bomb. As the cold war and McCarthyism intensified in the United States, the American government pressured the French to fire Fred.

Irène sympathized with the Communist party but never joined. Socialists and Communists had always been the biggest supporters of feminism in Europe. Her main interest in the party was Fred, according to her French biographer, Noelle Loriot. Irène was also too independent to join anything as disciplined as the Communist party. As she often told her daughter, her symbolic position as a Curie precluded her ever joining any particular party. On the other hand, she still spoke out frankly when she believed in something. It was she, not Fred, who testified in court against a book that revealed the brutal conditions in Soviet *gulags*. She stated that she had visited the U.S.S.R. and that such conditions did not exist. "She did it because she was *really* convinced it wasn't true," her daughter said. "She hadn't seen, hadn't understood."

After the war, Marie Curie's friend, Missy Meloney, sent Irène streptomycin, which cured her tuberculosis. Throughout the late 1940s and early 1950s, Irène's health was better than it had been in decades. Besides continuing her research and running the institute, she attended international conferences for peace, atomic weapon bans, and women's rights. The atomic bomb horrified the Joliot-Curies. It was based on their research, and they felt partly responsible for the destruction it had caused in Japan.

In 1948, Irène Curie arrived in New York to speak at fund-raisers for Republican refugees from Franco's Spain. In 1921, when she had arrived in the United States with her mother, she had been the toast of the nation. This time, the U.S. Immigration authorities detained her overnight in a cell on Ellis Island. The United States attorney general considered the group sponsoring her trip to be subversive. French authorities protested the next day; Irène Curie was a high government official, and her husband was one of the most powerful men in France.

Released that morning, she told the press that Ellis Island's accommodations were adequate; she had had some good coffee and darned her stockings. Explaining that she was not a Communist, she added bluntly, "I am not surprised to have been arrested and detained, because I am here to aid the anti-Fascists. And, in the United States, they prefer the Fascists and even the Nazis to the Communists. They think that the first and second [the Fascists and Nazis] have more respect for money." Despite its inauspicious beginning, her tour went pleasantly. Outside New York City the press was both hospitable and fair, she wrote Fred, and nine hundred people attended a Hotel Astor banquet in New York before she left.

The Joliot-Curies fell out of favor in later years because of their Communist leanings. The French government fired Fred as head of the French Atomic Energy Commission in 1950. His ouster ended the Curie family's half century of domination of French nuclear physics. Ostracized, he spent his remaining years working for peace organizations. Irène's term on the Atomic Energy Commission was not renewed after 1951. When she arrived later in Stockholm for a physics conference, city hotels refused to give her a room. The British declined to give her a visa to attend a scientific conference. Despite her Nobel Prize, the American Chemical Society rejected her application for membership. Her Sunday afternoon open houses attracted only a few old friends.

Unlike Fred, Irène still worked in her laboratory. She secured government approval for a large new nuclear-research center at Orsay, south of Paris; after her death, it replaced the Radium Institute. Throughout the mid-1950s, she worried about Fred's health. He was gravely ill with radiation-induced hepatitis. She was not at all worried about her own health and worked in the lab every day through January 1956.

In February, she went alone to their ski chalet in the Alps. Suddenly ill, she took a train back to Paris and checked herself into the Radium Institute's hospital. She never left it.

Although doctors diagnosed leukemia, Fred was so sick that he could see her for only a few moments. She told her childhood friend Aline Perrin, though, "I am not afraid of death. I have had such a thrilling life!"

To the end, she maintained her faith in science. During the last year of her life, she wrote, "Science is the foundation of all progress that improves human life and diminishes suffering." She died at the age of fifty-eight on March 17, 1956. When the French government staged a national funeral for her, her family requested that the military and religious portions of the ceremony be omitted.

Fred and Irène Joliot-Curie had lived and worked together for thirty years. Two years after her death, he died—also of what he called "our occupational disease." He, too, received a national funeral. Of the four Curie scientists, only Marie Curie had lived past sixty.

* * *

7

Barbara McClintock

June 16, 1902–September 2, 1992

GENETICIST

Nobel Prize in Medicine or Physiology 1983

When a Miss Barbara McClintock of St. Louis announced her 1936 engagement in the newspapers, the chairman of the University of Missouri's botany department was horrified. Mistaking his new thirty-four-year-old assistant professor for the woman in the newspaper, he summoned Dr. Barbara McClintock to his office. Then he threatened her, "If you get married, you'll be fired."

The University of Missouri was "awful, awful, awful," McClintock complained years later. "The situation for women was unbelievable, it was so bad."

Eventually, she marched into the dean's office and asked point-blank whether she would ever get on the university's permanent staff. He shook his head no. In fact, he confided, when her mentor left she would probably be fired.

McClintock retorted that she was taking an immediate leave of absence without pay and that she would never return. Then she packed her Model A Ford with all her belongings and drove off, without a job or even any prospect of a job. Toying with the idea of becoming a weather forecaster, she finally decided that she never wanted a job of any kind again. It was years before she changed her mind.

McClintock was at the top of American science when she quit it. She had revolutionized maize genetics; one of her early experiments still ranks among the twentieth century's most important biological experiments. She was the vice president of the Genetics Society of America and was about to become its president. She had not yet done her Nobel Prize–winning project, but she had already received an honorary doctorate from a well-known university and would soon be

Barbara McClintock at a press conference when her Nobel Prize was announced in October 1983.

Barbara McClintock at Cornell in 1929, with (from left to right) Charles Burnham, Marcus Rhoades, and Rollins Emerson. Kneeling, George Beadle.

Barbara McClintock at the far right of the first row in her first grade class, P.S. 139 in Brooklyn, N.Y. Her name is on the blackboard on the right with four other girls who made the honor roll.

Barbara McClintock at her parents' home in Brooklyn while she was attending Cornell University.

elected to the National Academy of Sciences, then the nation's highest scientific honor.

But McClintock was a woman who wanted to do research full-time—and she was a feisty woman at that. So, despite friends in high places, she had no permanent job. Universities were the chief sponsors of scientific research in the United States, and they reserved their research positions for men. Thus, McClintock's decision to leave academia meant giving up the passion of her life—genetics.

McClintock had wanted to be free and independent all her life. Born in Hartford, Connecticut, on June 16, 1902, she was the youngest of the three daughters of Dr. Thomas Henry McClintock and the former Sara Handy. "My parents were wonderful," McClintock recalled. "I didn't belong to that family, but I'm glad I was in it. I was an odd member."

Dr. and Mrs. McClintock had hoped for a boy and had chosen his name: Benjamin. "My mother took the blame because it was her fault she didn't deliver the right thing," McClintock noted dryly. Neither Mrs. McClintock nor Barbara could conceal their feelings from the other: Barbara knew that her mother was disappointed, and Mrs. McClintock knew that Barbara knew.

The two maintained a wary, arm's-length relationship. When Barbara was an infant, her mother often put her on a pillow on the floor and gave her a toy; Barbara played happily alone. Within four months of McClintock's birth, her parents decided that her name—Eleanor—was far too sweet and gentle. So, as McClintock enjoyed explaining, they changed Eleanor to Barbara because it sounded harsher.

When Barbara was two years old, the long-awaited boy was born. Barbara's mother was overwhelmed with caring for four small children. A Mayflower descendant and a Daughter of the American Revolution, she had lived in affluence until she defied her father to marry a homeopathic physician. To devote more time to her son and to relieve the strain between herself and Barbara, she periodically sent her daughter to stay with an aunt and uncle in rural Massachusetts. The uncle sold fish from a horse-drawn wagon, and Barbara enjoyed accompanying him on his rounds. From him, she learned to repair machinery and to love nature. Back home, she continued to rebuff her mother's hugs and kisses. "I didn't get approval, but I didn't get harsh treatment from her," McClintock admitted.

Barbara's father raised her as a boy, free from the conventional restraints placed on girls. When she was four years old, he gave her boxing gloves. "I didn't play with girls because they didn't play the way I did," McClintock said. "I liked athletics, ice skating, roller skating, and bicycling, just to throw a ball and enjoy the rhythm of pitch and catch; it has a very wonderful rhythm."

"My parents supported everything I wanted to do, even if it went against the mores of the women on the block. They wouldn't let anybody interfere," McClintock explained. When a neighbor tried to teach Barbara "womanly" things, Mrs. McClintock sternly told the housewife to mind her own affairs. When Barbara decided that her teacher was "emotionally ugly," her father let his daughter stay home from school.

Playing baseball with the boys on her block, however, did not make her one of the boys. Once her team was so embarrassed to have a girl catcher that they refused to let her play an away game. Luckily, the other team did not care and invited her to join its side. On the way home, her neighborhood buddies accused her of being a traitor. "So you couldn't win," Barbara realized, concluding reasonably enough that "you had to be alone. You couldn't be in a society you didn't belong to. You were only *tolerated* by the boys.... I knew I couldn't win—and that's a dreadful feeling as a child." As a result, some of Barbara's happiest childhood moments were spent reading or just thinking about things.

Far from being unhappy, Barbara felt a great sense of freedom and opportunity as she grew up. When she was eight years old, the McClintocks moved to Flatbush, a rural neighborhood in Brooklyn, New York. With telephones and Morse code speeding the latest world news into the little community, "You felt as if you were branching out," McClintock said. "There was change. Everything was changing." Barbara was ready to take on the world. When she learned that the Statue of Liberty was 152 feet high, she announced with confidence, "That's no problem! I can shinny up!"

"When I reached adolescence, my mother panicked." Barbara wanted to attend Cornell University, but Mrs. McClintock thought that higher education would make her daughters "strange" and unmarriageable. She had convinced Barbara's older sister to reject a scholarship to Vassar College. "My father was an M.D. though. He sensed from the beginning that I would be going into graduate work. He didn't want me to be an M.D. He thought I would be treated so badly. Women got such nasty treatment. But he *warned* me; he didn't coerce me. He was very supportive with me. He had great faith I'd come out all right."

Unfortunately, when Barbara graduated from high school during World War I, Dr. McClintock was in France in the army medical corps. Acting on her own, Barbara's mother flatly vetoed her daughter's Cornell plans. Instead, Barbara found a job as an employment agency interviewer and spent her evenings and weekends studying frantically in the public library.

When her father returned from France in 1919, he supported Barbara immediately. Within days, she was enrolled in Cornell's College of Agriculture, where the tuition was free. McClintock remained grateful to her father for the rest of her life, emphasizing, "I just knew what I wanted to do. It was easy because it was so clear and because I had the support of my father, the complete support. My mother—if she could have done it without raising trouble—she'd have stopped it." After her children were grown, Mrs. McClintock took summer courses in art and writing at Cornell and finally understood what education had meant to Barbara. The revelation came too late; Barbara was the only McClintock child to attend college.

Cornell thrilled McClintock from beginning to end. Sometimes she was so immersed in her work that she could not remember her own name. A shade over five feet tall and wiry slim at ninety pounds, McClintock had a belly laugh like a child and loved jokes. Years later a photographer took so long setting up a picture of her in her laboratory that, just as he pressed the button, she popped a microscope cover over her head, and that is the picture he took. Sometimes her dreams seemed so funny that she woke up laughing. She was president of the freshmen women and played tenor banjo in a jazz group around town, until she decided that late hours interfered with her work.

McClintock was a modern woman who smoked, bobbed her hair, and wore golf knickers—plus fours—for field work. Although she relished scandalizing Cornell with her haircut and pants, she may not have succeeded. Bobbed hair was de rigueur for fashionable young women from 1921 on, and knickers were standard wear for both men and women throughout the twenties, much like blue jeans today. Her choice of friends was *avant-garde* though. The social gap between Jews and Gentiles was enormous at Cornell, but most of the women in McClintock's circle were Jews. She studied Yiddish, and when her friends were not invited to join sororities, she rejected her own bids.

McClintock's vigor, intensity, and enthusiasm marked her as special, and by the time she graduated in 1923, she was already deep in graduate work. So were many other young American women. Between 30 and 40 percent of all graduate students in the United States during the 1920s were women. In fact, women accounted for approximately 12 percent of the science and engineering Ph.D.s awarded in the United States—a proportion they would not reach again until the 1970s. Most studied biology, and almost one in five was a botanist. A goodly number of them specialized in genetics. Most of the rest were in zoology and psychology, which required little mathematics.

Getting a good science education, however, was much easier than getting a research job. Industry, government, and most colleges and universities refused to hire women. Most women scientists taught in women's colleges, where teaching loads were heavy and research time short. Only four percent of women scientists in the United States were employed by coeducational colleges and universities, and they were concentrated in home economics and physical education and in low-ranking positions as assistants, instructors, and assistant professors.

Genetics, however, was a wide-open field. McClintock and genetics were born and raised together. Gregor Mendel's studies of heredity in garden peas were rediscovered in 1900, just two years before McClintock's birth. By the 1920s, genetics was America's first world-class science and biology's most abstract specialty. When McClintock entered graduate school in 1923, many biologists still did not accept Mendelian genetics. The word *gene* had been coined but it had no clear definition or physical reality. It was just an abstract concept and controversial theory describing the way inherited traits are passed from one generation to another. As Thomas Hunt Morgan put it, geneticists assumed "there is something in the egg that is responsible for every detail of character that later develops out of the egg."

Chromosomes were known to carry hereditary elements inside the nucleus of the cell. Before a cell divides, its chromosomes double in number. Then half move to one end of the cell and half to the opposite end. As the cell stretches out, it divides into two halves, each the exact duplicate of the other and each containing the same amount of chromosomal matter. Geneticists also knew that each species has a characteristic number of chromosomes, ninety-four for goldfish, forty-six for human beings, ten for corn, and so on. The discovery that DNA is the chemical basis of genes was decades in the future.

As McClintock began her career, fruit flies and corn were vying as genetics' leading research tool. Morgan, who was studying fruit flies at Columbia University, had shown that many of the fly's physical traits are inherited as a package deal, like red hair and freckles in human beings. He correlated their visible characteristics—long and short wings, gray and black bodies, and so on—with changes in their chromosomes. Traits that are "linked" or inherited together correspond to genes residing on the same chromosome. In fact, the probability of those traits being inherited together increases the closer they are together on the chromosome. According to how often the traits were inherited together or separately, Morgan and his followers mapped the relative locations of the fruit fly's genes on its chromosomes.

Cornell geneticists worked with corn, however. Scientifically known as *Zea mays,* corn is an economically important crop. It was an ideal research tool, too. The variegated colors of its kernels functioned like a technicolor spread sheet of genetic data; genetic changes were as plain as the kernels on the cob. Furthermore, maize could be self-fertilized, inbred to produce tightly controlled extremes of genetic behavior. Each maize plant produces both male and female flowers: female flowers borne on the ear contain egg cells; male flowers produced in the tassel at the top of the stalk contain sperm cells, known as pollen.

When spring-planted corn reached sexual maturity in July, Cornell's geneticists began working from dawn to dark seven days a week to control the mating. Normally, wind wafts pollen from the tassel of one plant to the silk of another. There a pollen grain germinates, growing a long phallic tube down through a silk to carry the sperm to the egg at the bottom of the cob. Sperm and egg cells fuse, starting the next generation's seed, a kernel on the cob. Each fertilization produces one kernel.

To prevent random promiscuity, geneticists cover the ears and tassels with paper bags and transfer the pollen to the silk by hand. To self-fertilize a plant and inbreed exotic strains, they fertilize the silk of one plant with its own pollen.

Despite the attractions of maize as a research tool, Cornell's geneticists had not studied its chromosomes. They had no way to

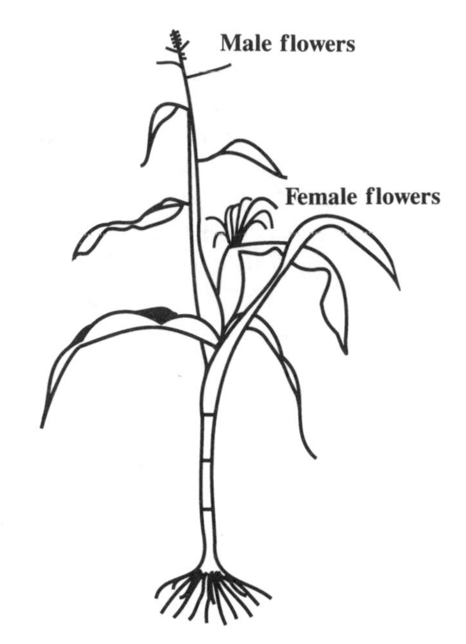

Figure 7.1. Corn Plant.
A corn plant produces both male and female flowers.

identify which chromosomes carried which inherited traits. Working in the botany department because Cornell's plant breeders refused to have women in their department, McClintock devised a system.

Using new staining techniques, she discovered that each of the ten chromosomes in maize could be distinguished under a microscope by their tiny knobs, extensions, and constrictions. Then she went on to identify each chromosome with a group of visible traits that are generally inherited together. By plotting the probability of these traits appearing together, she mapped the position of the genes on the chromosomes, just as Morgan had done with the fruit fly.

At first, none of her Cornell colleagues understood her project. Then Marcus Rhoades, who had earned his Ph.D. with Morgan, came to Cornell. Rhoades immediately realized how good McClintock was. "Hell," he said, "It was so damn obvious. She was something special." Immediately, he asked McClintock, "Can I join you?" Then Rhoades—her champion, interpreter, and soul mate for decades to come—explained the importance of McClintock's work to Cornell.

From then on, McClintock was the enthusiastic leader of a little band of professors and young men who already had their Ph.D.'s. "It was quite a remarkable thing that this woman who hadn't gotten her Ph.D. yet, or probably even her master's, had these postdocs trailing around after her, just lapping up the stimulation that she provided," recalled Ernest Abbe, later a University of Minnesota professor. "Lester G. Sharp was a prominent geneticist, but she was telling him what the answers were. It was very cute," Abbe laughed. McClintock even interviewed prospective graduate students for faculty members because she noticed so much more than anyone else. Later, during the late 1920s and early 1930s, Sharp propagated McClintock's research to the genetics community at large in his authoritative textbook *An Introduction to Cytology*. "His textbook was very important in getting her recognition early on," Abbe emphasized.

As she recently recalled, McClintock and her little band "did very powerful work with chromosomes. It began to put cytogenetics, working with chromosomes, on the map in the late 1920s–early 1930s.... It was just a little group of young people. The older people couldn't join; they just didn't understand. The young people were the ones who really got the subject going because they worked intensely with each other. It was group activity because they discussed everything and were constantly thinking about what they could do to show this, that, or the other thing." Two members of the group, McClintock and George Beadle, would later win Nobel Prizes, Beadle for his "one gene, one enzyme" hypothesis. Following McClintock's lead, the Cornell maize group entered its golden age.

McClintock's enthusiasm and intensity swept her ahead of the others. She worked in spurts, night and day for weeks, to solve a problem. During a long drought, she saved her corn by laying water pipes up to her hilltop patch; standing in the hot sun, she watered her plants with tears of fatigue coursing down her cheeks. During a late-night flood, she replanted her washed-out corn by the light of car headlights. To Beadle's dismay, McClintock could interpret his experimental data faster than he. He complained to the department chairman, the eminent geneticist Rollins A. Emerson. "Emerson told him that he should be grateful there was someone around who could explain it," McClintock commented dryly. "The fun was solving problems, like a game. It was entertaining."

McClintock earned her Ph.D. degree in 1927 at age twenty-five and stayed on as botany instructor. Over the next few years, she published nine papers on maize chromosomes. Rhoades considered each one a milestone in genetics and thought that she already deserved a Nobel Prize.

In the meantime, McClintock's mother still hoped her daughter would quit work and get married. "Every time I went home at vacation time, she'd try to persuade me to let somebody go up and get my things and not go back. It was a real fear on her part that I'd be a professor." But McClintock finally decided that she was too independent for close, emotional relationships. She had a faithful beau, her undergraduate chemistry instructor Arthur Sherburne, but she concluded that "marriage would have been a disaster. Men weren't strong enough...and I knew I was a dominant person. I *knew* they would want to lean against you.... They're not decisive. They may be very sweet and gentle, and I knew that I'd become very intolerant, that I'd make their lives miserable." Eventually, she told Sherburne "not to stay in touch with me."

Instead of marrying, she managed her life with "a fastidious spareness, an aesthetic of order and functionality," as her biographer Evelyn Fox Keller expressed it. Highly organized, McClintock arranged her data on cards, the cobs neatly tagged and cross-referenced to tables. She scheduled her time so that she could play a fast tennis game each day at five o'clock and still drive to her friend Dr. Esther Parker's cottage for supper before dark. Dr. Parker, a physician, had been an ambulance driver during World War I for the American Friends Service Committee. Her house was McClintock's home away from home.

Late in the summer of 1929, Harriet Creighton came to Cornell as a botany graduate student from Wellesley College. Within minutes of meeting, McClintock had organized Creighton's academic career, too, steering her to the right courses and advisers. Technically, instructors were too low-level to advise graduate students, but practically speaking, McClintock was in charge. McClintock gave Creighton her best research project as a thesis topic. In the late 1920s, there was circumstantial evidence, but no hard proof, that chromosomes carried and exchanged genetic information to produce new combinations of physical traits. McClintock wanted the proof.

She had bred a special strain of corn with an easily identifiable ninth chromosome that usually produced waxy, purple kernels. Under her microscope, she could see an elongated tip on one end of the ninth chromosome and a knob that readily absorbed stain at the other end. According to her mathematical analysis, the elongated tip was located near the region of the chromosome that determined whether the plant would produce waxy kernels. She suspected that the region near the knob was responsible for supplying purple pigment.

That spring, Creighton and McClintock planted waxy, purple kernels from the strain. In July, they fertilized the silks with pollen from a plant of the same strain whose kernels were exactly opposite types, that is, they were neither waxy nor purple.

That fall, when McClintock and Creighton harvested the ears, some of them had the usual waxy, purple kernels and some kernels were the opposite, neither waxy nor purple. But some ears were different: they had inherited one trait—but not both. Thus, they

Figure 7.2. Crossing over.
Step One. Barbara McClintock specially bred corn to produce many waxy, purple kernels. The chromosome responsible had an elongated tip on one end and a knob at the other.

were either waxy or purple, but not both. When McClintock and Creighton examined the chromosomes of these new kernels under their microscopes, they could see that their structure had changed markedly. Physical bits of the ninth chromosome—either the knob or the elongated tip—had actually exchanged places. Whereas every elongated chromosome in the parent plants had a knob, they now found a mix: elongated chromosomes without knobs and knobby chromosomes without tips.

McClintock and Creighton had proved that genes for physical traits are carried on the chromosomes. They had produced the first physical proof that exchanging chromosomal parts helps create the amazing variety of forms present in the biological universe.

Normally, McClintock liked to publish enormous amounts of supporting data in her papers; today each one of her reports would make several separate articles. So she was waiting for a second crop before publishing the data. Luckily, Thomas Hunt Morgan visited Cornell and heard about the experiment. He urged them to publish

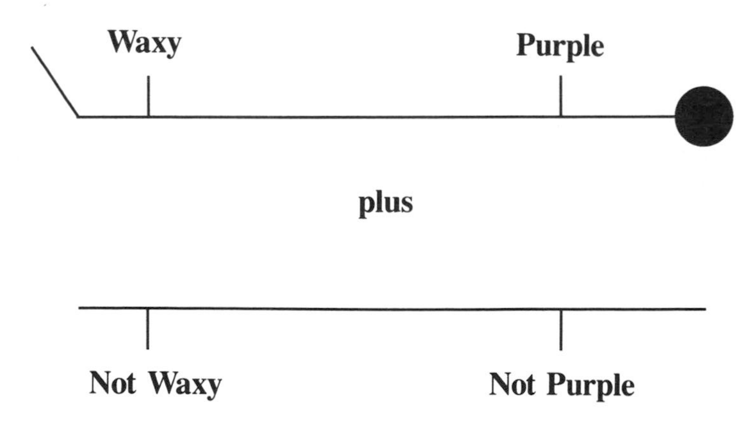

Figure 7.3. Crossing over.
Step Two. Barbara McClintock fertilized a plant with waxy, purple kernels using pollen from a plant with kernels that were neither waxy nor purple.

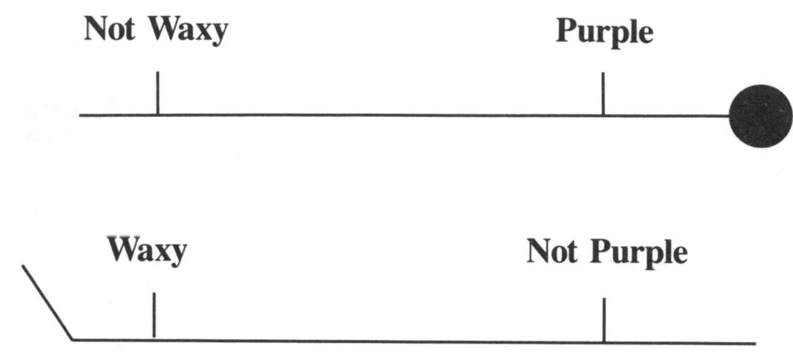

Figure 7.4. Crossing over.
Step Three. Some of the kernels produced had mixed characteristics, and McClintock could see through a microscope that bits of the responsible chromosome had exchanged places.

immediately. In his excitement, he wrote a journal editor that an important article would arrive in two weeks. Thanks to Morgan, McClintock's article was published in August 1931. A few months later, a German geneticist, Curt Stern, published parallel data on fruit flies. Had McClintock waited for another crop, Stern would have been first.

The paper made McClintock's reputation. "Beyond any question, this is one of the truly great experiments of modern biology," Mordecai L. Gabriel and Seymour Fogel declared in their book *Great Experiments in Biology*. James A. Peters, editor of *Classic Papers in Genetics*, wrote, "This paper has been called a landmark in experimental genetics. It is more than that—it is a cornerstone." Then he warned, "It is not an easy paper to follow, for the items that require retention throughout the analysis are many and it is fatal to one's understanding to lose track of any of them. Mastery of this paper, however, can give one the strong feeling of being able to master almost anything else he might have to wrestle with in biology." James Shapiro, a University of Chicago microbiologist, told the *New Scientist* magazine that the experiment should have won a Nobel prize by itself.

When Marcus Rhoades asked McClintock how she learned so much from a microscope, she replied, "Well, you know, when I look at a cell, I get down in that cell and look around." Explaining the remark later, she said, "You're not conscious of anything else.... You are so absorbed that even small things get big.... Nothing else matters. You're noticing more and more things that most people couldn't see because they didn't go intently over each part, slowly but with great intensity.... It's the intensity of your absorption. I'm sure painters have the same thing happen right along."

"When you're doing something like this, the depth of your thinking is very penetrating. You can feel the intensity of it," she added. Many scientists believe that the thrill of discovery is unique to science. But McClintock contended that engineers, historians, and writers—"anyone who must think intensely and integrate vast amounts of information to solve a problem"—must feel it too. "The thrill comes from being intensely absorbed in the material."

By this time, McClintock knew she would have to leave Cornell. Emerson, the department chairman, was one of her greatest fans but he could not override the faculty, who were strongly opposed to giving permanent faculty positions to women.

For the next five years, from 1931 to 1936, McClintock crisscrossed the country in her beloved Model A Ford. At the top of her profession, she was at the bottom of the career ladder. While her friends worked frantically to find her a permanent job, she won a series of short-term fellowships to do research at various universities. The fellowships were highly prestigious stepping-stones for men on the way to professorships. For the few women who received them, however, they were stopgaps intended to tide them over. Nevertheless, McClintock was happy to use grants from the National Research Council, the Rockefeller Foundation, and the John Simon Guggenheim Memorial Foundation to work at Cornell, the California Institute of Technology, and the University of Missouri. As she confessed, "I couldn't wait to get to the laboratory in the morning, and I just hated sleeping.

Years later, McClintock explained in a speech to the American Association of University Women what those fellowships meant to her: "For the young person, fellowships are of the greatest importance. The freedom they allow for concentrated study and research cannot be duplicated by any other known method. They come at a time when one's energies are greatest and when one's courage and capacity to enter new fields and utilize new techniques are at their height."

Of all the advances in genetics during the 1920s, one of the greatest was the discovery that X rays enormously speed up the rate of mutations, fifteen-hundred-fold in fruit flies, for example. Instead of waiting for spontaneous mutations, scientists now could produce them at will. Lewis Stadler had a Rockefeller grant to build a genetics center at the University of Missouri to study X-ray-induced mutations. Stadler planted a field with kernels from X-ray-irradiated pollen and asked McClintock to figure out how the mutations had occurred.

Studying Stadler's fields, McClintock discovered that X rays actually break a plant's chromosomes and leave them with damaged, frayed ends. Then, she was surprised to see the chromosomes mend themselves: their frayed ends fuse with the frayed ends of other damaged chromosomes. She even found that some damaged chromosomes fuse together in rings. Often, two fragments fuse in such a way that the ends of the repaired chromosome pull in opposite directions during cell division and make the chromosome break *again*. As a chromosome breaks, repairs itself, and rebreaks, its ends lose more and more genetic material. She called the entire process the breakage-fusion-bridge cycle.

Many scientists would have been content to have discovered ring chromosomes, but McClintock was always interested in maize for the clues it offered to nature as a whole. She constantly tried to integrate her specialized studies with broad questions regarding heredity in other species. Thus, when she discovered ring chromosomes, she immediately asked how the frayed ends of the damaged chromosomes find each other and repair themselves. If the genetic process includes emergency repairs, it must be able to recognize and process information. As she pointed out, "The conclusion seems inescapable that cells are able to sense the presence in their nuclei of ruptured ends of chromosomes and then to activate a mechanism that will bring together and then unite these ends, one with another.... The ability of a cell to sense these broken ends, to direct them toward each other, and then to unite them so that the union of two DNA strands is correctly oriented is a particularly revealing example of the sensitivity of cells to all that is going on within them."

McClintock's insight came a good fifteen years before other scientists like Evelyn Witkin began work on DNA repair processes in the 1950s. McClintock was already poking holes in the standard picture of the chromosome as a rigid string of stable genes, arranged like pearls along a necklace chromosome. She was starting to think of the genetic process as responsive to signals, processing information, and receiving and interpreting signals from inside and outside the cell. She was looking at nature afresh, free of the conceptual constraints that most scientists work within, observed Witkin, who, until her retirement, was the Barbara McClintock Professor of Genetics at Rutgers University. Eventually, McClintock's unbiased approach would meet head-on with those who still believed in the stable chromosome.

When McClintock and her Model A Ford moved on to Caltech in 1931, she was the first woman postdoctoral fellow to work at the men's school. Although McClintock was paying her own way with her fellowship, Caltech's board of trustees had to give its approval before she could come. Her first day there, a colleague took her to lunch at

Caltech's elegant faculty club. As she walked the length of the dining room to an empty table, everyone stopped eating and stared at the tiny thirty-year-old woman with her boyish figure, tousled hair, and practical clothes. To Warren Weaver of the Rockefeller Foundation, she seemed "more boy than girl."

Alarmed at the stares, McClintock demanded, "What's wrong with me?"

"Oh, everyone's heard about the trustees' meeting, and they're looking you over," her host replied cheerily.

Caltech's practice was to make visiting researchers with fellowships automatic members of the faculty club, but McClintock was never allowed in the building again. Nor did she visit any labs other than her own and that of Linus Pauling, the politically liberal chemist who later won two Nobel prizes. Scientifically, however, her visits to Caltech were productive. Two summers later, she discovered the nucleolar organizer there. The nucleolar organizer region of the chromosome helps form the nucleolus, the cell's factory for synthesizing ribosomes. Although Caltech would not hire her full-time, she did not mind helping men who were hired there. When Charles Burnham, one of her old gang at Cornell, asked her what he should teach in his cytology techniques class at Caltech, she laid the course out for him. It was 1971 before Caltech hired its first woman professor, Olga Taussky Todd, a protégé of Emmy Noether.

Using her Guggenheim fellowship, McClintock visited Germany in 1933, the traumatic year in which Hitler became chancellor and fired the Jews from German universities. Science laboratories were in chaos, and her student residence was empty except for herself and a Chinese gentleman, who dined in silence. Loneliness, the politicizing of genetics, and the persecution of Jews appalled her. In December, she fled back to Cornell.

She returned at a bad time. The Depression was worsening and universities were cutting back. Few could afford a pure researcher. As Warren Weaver observed at Cornell, "The Dept. of Botany does not wish to reappoint her, chiefly because they realize that her interest is entirely in research and that she will leave Ithaca as soon as she can obtain suitable employment elsewhere; and partly because she is not entirely successful as a teacher of undergraduate work. The Botany Dept. obviously prefers a less gifted person who will be content to accept a large amount of routine duty."

Friends interceded with the Rockefeller Foundation, however, and arranged $1,800 a year for McClintock to spend two more years at Cornell. Morgan wrote the foundation that "she is highly specialized, her genius being restricted to the cytology of maize genetics, but she is definitely the best person in the world in this narrow category." The $1,800 was the largest income McClintock had ever earned.

Testifying on her behalf, Morgan also confided that "she is sore at the world because of her conviction that she would have a much freer scientific opportunity if she were a man." But McClintock denies that she was ever bitter. Realistic at recognizing prejudice, yes, but never bitter. "If you want to do something, you have to pay the price and never take it seriously. I never worried. I couldn't compete with men, so I didn't try."

When McClintock left Cornell for good in 1936, Cornell's golden age of maize genetics ended. After years of trying to get her a permanent position, friends had finally found her a job with Lewis Stadler at the University of Missouri starting in 1936. She would be only an assistant professor—far below the rank and pay of a man with comparable attainments—but it was her first faculty position. Her wandering years were over. Or so she thought.

For several years, McClintock worked in Columbia, Missouri, during the winter and raised her corn plants at Cornell during the summer. She grew only a few thousand plants each year, but they were highly selected, so she had no waste. "I wanted to know each plant well, so I carefully organized what I was going to need and why, and how many samples I needed in each case. I was highly organized... so that it was manageable. It had to be manageable. The recording was equally foolproof. I didn't want to have anything come up that seemed irrational and not right, and if I did it myself I would know, because my memory would tell me where to look... and how to find the error."

Helen Crouse, who had read McClintock's nucleolus paper as an undergraduate at Goucher College, visited Ithaca the summer of 1938. When she asked a timid young man how to find McClintock's lab, he replied, "Oh, well, she's up under the roof, and she doesn't want to see anybody." But he took Crouse up anyway. McClintock came to the door with a green, opaque visor over her eyes and a cigarette in a long filter holder in her hand. "What do you want?" she demanded. Crouse turned around, but her companion had vanished. After Crouse introduced herself, McClintock answered, "I heard you were coming. I was expecting you. Let's go home for lunch."

Home was Dr. Parker's house. When they got to the porch, McClintock sprayed their ankles well with flea repellent because Parker kept three large Irish setters. "We had a great lunch with Dr. Parker, who never knew whether her dress was right- or wrong-side out and didn't care. She was a wonderful vigorous sort of person.

And I must have stayed a week," Crouse said. A few weeks later, McClintock invited Crouse to the Genetics Society meeting in Woods Hole, Massachusetts. "I didn't have fifty cents; but she said she'd pay all my expenses to go, that she'd like to have someone to go with her," Crouse said. "I had a glorious time."

After Crouse's sun-filled visits in Cornell and Woods Hole, she started graduate studies at the University of Missouri. There she was surprised to discover McClintock's position was not only clouded over but downright stormy. As a teacher, McClintock was intense, inspiring, and so full of ideas and fast talk that it was hard to keep up. She had insisted on proper equipment, and the university bought her new microscopes for a lab course. She installed them late one Friday night, putting a slide in each and delicately adjusting their lights and lenses to highlight the important feature in each demonstration. The next morning, the students gave a passing glance to the demonstrations on their way to pollinate their fields. McClintock was crushed. On the way to lunch with Crouse, she burst into tears—because the "corn boys" had skipped some of the slides. "She took it all so intensely, " Crouse realized.

As usual, McClintock was way ahead of everyone else. Taking a quick look through Crouse's microscope one day, she discovered more than Crouse had found in her own material. Crouse had not adjusted her microscope's light and lens properly, and McClintock stalked out of the lab, slamming the door behind her. "You had to have a pretty sturdy constitution to survive," Crouse decided. McClintock was not about to waste her time on inept students, especially when jobs were scarce for even the best.

McClintock reigned over a spacious third-floor lab like "the Queen Bee. Everyone was scared of her," according to Crouse. Technically, Crouse was not McClintock's graduate student, so there was little tension between them. But McClintock's sharp tongue so terrified one of her official graduate students that he fled by the back greenhouse door whenever she entered the front. Another young man escaped to Berkeley.

Although the Rockefeller Foundation regarded Stadler and McClintock as the leaders of the genetics center at Missouri, university administrators thought that McClintock was a troublemaker and hoped she would leave. While everyone wore knickers for field research, McClintock wore pants *all* the time. She even let her students work in the lab past the eleven P.M. campus curfew. Then one Sunday she forgot her keys, climbed into her lab through a ground-floor window, and totally scandalized the locals. The culture shock was reciprocal. Crouse was appalled that agriculture students practiced their hog calling on campus. She was even more upset to learn that wildlife students hunted at night by blinding animals with their car headlights before they shot them.

Whatever the reason, McClintock was in a no-win situation. Excluded from faculty meetings, she was not part of the department. The authorities would not accommodate her research needs; she arranged for substitute lecturers each fall so that she could harvest her plants in Cornell, but the administration disapproved. At the same time, she could not get another job. She was expected to recommend male colleagues for the likes of Yale, Harvard, and elsewhere—"jobs that would have been just right for me, with my experience"—but she was never considered for those jobs herself.

"Missouri was very conventional, and there was no hope. And also, you get tired of being always the lowest one on the ladder," she thought. Crouse thought she was "absolutely furious that no one paid her any attention."

McClintock had a wry sense of humor. When the University of Rochester gave her an honorary degree, for example, she called it "getting my shirt stuffed." But the bite of her wit grew sharper at Missouri. She and some of her students ate supper at Mrs. Pyles's boardinghouse, which was extensively decorated with religious objects. One day Mrs. Pyles rearranged her pictures, and McClintock joked about "creeping Jesus." Bible Belt Missouri was unamused. Crouse had a professor who could not remember her name, until a friend pointed out that Crouse rhymes with "mouse." McClintock laughed and took to calling Crouse "Miss Louse." But when Crouse retaliated and called McClintock "Babs," McClintock was irritated.

"I hadn't known she was such a tiger," Crouse conceded. But then, Crouse realized, women who succeeded in science were "the ones with the strength to abide in a world where they weren't wanted. They had to have stamina and brains and nerve and gall to survive. You're not going to find any weeping willow making it." As for McClintock, she claimed that when she was nervous or upset, she talked too much. Then she would blow off verbally and afterward not remember why. "I don't remember bad things."

McClintock was searching for ways out of the trap. On the way to lunch at Jack's Latch cafeteria, she often stopped in the post office to chat with the federal meteorologists. She was trying to teach them new forecasting methods, and as the University of Missouri became more intolerable, she toyed with the idea of becoming a weather forecaster, too. Finally, in 1941, she asked Missouri's dean if she would ever get promoted to a permanent position. "If Stadler leaves," the dean answered, "you'll probably be fired."

"I want a leave of absence—and I won't be back," McClintock snapped back.

"I thought you were going to say that" was his only response.

"There was no use staying there," McClintock thought. "Though it was good for the work, it was bad for the morale and too hard to take....I didn't want a job. I just didn't want one anymore, and I decided I'd never go back to a university. That was out."

"I just quit the whole business," McClintock declared. She had no job, no means of support, no place to work, and no prospects.

She did not care about her career, but she did care about her corn. Writing Marcus Rhoades, then at Columbia University, she inquired where he grew his plants. "Cold Spring Harbor" was the reply.

Cold Spring Harbor had been founded on rural Long Island in 1890 as a summer center for the study of Darwin's evolution theory. In 1941, a handful of researchers worked there year-round, financed by the Carnegie Institution of Washington. In summertime, as many as sixty geneticists, including Harriet Creighton, Marcus Rhoades, Max Delbrück, and Salvador Luria, flocked there. Today, Cold Spring Harbor is a large, privately funded research center for basic biological research and is financed by federal and private grants.

McClintock wangled an invitation to plant her corn at Cold Spring Harbor that summer. In the fall, she stayed on in a summer house until the weather turned cold and Marcus Rhoades lent her a spare room in his New York apartment. Finally, a friend, Milislav Demerec, became genetics director at the lab and offered her a temporary position.

Before she could get permanent status from the Carnegie Institution, she had to go to Washington, D.C., to be interviewed by its president, Vannevar Bush. Demerec nagged McClintock to go, but she kept postponing the trip. Finally, he ordered her to take a plane.

Not caring whether she was hired or fired, McClintock went to see Bush "with complete freedom from any nervousness. And, as a consequence, we had a very good time talking, because I simply didn't care what his opinion would be. It took three or four years before I realized that I could stay in a job, that this was more like no job at all. I had complete freedom....I could do what I wanted to do, and there were no comments. It was simply perfect. You couldn't mention a better job. It was really no job at all."

The decade that had started so disastrously in Missouri ended gloriously at Cold Spring Harbor. It was Barbara McClintock's kind of place. Everyone wore blue jeans, worked seventy to eighty hours a week, and loved biological research. Teaching was not required, and there were no restrictions on research. Thanks to support from the Carnegie Institution, McClintock was free and independent of any changing administrations at Cold Spring Harbor.

McClintock settled into a routine undisturbed by passing decades. She alternated quiet winters analyzing data with busy summers filled with visitors and corn growing. For exercise, she ran, swam, and played tennis. Loaded with field guides, she took long nature walks, gathering black walnuts for brownies or checking the spots on ladybug beetles.

In addition to her cornfield, she had a spacious laboratory within a stone's throw of Long Island Sound. Seven days a week she worked from early morning until late evening on a long surface made of several desks pushed together. In a small side room she stored boxes of dried corncobs, each carefully tagged and cross-referenced so that when colleagues asked for seed of a particular strain she could explain its lineage. When she entertained friends, they met in the lab.

Across the road she kept an unheated, two-room pied-à-terre in a converted garage. Her real home was her lab, so she kept no telephone in the apartment; lab employees relayed night-time emergency messages. The apartment was as meticulously organized as her work. All the hangers in her closets faced the same direction and none touched another. Each sheet in her linen closet was enclosed in a plastic bag and tagged for size. "She was totally dedicated to efficiency," Crouse observed on visits.

McClintock enjoyed quality equipment. Although she ate most meals in the lab dining room, she bought a spectacular electric range with purple, green, and red lights and a complete set of copper-bottomed Revereware pots. She cared for her cars and, until she was eighty, changed their tires. She stripped and reassembled her microscopes. When she found a piece of machinery she liked—an electric fan or a tabletop vacuum cleaner to remove corn chaff—she often bought three of each.

Life at Cold Spring Harbor became both McClintock's strength and her weakness. Thanks to the support of the Carnegie Institution, she could work without interruption, even on unpopular projects. But isolation also left her without colleagues to popularize her research to the scientific community at large. For the first time in her career, McClintock would have to explain her own work.

She began reaping the benefits of her international reputation during her early years at Cold Spring. In 1944, she was elected the first woman president of the Genetics Society of America. That same year, she was named to the prestigious National Academy of Sciences, which had admitted only two other women in eighty-one years. Surprisingly, when McClintock heard about the honor, she burst into

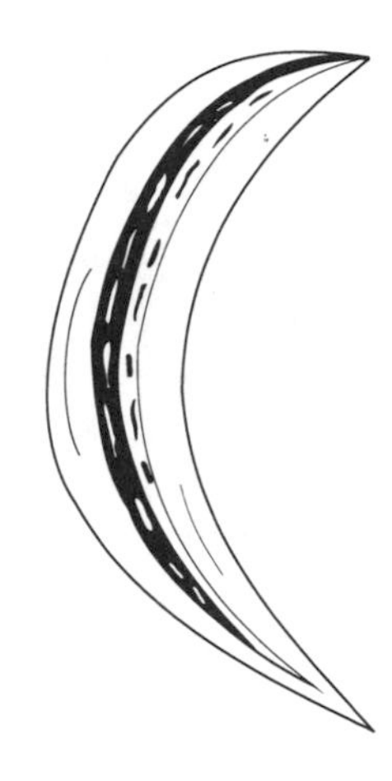

Figure 7.5. Breaking.
A plant with a long history of inbreeding and self-fertilization produced leaves with bizarrely colored twin splotches.

bitter tears. Had she been a man, she said, she would have been delighted by the honor. But as a woman, she felt trapped. She wanted to be free to walk out on genetics if she ever got bored. Now she would never be able to leave it. "It was awful because of the responsibility to women," she explained. "I couldn't let them down." As she wrote a friend, "Jews, women, and Negroes are accustomed to discrimination and don't expect much. I am not a feminist, but I am always gratified when illogical barriers are broken—for Jews, women, Negroes, etc. It helps all of us."

World War II had put women to work in unprecedented numbers. In its wake, McClintock felt buoyant and self-confident enough in 1947 to declare, "Opportunities for women have never been greater than they are at the present time. There is no question in my mind that these opportunities will become increasingly better and at a very rapid rate. The restrictions in opportunity...are being steadily removed."

Challenging her maize plants with broken chromosome problems at Cold Spring Harbor, McClintock was fascinated by their response. During the winter of 1944–1945, she planted a greenhouse with self-pollinated kernels. Each was the heir to a long traumatic history of inbreeding and self-fertilization that had resulted in broken arms at the end of their ninth chromosome. When the seedlings sprouted, she was astounded. The leaves had broken out with quirky patches of curiously colored patterns. Moreover, the bizarre patches occurred in pairs. The leaf of one plant, for example, had two albino splotches of similar size side by side: one patch contained many fine green streaks while its complementary twin patch contained only a few green streaks. The results, McClintock thought, were startlingly conspicuous and totally unexpected. Generations of breaking, healing, and rebreaking the chromosomes had created a crisis in the plant's genetic system. Every time a cell divided, chromosomes broke and some genes were lost.

Because the complementary patches sat side by side, McClintock immediately realized that some bizarre event had struck the plant's cells as they had divided. "One cell had gained something that the other cell had lost," she told herself. "I set about to find out what it was." Eventually she realized that when a chromosome that has broken and refused breaks in two again, one of its parts may gain some genetic material while the other part may lose some.

McClintock was fascinated by everything around her, including her own mind, and she described its functions as objectively and precisely as she did her plants. Hence, she described her reaction to the strangely spotted plants by saying, "My mind went straight on it and worked quite hard on thinking about it, and it seemed all logical that we'd just missed the idea. So I had a pretty good feeling for it, and I had a pretty good feeling."

From the beginning, she knew she had discovered a basic genetic phenomenon, not just an event unique to maize. Long before scientists knew that genes are made of DNA, she asked the next question: How are genes controlled?

Comparing chromosomes of both the plants and their parents under her microscope, she deduced that parts of their chromosomes had changed positions. Six years of painstaking research later, she would be able to prove that a gene need not have a fixed position on a chromosome. She would conclude that genes are not stable pearls laid out along a chromosome string. Instead, they can move around and turn on and off at various times during a cell's development.

Eventually, McClintock described and characterized two new kinds of genetic elements: The first is a controlling element, a switch to turn on and off the genes that express physical characteristics like color or size. The second type is an activator that can make the on-and-off switch jump around from one part of a chromosome to another. Today, McClintock's discovery is called genetic transposition, and the moving chromosome parts are called transposable elements, transposons, or "jumping" genes.

Thus, an activator gene can cause the off-switch gene to jump next to a pigment gene and turn off the color. If the off-switch turns off the pigment gene early in a plant's development, a large region of the plant gets no pigment. If the pigment gene is turned off partway through development, parts of the plant are streaked or spotted with color. When the activator makes the off-switch turn back on, the pigment gene resumes work.

As a result, not only are genes unstable, but their mutation effects

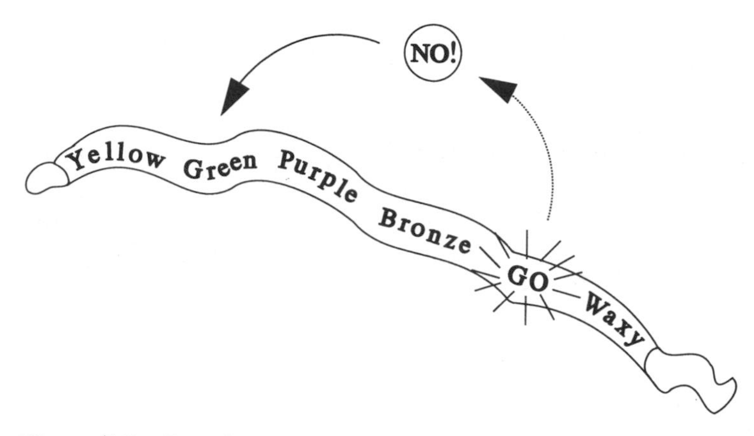

Figure 7.6. Jumping genes.
The genetic activator makes the switching gene move from one part of the chromosome to another to turn genes on and off.

are too. Geneticsts had assumed that a mutated gene was dead and could not be reactivated. But McClintock showed that environmental conditions could reverse some mutations and turn the genes back on. Her experiments provided a radically fluid picture of genetics, in contrast to the old view of stable mutations and immovable genes.

The implications of transposable elements fascinated McClintock even more than the discovery itself. She saw immediately that transposons are a fundamental phenomenon that helps explain the incredible variety of organisms produced by nature. In 1951 she noted, "The same mechanisms may well be responsible for the origins of many of the observed mutations in plants and animals." In a famous 1955 statement, McClintock prophesied that it "would be surprising indeed if controlling elements were not found in other organisms."

For six years, McClintock collected evidence, stuffing cards, tables, filing cabinets, and shelves with data. She was so excited that she often called Evelyn Witkin down from her lab to see the latest wonder. "It was a great thing to see. She was getting such really intense joy out of it," Witkin remembered. "She was so very sure of what she was seeing, and her evidence was absolutely convincing."

While McClintock was studying transposons, the world of genetics was changing. Chemists and physicists had joined the hunt for the physical basis of hereditary. Trained in Cold Spring Harbor summer schools, they applied the principles of physics to biological problems. In their excitement, these new molecular biologists ignored previous work by crystallographers, biochemists, bacterial experts, chemists, and geneticists, including McClintock. The molecular biologists' softball games became a symbol of their disregard. As the codiscoverer of DNA's structure James Watson told the story, the softball "all too often" wound up in McClintock's cornfield.

In an hour-long talk at a major Cold Spring Harbor symposium in 1951, McClintock summarized her findings before a group of leading scientists. The report was long, complicated, and dense with statistics and proofs. When she finished, there was dead silence, Witkin remembered. "It fell like a lead balloon," recalled Harriet Creighton. McClintock felt as if she had "collided with the stable chromosome."

Scientists scrambling to learn molecular biology wanted it simple; they did not like a genetic system that was fluid, moving, changing, and intricately regulated. They reacted with puzzlement, frustration, even hostility. "I don't want to hear a thing about what you're doing. It may be interesting, but I understand it's kind of mad," a biologist told her. A leading molecular biologist called her "just an old bag who'd been hanging around Cold Spring Harbor for years."

Understandably, McClintock was upset and disappointed. She summarized her work in a longer article published in 1953. Maize geneticists understood and accepted the data, but she wanted the science community at large to realize the wider significance of her work. Only three scientists outside her field, however, requested copies of the article. McClintock concluded that publishing was a waste of time. From then on, she wrote up her work in large notebooks, all tabulated, documented, and analyzed, and filed the notebooks on a shelf. She submitted only brief summaries of her work for publication in the annual reports of the Carnegie Institution of Washington—which only a few libraries purchased. "I don't know of any other scientist who would have had the discipline or self-confidence to do that," observed a friend of her later years, molecular biologist Bruce Alberts of the University of California at San Francisco. She stopped giving seminars at Cold Spring Harbor, too. Twenty years ahead of her time, McClintock went into "internal exile" at the lab, waiting for the scientific community to catch up with her.

McClintock so enjoyed ideas and thinking that the pain of being ignored soon slipped away. "I was startled when I found they didn't understand it, didn't take it seriously," she explained. "But it didn't bother me. I just knew I was right. People get the idea that your ego gets in the way a lot of time—ego in the sense of wanting returns. But you don't care about those returns. You have the enormous pleasure of working on it. The returns are not what you're after."

Being ignored gave McClintock more time to work and learn about other fields of biology. She was one of the few nonmolecular specialists who kept up with molecular biology. "Despite her age and her coming from a very specialized area of biology, she's on top of everything," as Alberts noted while McClintock was still alive. She devoured nonfiction—from biographies to monographs on offbeat biological subjects. Keeping an open mind about anything she could not understand, she viewed nature's oddities as windows onto fundamental phenomena in nature. She read up on stick insects, animal mimicry, plant galls, midwife toads, extrasensory perception, and the methods by which Tibetan Buddhists control their body temperature. She regularly scanned twenty biological journals of widely differing specialties; one year she spent a month reading all the literature on insect evolution.

Finding transposable elements everywhere in nature, McClintock photographed them for her own pleasure and for teaching her friends. Driving past a field of Queen Anne's lace, she would stop her car to walk through the field. Each flower of Queen Anne's lace consists of a cluster of florets, each formed from the progeny of a single cell. Normally, the white florets on the outside rim of the blossom open first and the center floret opens last to reveal a spot of pink, green, or purple pigment. But on closer examination, McClintock found blossoms where the colored floret was not confined to the center. The activator gene had turned the pigment gene on too soon. "It was the right pattern in the wrong place at the wrong time," she realized. Her face lighting up at the memory, she insisted, "You can see the pleasure. The pleasure is very great....I love the springs, summers, and falls for all they can entertain you with."

As legend has it, McClintock was ignored because she was a woman and because scientists thought her "crazy" and "mad." But this is incorrect; most geneticists did not think she was crazy. McClintock had been famous and highly respected for years, Witkin emphasized. "Most geneticists didn't think she was crazy. It was just extremely difficult both to understand her experiments and to reconcile her conclusions on transposable elements with the prevailing belief in the stability of genes on the chromosomes." Asked about McClintock's work in 1951, the great geneticist Alfred H. Sturtevant replied, "I didn't understand one word she said, but if she says it is so, it must be so!"

Corn and fruit fly geneticists quickly incorporated her ideas into their graduate courses and conducted follow-up experiments. Her work was included during the 1950s and 1960s in authoritative books like James A. Peters's *Classic Papers in Genetics* (1959) and L. C. Dunn's influential *Short History of Genetics* (1965). Nobel Prize–winning biologist David Baltimore said, "I remember growing up as a student in the sixties; one of the things all of us tried to read were Barbara McClintock's papers in the Cold Spring Harbor symposia from the

1950s. But a lot of us gave up." Her results were complex and possibly irrelevant to molecular biology in other organisms.

Nevertheless, the fact remains that the scientific community at large ignored transposable elements for years. "Transposable elements are an example of how new ideas are accepted coldly by the scientific community," a much younger geneticist, James Shapiro, declared. "If she says something has happened, she has seen it in dozens and hundreds of cases. One reason that people don't read her papers is because the documentation is so dense. So first they said she's crazy; then they said it's peculiar to maize; then they said it's everywhere but has no significance; and then finally they woke up to its significance."

McClintock became discouraged enough to write Marcus Rhoades and Helen Crouse during the 1960s and 1970s to ask about jobs elsewhere. For two winters in the late 1950s, she even suspended her research entirely and trained Latin American cytologists to identify maize strains for the National Academy of Sciences. The adoption of modern seed was destroying indigenous maize strains. Studying the geographic distribution of particular chromosomes, McClintock realized that they revealed ancient migration and trade routes. Corn seeds are so tightly enclosed in their husks that the plants cannot travel without people. Her insights led to a major study of ancient migrations based on the chromosomes of present-day maize plants. Thanks to her Latin American visits, McClintock mastered the Spanish language, which she kept up by watching Spanish television stations.

During the 1960s, when McClintock could have considered retirement, she collected awards from Cornell University, the National Academy of Sciences, and the National Science Foundation. None of these honors was given for her transposable element work. Nevertheless, a parade of pilgrims began to line up outside her door to learn from her. Many remained her friends. As always, to save time for activities she loved, she concentrated on her family and on close friends who interested her; she ignored casual acquaintances who bored her.

With friends, she was warm, charming, and open—far from the recluse that the media made her out to be. In fact, she studied human nature the way she studied corn—carefully, precisely, and with absorbing interest. An enthusiastic teacher one-on-one, she moved instinctively to the age and intellectual level of the person she was talking to, Guenter Albrecht-Buhler discovered. Speaking before McClintock's death he said, "She's far ahead of her time and tries not to startle you with it. I think it's a defense mechanism from the time when it was important for women not to be brighter than others....She enjoys making things clearer. She's a passionate teacher. The passion of her existence is removing the fog...."

Often, the highlight of a visit with McClintock was a nature walk, during which she showed these professional biologists things they had never seen before. For example, "I'm *very* interested in galls. When an insect injects a chemical into a plant, the plant grows an elaborate, highly specific home that fits that particular kind of insect perfectly. And one grape plant may have many different types of galls.This tells me that organisms have all the necessary machinery, the potential, to make any kind of organism. All around you, there is so much pleasure, if you think about it."

Molecular biology finally caught up with McClintock during the late 1960s when James Shapiro and others discovered transposable elements in bacteria. Suddenly, molecular biologists started finding mobile genetic elements in all kinds of organisms, including people. Transposable elements are used in much of today's genetic engineering. They are responsible for many mutations and play an important role in evolution, inherited birth defects, resistance to antibiotics, and perhaps the incidence of cancer. The movement of genes and gene segments on chromosomes helps to explain how cells produce antibodies to combat a host of different viral and bacterial threats, how bacteria retaliate by acquiring immunities to human defenses, and how certain cancer cells develop. These genetic elements, cloned by recombinant DNA techniques, are used to carry desired genes to new hosts. Scientists today make mutations with transposable elements, instead of with chemicals and X rays. Watching the discoveries multiply, McClintock wrote a friend, "All the surprises...revealed recently give so much fun. I am thoroughly enjoying the stimulus they provide."

Contemporary scientists regard the inheritance process as a fluid information-processing system, much like a computer. "We now think of a dynamic storage system subject to constant monitoring, correction, and change by dedicated biochemical complexes," Shapiro explained in an article in *Genetica*. "We can now think about integrated, multigenic systems that can be turned on and off in a coordinated fashion according to the needs of the organism."

By the late 1970s, McClintock's honors were piling up in glorious profusion, this time for transposable elements. In 1980–1981, she received eight major awards, three of them in one week: the Albert Lasker Basic Medical Research Award, the $100,000 Wolf Prize in Medicine from the Wolf Foundation in Israel, and the MacArthur Foundation Fellowship, $60,000 a year tax-free for life. As McClintock noted, she made her money late in life.

Her reaction? "Rather upset. I'm not a person who likes to

accumulate things," she explained, squirming miserably in her chair during a press conference. "I don't like publicity at all....It's too much at once." Her biographer, Evelyn Fox Keller, conducted five interviews before McClintock broke off discussions. Keller wrote about McClintock as a brilliant recluse, a mystic whose "passion is for the individual, for the difference," not in broad fundamental issues common to all of biology. When *A Feeling for the Organism* was published in 1983, McClintock announced tersely, "I want nothing to do with a book about me. I do not like publicity." She never read the book. She even refused to autograph it for a colleague.

McClintock's friends reacted to the book in a variety of ways. But virtually all stressed that McClintock was neither a recluse nor a mystic. And they argued that she had always been interested in maize as a window on fundamental biological phenomena and not just as a study in and for itself. McClintock herself denied that she was a mystic, if being a mystic meant believing in something she knew little about. She said she did not dismiss phenomena that she did not understand, but she did not believe in them either. "You just don't know," she declared flatly.

Early in the morning of October 10, 1983, McClintock was listening to her apartment radio when she learned that she had been awarded the Nobel Prize for Medicine and Physiology. The Nobel Committee called her work "one of the two great discoveries of our times in genetics," the other being the structure of DNA. The prize was remarkable in many respects. Only once before had the Nobel Committee waited so long to award a researcher. She shared the award with no one; in the past several decades, all but a handful of the medical and physiology prizes have been shared by two or three winners. She was the seventh woman to receive a science Nobel. And finally, the prize, which is generally given for medical or animal biology, had never been awarded for studies of higher plants. McClintock won only after it was clear that her work had implications beyond botany.

Overwhelmed at the news, McClintock took a walk in the institute woods, collecting black walnuts and her thoughts. "I knew I was going to be in for something," she explained. "I had to psych myself up. I had to think of the significance of it all; to react. I had to know what approach I would take."

Then she told the lab's administrative director, "I will do what I have to do." She issued a press release noting how unfair it seemed "to reward a person for having so much pleasure, over the years, asking the maize plant to solve specific problems and then watching its responses." Then she held a press conference, sitting on a stool in her carefully pressed dungarees and shirt, whispering courteously.

At eighty-three, her brown hair was graying, her skin was sun-wrinkled, and her eyes were bright.

"I don't even know what the award brings in," she admitted.

"It's approximately $190,000," a reporter replied.

"Oh, it is," she whispered. The reporters laughed. Then, with characteristic objectivity, she spelled out how her mind was working. "No, I didn't know, and I'll just have to get to one side and think about this."

Thanking the Carnegie Institution of Washington, she said, "I don't think there could be a finer institution for allowing you to do what you want to do. Now, if I had been at some other place, I'm sure that I would have been fired for what I was doing, because nobody was accepting it, but the Carnegie Institution never once told me that I shouldn't be doing it. They never once said I should publish when I wasn't publishing."

Asked if she was bitter at having to wait so long for recognition, she took pains to explain, "No, no, no. You're having a good time. You don't need public recognition, and I mean this quite seriously. You don't need it. You need the respect of your colleagues.... When you know you're right, you don't care. You can't be hurt. You just know, sooner or later, it will come out in the wash, but you may have to wait some time. But...anybody who had had that evidence thrown at them with such abandon couldn't help but come to the conclusions I did about it."

Furthermore, she reiterated, "It's such a pleasure to carry out an experiment when you think of something—carry it out and watch it go—it's a great, great pleasure. It couldn't be nicer.... I just have been so interested in what I was doing, and it's been such a pleasure, such a deep pleasure, that I never thought of stopping.... I've had such a good time, I can't imagine having a better one.... I've had a very, very satisfying and interesting life."

The announcement that Barbara McClintock had won the Nobel Prize electrified the scientific community like no other recent prize—as much for the beauty of her motivation and dedication as for her scientific tour de force. When McClintock accepted her award from King Carl Gustaf in Stockholm, the ovation from the normally reserved and formal audience was so loud that it made the concert hall floor vibrate. Her solitary excellence, her quiet thoughtfulness, and her perseverance in the face of male prejudice and scientific rejection had captured their imaginations. Talking briefly with a Carnegie trustee afterward, McClintock parted with the words, "We women have to stick together."

The Nobel Prize with its competition, publicity, fawning hangers-on, and name-droppers was a burden for McClintock. "You put up with it," she remarked tersely. "It's a good thing that it happened so late in life," she told a friend. Otherwise, it would have interfered with her work. Overall, she said, "It's been very, very difficult on a person. It hasn't been easy or pleasant."

Despite the Nobel, McClintock continued with her research. In her eighties, she switched her exercise program from running to aerobic dancing. She ate a chocolate a day, traveled twice yearly to South America where much of today's maize research is conducted, and worked twelve-hour days. Her reading was as encyclopedic as ever. Her work table was covered with neat piles of reading material carefully underlined with a ruler with coded red, blue, and green ink. She read thoroughly and in an organized manner on multiple levels. Pencil notes filled the margins: "imp" beside each important point and "exp" for possible experiments. She spent much of her time helping molecular scientists analyze her material.

The tiger in McClintock mellowed, and there were fewer blasts of impatience. As McClintock neared ninety, she began to slow down to an eight- or nine-hour work day. Minor health problems irritated her. "I'm almost ninety," she told a caller. "And in my family ninety is the end, and I'm beginning to feel it."

She still passionately resisted anything that bored or distracted her from the main joys of life. As she protested, "I want to be free."

On September 2, 1992, Barbara McClintock died. At age ninety, she was free.

* * *

8

Maria Goeppert Mayer

June 28, 1906–February 20, 1972.

MATHEMATICAL PHYSICIST

Nobel Prize in Physics 1963

"NEVER BECOME *just* a woman," her father insisted.

Maria understood instantly. Someone who was "just a woman" was a housewife interested only in her children. "This he didn't like," she explained late in life. "My father always said I should have been a boy."

To follow in her father's footsteps, she vowed to become a scientist and the seventh-generation university professor in the Goeppert family. To that end, Maria Goeppert Mayer patched together a career of volunteer work unique in the annals of the Nobel Prize. In all, she worked for thirty years in three different fields for three American universities—as an unpaid volunteer. She taught, supervised graduate students, served on university committees, and published articles—but she did not receive a university salary until ten years *after* her Nobel Prize–winning work.

Maria Gertrud Käte Goeppert was the only child of Friedrich and Maria Wolff Goeppert. She was born on June 28, 1906, in Upper Silesia, which was then a part of Germany but is now in Poland. When she was four years old, her family moved to the small university town of Göttingen in central Germany. Göttingen treated Maria like a little princess, and both the town and its university profoundly influenced her life.

Friedrich Goeppert was the professor of pediatrics at the university. He also directed a children's hospital and founded a day-care center for the children of working mothers. As a German university professor, he occupied a position of great prestige. German professors were on a par financially with successful American doctors, lawyers, and businessmen. Socially, they were the equals of high government officials. When Frau Professor Goeppert shopped, she was served first—unless another "Frau Professor" who outranked her in some subtle way was also present. Göttingen's social life revolved around the parties hosted by professors' wives, and their children were Göttingen's socialites.

Competition to be Göttingen's leading hostess was fierce, but Frau Goeppert's parties set the standard for hospitality. When she entertained, she opened every room of the house to guests. If the dance band quit at midnight, she played the piano for singing until four A.M. When she bought a Christmas tree and discovered a taller one elsewhere, she returned the inferior tree for the better one.

Although Frau Goeppert set her daughter's entertaining and mothering standards, Maria's father was her favorite parent. "He was, after all, a scientist," she explained. She called him a gentle bear of a man, because he doted on children and they followed him Pied Piper–like through the streets. He took his daughter on science walks, hunting for quarry fossils and studying forest plants. When she was three and a half years old, she asked for—and he gave—an accurate description of a half-moon; when she was seven, he made her dark lenses for watching a solar eclipse.

As a pediatrician, Professor Goeppert liked children to feel brave and self-confident. He was famous for rousing his little surgical patients from bed the day after their operations; at the time, other doctors insisted on weeks or months of bed rest. He thought that mothers were the natural enemies of their children: women stifled inquisitiveness and daring. When Maria climbed a tree, her father led Frau Goeppert away, lest she inhibit her daughter's spirit.

An only child, Maria was thin and pale, with almost translucent skin—definitely not an "outdoor" girl. She suffered from frequent and throbbing headaches all her life, but her father told her, "We've done all we can for them. You can either be an invalid, or you can ignore them and go on as best you can." She decided to ignore them.

On Maria's eighth birthday in 1914, Archduke Ferdinand of Austria was assassinated. World War I soon followed. Toward the end of the war and during the runaway inflation that followed, the Goepperts dined on turnip soup and pigs' ears to save food for the children in Dr. Goeppert's clinic. A Quaker organization fed Maria hot lunches at school, and when a *New York Times* reporter wrote that "the professor who is feeding the children is suffering from malnutrition himself," food packages poured in from the United States. By the time Maria was in her late teens, however, conditions had returned to normal and Mrs. Goeppert had resumed her fabulous entertaining schedule. Maria frequently voiced her bitterness, however, about the postwar transfer of Upper Silesia, the region where she was born, to Poland.

Maria Goeppert Mayer.

Maria Goeppert Mayer with Joseph Mayer about the time of their marriage.

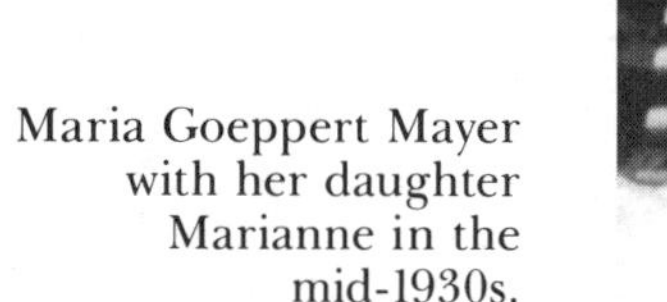

Maria Goeppert Mayer with her daughter Marianne in the mid-1930s.

Marie Goeppert Mayer with her thesis adviser Max Born.

"When I grew older," she explained later in life, "there was never any doubt in the minds of my parents or myself that I would study at the university.... Ever since I was a very small child, I knew that when I grew up I was expected to acquire some training or education which would enable me to earn a living so that I was not dependent on marriage." Getting enough education to pass a university's rigorous entrance examination seemed impossible, though.

Even as late as 1924, Göttingen had two large schools that prepared hundreds of boys for the university every year while the city had nothing comparable for girls. For several years, Mayer attended a small, private school endowed by suffragettes. When it collapsed during the post–World War I inflation, she decided to take the university examination a year early. Her former teachers were horrified.

"You won't be able to do it," they scolded her.

"I will," she replied resolutely.

"You won't be able to do it, *and* you're too young. You won't be admitted," they warned.

"All right, I'll take my chance on that too," she announced firmly. And then she pulled strings until she was admitted to the examination.

She and four other girls from the suffragettes' school took and passed the exam. "Imagine," Mayer said indignantly years later, "Five girls compared to hundreds of men!" When Mayer entered the university at Göttingen in 1924 to study mathematics, fewer than one in ten German university students was a woman. In the United States, women accounted for roughly one out of three students.

Mayer attended Göttingen at the peak of its prestige in mathematics and physics, and she became the personal friend of many of its leading stars. Professor David Hilbert, Emmy Noether's mentor, was regarded as the greatest mathematician since Karl Friedrich Gauss and the greatest geometrician since Euclid. Hilbert's famous garden, with its long sheltered blackboard for outdoor seminars, bordered the Goepperts' yard. Hilbert always invited his latest lady friend to attend the public lectures he gave each Saturday morning. One Saturday when he was fresh out of ladies, he invited Maria to substitute.

"Won't you come?" Hilbert asked her. Arranging an excuse from school, she got her first glimpse of atomic physics. Later, when Hilbert was dying of pernicious anemia, Maria waited anxiously for the first liver extract to arrive in Germany. Daily doses of the vitamin B_{12} extract kept Hilbert alive for years.

Two of Göttingen's leading physicists, Max Born and James Franck, became Maria's lifelong admirers as well. Born regarded her as his daughter and Franck considered her "at least a niece." Thanks to them, most of the physicists who developed quantum mechanics during the mid-1920s and early 1930s visited Göttingen. She was friendly with many of them and, as a result, was among the first practitioners of the new physics.

Quantum mechanics—perhaps the greatest intellectual achievement of the twentieth century—describes the behavior of atoms, nuclei, and their components. The key to understanding all matter, quantum mechanics united physics, chemistry, astronomy, and much of biology. It was developed under the loose leadership of Niels Bohr in Copenhagen by a small group of Europeans, particularly Maria's teacher Max Born and Werner Heisenberg in Göttingen, Erwin Schroedinger and Wolfgang Pauli in Austria, and Paul Dirac in England.

"Maria was a lovely and lively young girl and, when she appeared in my class, I was rather astonished," Max Born wrote. "She went through all my courses with great industry and conscientiousness, yet remained at the same time a gay and witty member of 'Göttingen society,' fond of parties, of laughing, dancing, joking. We became great friends." Even before Maria joined his seminar, Born asked her to join him and his students in walks through the hills surrounding Göttingen to nearby village inns for supper. Eventually, it became obvious that she was Born's favorite student, and Göttingen's small-town rumor mill, still unaccustomed to women students, concluded that there was more than physics between Maria and her married professor.

Maria Goeppert entered the university expecting to study mathematics, but her first taste of quantum mechanics turned her into a physicist. "This was wonderful. I liked the mathematics in it.... Mathematics began to seem too much like puzzle-solving.... Physics is puzzle-solving, too, but of puzzles created by nature, not by the mind of man.... Physics was the challenge." Furthermore, she observed, "quantum mechanics was young and exciting."

"The spirit in Göttingen was so different at that time than anywhere else," Maria thought. Born and Franck ran a joint seminar in which participants were required to interrupt the speaker and criticize ruthlessly. Born and Heisenberg were developing quantum mechanics, and a friend in Heisenberg's class reported to Maria, "It's very exciting. Heisenberg tells us what he thought about last night."

Born's quantum mechanics students sometimes overwhelmed him, but they did not scare Maria Goeppert. Born believed that his advanced theory seminar attracted "the most brilliant gathering of young talent then to be found anywhere." Among its stars were Paul Dirac, Robert Oppenheimer from the United States, Enrico Fermi from Italy, and Eugene Wigner and John von Neumann from Hungary. Oppenheimer, the Harvard wunderkind, interrupted Born so often that Maria Goeppert finally organized a student petition demanding that he keep quiet. Born casually left the petition where Oppenheimer would read it and was relieved when the interruptions ceased. "But I am afraid he was deadly offended," Born reported.

Maria Goeppert's closest friends were fellow students Victor Weisskopf and Max Delbrück, who became Lise Meitner's assistant in Berlin and Barbara McClintock's colleague at Cold Spring Harbor. Maria often ate two dinners a day—one with her mother at home and one with her friends in town—so she could talk quantum mechanics with them. As Weisskopf, one of the great men of modern physics, explained, "We both were students of Max Born. We actually worked on rather similar problems and worked together and, if I may be a little personal, I fell in love with her." Delbrück, on the other hand, flunked his final examination the first time around, switched fields and eventually became a leading molecular biologist. Other friends who became science luminaries were Linus Pauling, Leo Szilard, and Arthur Holly Compton. As Maria Goeppert herself remarked, "I have to confess, I never much associated with the women. I always was with the boys...and that was enjoyable, you see."

When Professor Goeppert died in 1927, his wife followed Göttingen's time-honored custom and took in student boarders. The university did not provide student housing. Joseph Mayer, a good-looking Californian who had come to Göttingen to study quantum mechanics with Franck, knocked on Frau Goeppert's door and asked for a room. "To my enquiry, the maid brought a pretty little blond girl who, in spite of her obvious attractiveness, annoyed me by ignoring my painfully halting German and talking with me in faultless Cambridge English," Joe said. "I soon found that Maria, in addition to her pulchritude and linguistic accomplishment, was a student of Max Born's."

Joe was smitten. She was "a terrible flirt—but lovely, and brighter than any girl I had ever met." In fact, a chemist who knew them both concluded that she was brighter than Joe—and that he knew it. She liked to ski, swim, and play tennis, and wanted to go dancing almost every night. According to Joe, "She didn't do any of those things well, but she was a joy to be with."

Small and slim with blond hair and blue eyes, Maria Goeppert was called "The Beauty of Göttingen." Male scientists found her a delightful blend of intelligence and femininity. "She was nothing else if not feminine," her daughter Marianne Wentzel mused later. "For most of those scientists, they'd never met a woman who was as intelligent as they were, and then to have a woman who was very definitely a woman—it was an unbeatable combination."

"I wasn't beautiful at all," Maria Goeppert protested. "Göttingen was the classic European university town in which professors' daughters were tops in society and were terribly spoiled and popular."

Weisskopf and Joe Mayer, however, were not the only young men infatuated with her. "Everybody who boarded in that house apparently was taken with her, and half those guys proposed to her," according to Robert G. Sachs, who became her student in the United States. Years later, the American Nobel Prize–winner Robert S. Mulliken, who had boarded with the Goepperts, wondered, "What if I had married Maria Goeppert?" And another American student confessed, "We were all in love with her—not that I ever said a word to her. I didn't even know she spoke beautiful English." As Maria's daughter explained, "Men fell in love with her all the time. It wasn't important to her. It always seemed to me so amazing that she could just bend men around her finger." The skill proved useful in the years ahead.

Joe Mayer was a tanned, jazz-age Californian who had graduated from the California Institute of Technology and earned his Ph.D. at the University of California at Berkeley. He had pizzazz and could argue with verve and style about almost anything. A child of the Roaring Twenties and Prohibition, he laid in a supply of whisky and gin as soon as he arrived in Göttingen. In a town where Fermi felt like a millionaire because he owned a bike, Joe bought a convertible for cash. Gossip had it that Maria chose Joe from all her suitors because he had a car. Joe had another advantage, though. Like her father, he wanted Maria to become a professor.

When Maria and Joe became engaged, she thought about quitting physics, but Joe convinced her to continue. He was the feminist of the two. As a chemistry professor, he would have several women students; she never had any. By far the more famous of the two for most of their marriage, Joe pushed, prodded, cajoled, and supported Maria into doing physics all her life.

That fall, while she was still procrastinating with her thesis, he drove her to the Netherlands to visit one of Albert Einstein's closest friends, Paul Ehrenfest, a physicist. Ehrenfest and his wealthy Russian wife had built a beautiful house just before the Bolshevik Revolution intervened and left them with no money for decorating the walls. So Ehrenfest simply hung up his pocket watch and asked visitors to sign the guest room wall.

During their visit, Maria Mayer chatted happily about her thesis until suddenly Ehrenfest interrupted, "You've talked enough: now write." Locking her into the guest room, he told her not to emerge without an outline.

"It was a marvelous room, with one wall covered by the largest collection of detective novels I have ever seen and the other walls covered by autographs of famous scientists," Maria Mayer recalled. "Einstein's autograph was over my bed. I solved my problem within an hour, and it was the basis for my doctor's thesis." Then she signed the wall and departed.

Joe insisted that Maria finish her thesis. The day after Christmas was the maid's day off, and Maria decided to cook an elaborate dinner of deer loin and trimmings. Joe, who knew how to cook, soon realized that Maria did not. Waiting until she became thoroughly distraught, Joe struck a deal: "If you go on in science, I will always keep a maid for you. If you stop, you'll have to learn to do your own cooking. I won't be able to afford a maid."

Maria and Joe were married on January 18, 1930, and by March she had finished her thesis and passed her final examination. Three Nobel Prize winners sat on her examining committee. She had calculated the probability that an electron orbiting an atom's nucleus would emit not one, but two, photons or quantum units of light as it jumps to an orbit closer to the nucleus. Eugene P. Wigner, who later shared the Nobel Prize with Maria, called her thesis "a masterpiece of clarity and concreteness." Her solution was confirmed by lasers in the 1960s and is still quoted in papers on optics, atomic physics, and molecular physics. She also contributed an eight-page section to a quantum mechanics book that Born wrote with Pascual Jordan.

When Joe and Maria left Göttingen, she had "more first-class mentors than anyone I ever knew in the world," Bob Sachs said. "She had everybody—David Hilbert, James Franck, Max Born, Karl Herzfeld—just everybody was enthusiastic about her. She had a tremendous support structure. Everybody was helping her." She would need them all. She arrived in the United States on April Fool's Day, 1930.

The Mayers went straight to Baltimore, Maryland, where Joe had an assistant professorship at Johns Hopkins University. Maria loved Baltimore all her life and often said that her ten years there were her happiest. The bitterness came only at the end of their stay.

Initially, however, she was homesick. America seemed wild and woolly. When she ordered wine in a Prohibition-era restaurant, it came disguised in a coffeepot. When a large wooden box from a German arms manufacturer arrived for Joe, she thought the worst: "My God, is he buying weapons? Is he a gangster?" But his response was reassuring: "Oh good, my razor blades have arrived."

She eased into American life by writing and cabling her mother frequently and by spending her first four summers of married life among Europeans. The University of Michigan held a summer school in Ann Arbor during the early 1930s to teach European quantum mechanics to American physicists. Her old friends Fermi and Ehrenfest sat in the front row at each other's lectures, gleefully correcting one another's fractured English. So many of the visitors spoke German that Maria felt sorry for the Americans. The next three summers, she returned to Göttingen to work with Max Born.

Maria's golden youth in Göttingen had convinced her that she could have it all: children, an elaborate social life like her mother, and a career like her father. Joe agreed; he calculated that a working mother with hired help could handle the household chores easily in two hours a day. The social part of the equation was fun in Baltimore. The Mayers sailed Chesapeake Bay, hiked Maryland's hills, and crushed enough grapes in an old washing machine to make 100-liter batches of wine. A faculty wife observed a trifle sourly, "Whenever there's a party, the men seem to collect around Maria." And when an innocent young scientist asked how to emulate her gracious lifestyle on an academic salary, Maria advised him simply, "Live beyond your means!"

When her daughter Marianne was born in 1933 and her son Peter in 1938, her life became complicated. Raised an only child by a full-time housewife, Maria Mayer confessed, "I never stopped feeling guilty, thinking that I should have been home more. Marianne and

Peter are wonderful, but I will feel forever that they have missed something....The combination of children and professional work is not quite easy....There is an emotional strain due to the conflicting allegiances, that to science and that to the children who, after all, need a mother. I have had this experience in full measure." Joe still wanted her to continue in science, though, and she knew she would be miserable if she gave it up. She hated staying home, even when the children were sick.

The biggest complication of all was professional. Mayer wanted to be one of the world's best physicists, not just one of the best *women* physicists. Göttingen had proved she could hold her own at the top. But for her, being a top scientist "did not mean doing science at home," her friend and collaborator Jacob Bigeleisen noted. "Going to the office and doing science was important because the office was an indication of her status as a professor."

Maria Mayer realized that she could never have a university career in Germany. Emmy Noether, Lise Meitner and her friend Hertha Sponer—none of them ever became a regular professor in Germany. But she had thought America might be different. During the 1920s, women earned 15 percent of all American Ph.D.'s. Johns Hopkins itself had a famous coeducational medical school, though the university proper was adamantly opposed to admitting women as regular undergraduates. Johns Hopkins had even appointed two women as professors of psychology and education, fields that were popular among women; one of them was a department chairman.

The Mayers had not counted on the Depression. "Nobody, especially no university, would give a paid position to the wife of a professor," Mayer soon learned. Hopkins, like most U.S. colleges and universities, had strict antinepotism rules prohibiting the employment of relatives (generally wives) of university employees. The rules are often described as an outgrowth of the Depression, when the public opposed two-paycheck families. But the practice began in the 1920s before the Depression and continued long after, into the 1970s. Nepotism rules trapped Maria Mayer for decades.

Johns Hopkins adorned her with a potpourri of titles like "fellow by courtesy," "voluntary assistant," "associate," and "research associate." The university did not even consider Mayer a regular "research associate"; she was never listed with the twenty or thirty male research associates in its catalog. When Mayer asked if she could use an empty office on the main floor, she was sent to the attic.

"So I worked for a number of years without pay, just for the fun of doing physics," Maria Mayer explained. To keep her skills up-to-date in case she ever had to earn her own living, "we considered the added expenses of part-time household help as insurance—insurance for me in case of my husband's death.... I sensed the resentment very early, so I simply learned to be inconspicuous. I *never* asked for anything, and I *never* complained."

She would not let Joe complain, either. She knew she was good, but she did not like to be angry. She was having fun doing physics, and she had a nice husband and two babies, a lovely home, and friends. "I would have fought if I had *had* to, but I only wanted to learn, to teach, and to work.... I just didn't feel put upon in the least."

Curiously, Hopkins did not seem to want a quantum mechanics expert anymore than it wanted a woman physicist. American theoretical physics lagged far behind Europe in 1930. Its physicists were oriented toward engineering and nineteen-century classical mechanics; they were uninterested in theory, especially the new quantum theory of atomic particles and waves.

Luckily, Hopkins was like Göttingen in one vital respect. The collaboration between chemists, physicists, and mathematicians was unusually close, and Maria could piece together a research program by linking the different disciplines. She produced ten papers and a textbook in nine years at Hopkins—what Joe proudly called "quite a bit of scientific work."

She collaborated with Joe and a kindly Austrian, Karl Herzfeld, a devout Catholic whose parents had converted from Judaism. Although physics at Hopkins was almost exclusively experimental, Herzfeld was an outstanding theorist. Max Born asked him to look out for Maria, and he did. Since Herzfeld was interested in chemical physics and Joe was a chemist, Maria began applying quantum mechanics to chemical problems, such as the structure of organic compounds.

Quantum mechanics was just beginning to have a profound effect on chemistry, and the range of available problems was broad, deep, and seemingly boundless. One of her papers, written with graduate student Alfred Sklar, was a milestone in quantum chemistry. Herzfeld paid her two hundred dollars a year for help with his German correspondence, and Maria and Joe wrote *Statistical Mechanics,* such a popular chemistry textbook about molecular systems that its various editions sold for forty-four years.

When Adolf Hitler came to power in Germany, Mayer's old friend James Franck fled to Hopkins and she found him a house in the well-to-do neighborhood of Roland Park. Franck's presence made Baltimore feel like Göttingen, and he turned Hopkins into a world center for atomic physics. Maryland was extremely anti-Semitic at the time, however. Beach billboards screamed "Gentiles Only" in huge letters, and Roland Park enforced strict real estate covenants against selling houses to Jews. Relishing the prospect of playing a joke on

Baltimore's snobs, Maria Mayer signed Franck's real estate contract as his proxy even though it included a clause against selling to Jews.

By then Hopkins was treating Maria Mayer quite graciously. She had a pleasant office in the physics department and was teaching and supervising her student Bob Sachs's Ph.D. thesis. But she still had neither job nor salary. Herzfeld asked the dean at least to put Maria Mayer's name on the department's stationery. The dean became so angry that he dropped every name but his own from the letterhead, Maria recalled.

Outraged, Herzfeld wrote to Hopkins' president: "Dr. Goeppert-Mayer does at least one-third of the work of a full-time associate, both as a teacher and in research....As so often happens in women, she does not often originate a paper, but she is always instrumental in reaching the result. It has often happened that I could not proceed and she saw the way out. Her mind is really brilliant and penetrating, although perhaps not very original....The adequate amount of remuneration would be $1,000." Nothing happened.

While most students took a romantic view of "Joe and Maria," most were overwhelmed by her extremely technical and highly condensed lecturing style and by her mathematical virtuosity. Before strangers, the self-confident flirt from Göttingen was "*very* shy. She was at her best when she was with people she knew well," Sachs said. Uncomfortable lecturing, she spoke fast and almost inaudibly in abrupt, clipped sentences. To calm her nerves, she puffed one cigarette after another.

Top students like John Wheeler and Sachs were inspired, though. In fact, she was one of Wheeler's favorite professors at Hopkins. Wheeler, who became an eminent professor in Princeton, New Jersey, and Texas, liked "her quiet firmness, the rising emphasis at the end of her last sentence, and her inability to leave a problem until it was clarified."

Sachs recalled that when he asked her for a thesis topic, she declared firmly, "Any young man—(and she said 'man')—starting out in theoretical physics at this time *must* work in nuclear physics where all the exciting new things are happening." In a remarkable series of events during the early 1930s, neutrons, positrons, deuterium, and artificial radioactivity were discovered and particle accelerators invented. Wheeler knew a man who had "traveled to the Arctic. He'd gone on mountain-climbing expeditions. He had carried a caravan across the Sahara. But now he had gone into physics because that's where the excitement lay."

Mayer was still unfamiliar with nuclear physics, she told Sachs. "Therefore, we'll go over to Washington to see Edward Teller. He'll tell us what you should do." Mayer loved to talk physics with her

Göttingen friend Teller. He, in turn, wrote her scores of letters over the years, mixing personal confidences and physics. Teller, who had emigrated to George Washington University, became the controversial father of the hydrogen bomb and of Star Wars. Together, Mayer and Sachs published a nuclear physics article that was her first and only foray into the field until shortly before her discovery of the shell model almost fifteen years later.

In the course of working with Mayer, Sachs discovered that "in a quiet way she was a driven person. She was *very* shy. Sometimes she would whisper very fast with great intensity. It was clear that she really wanted to get the solution to that problem.... Maria was very unassuming in demeanor but she was very competitive."

Sachs noticed something else about his advisor. "She wasn't very sensitive in the early stages to the implications of what was going on in Germany.... Before Born left Germany, she wanted me to go work with him in Göttingen." Sachs is Jewish and, "Well, I wasn't thinking very much of going to Germany in those days."

The prewar years were difficult for Maria and for many German-Americans proud of their homeland's culture. Mayer's attitudes, like theirs, changed slowly and not necessarily steadily. "She was very strongly oriented toward the Prussians," Sachs noted. "She was proud of that heritage, politically. She and Joe had a strong faith in the German way of doing things. In the early days of the Nazi regime, she was a strong believer that [then Reichstag President Hermann] Goering would be the savior, that he was biding his time, letting Hitler clean up 'the mess.' She was typical of many upper-class Germans who felt they could trust Goering. I heard her say it. I had a much different point of view." As it turned out, Sachs was right. Goering was one of Hitler's most loyal supporters and a prime architect of the Nazi police state and its military.

As time passed, Mayer became increasingly depressed about the Nazi takeover in Germany. As a student, Weisskopf had heard her denounce Nazi influence in student organizations. During her summer visits to Göttingen in the early 1930s, she was dismayed at the decay of its intellectual life. When government guards gave her a "Heil Hitler" greeting, she responded with a simple "Good morning."

In 1933, she became a U.S. citizen and with Herzfeld was temporary treasurer of a fund established by German professors in the United States who pledged a portion of their salaries to help German refugees. She opened her home to several exiles and signed affidavits of support for would-be immigrants. Nevertheless, the old feelings died hard, and in 1940, when Sachs saw her after the fall of Paris, she was torn between pride at Germany's quick victory and dismay at its aggression.

In 1938, when Maria was pregnant with Peter, Hopkins fired Joe. As Maria told the story, Hopkins president Isaiah Bowman decided that he could solve the university's financial plight by getting rid of the best—and hence most expensive—faculty members and by replacing them with junior faculty at a third the price. At the time, many universities, including Hopkins, did not give senior faculty members tenure after seven years of employment. Although Joe was by no means senior, he was thrown out during the housecleaning; even the *Baltimore Sunpapers* lamented the fact that young people like Joe were being let go. "My understanding is that the administration was prejudiced against Communists, against Jews, against Catholics, against women, and against foreigners," Sachs recalled. In the process of "cleansing" the atmosphere and balancing the budget, Hopkins lost almost a dozen of the world's leading scholars, among them Joe and Maria Mayer.

According to the Mayers, Donald Andrews, who chaired the chemistry department, was jealous of Joe's spectacularly successful theory of condensation. In a terse fifteen-minute interview, Andrews told Joe that his contract would not be renewed because he had not attracted enough graduate students. Maria Mayer detested Andrews and the Hopkins administration ever after. "I am sure that Joe was fired from the Johns Hopkins faculty in 1939 partly because I was around and was some trouble," she declared. "I was *very* careful after that."

"To use her expression, she felt she had to be a 'lady,'" explained Mayer's daughter Marianne. "She had to be absolutely fair, and if someone else did some work, she always let them publish it first and then she published her additions afterwards. She was always very conscious that she had to behave properly if she was not to be accused of being a conniving, abrasive woman."

On the other hand, "she never felt discriminated against particularly," reported her daughter. And, in some sense, Maria Mayer was right. The small and rarefied circle of top scientists—the only people she cared about—admired her; it was primarily outsiders like Andrews who did not.

Ironically, Joe's experience at Hopkins hurt Maria more than it did Joe. Joe wound up at Columbia University at twice his Hopkins salary. Harold Urey, who had won the Nobel Prize in chemistry in 1934 for discovering the deuteron and heavy water, chaired the chemistry department there and became Joe's mentor. From then on, wherever Urey went, the Mayers were sure to go. On the other hand, Maria Mayer's years at Columbia were among the most difficult of her life, even though they transformed her in the eyes of her colleagues into a full-fledged, professional physicist.

"She was shy, but she was absolutely driven. She was very competitive. She wanted to make her name on her own and to compete with the top people in theoretical physics in the world, no question about that," recalled Bigeleisen, who collaborated with her at Columbia. "Maria wanted to be recognized as one of the world's top scientists." So she applied for a job in Columbia's physics department.

Its chairperson, George B. Pegram, turned her down cold. He assigned her an office but made it clear she was unwelcome. From then on, Maria Mayer loathed Pegram. After the war, she even refused to publish her wartime research at Columbia if it meant getting clearance from him. "Over my dead body," she exploded. "This paper can rot in hell before I will ask Dean Pegram for anything." Eventually, she published it with no mention of Columbia. Years later, when she was ill, her biographer Joan Dash concluded that Maria Mayer had acquired such self-control that she never became angry. But Marianne said her mother had given that impression because she had felt "very uncomfortable" being interviewed. Among friends, Maria Mayer could erupt like the proverbial volcano.

Columbia's chemistry faculty refused to give her even an honorary job title to put on the title page of *Statistical Mechanics*. According to the scientific community, Maria had been Joe's editorial assistant, not his coauthor. However, Urey, the chemistry department chair, greatly admired Maria's abilities and gave her a minor teaching job. Thanks to him, she got an office in the chemistry building and the all-important job title. Resentments lingered, though. Someone in the department later told Maria and another scientist's wife, the only two women who attended the department's weekly seminars, to attend the lectures but not the dinners afterward.

"Naturally, I went to no more seminars and I made Joe promise not to complain," Maria said. "Later, they had a woman speaker and invited me, but I went to the opera with Edward Teller instead."

Soon after the invasion of Poland in the winter of 1939–1940, the Mayers moved to a modest clapboard house in Leonia, New Jersey, twenty minutes from Columbia. There they formed a little colony of past and future Nobel Prize winners, including the Fermis, the Ureys and a few years later the Willard Libbys. The Mayers' fierce bridge-playing, fervent partying, and heavy smoking were considered wildly dashing and sophisticated.

The Mayer children remember the Leonia years warmly. In the evenings, their mother read them German editions of French stories and of Kipling's *Jungle Book*, translating as she went. She sang so often that Peter thought, "She must know every song Schubert ever wrote." Joe, who was more talkative than Maria, generally answered the children's science questions and took them on beach walks and camping expeditions. Maria ran the house with lists and elaborate filing systems for everything from science to personal correspondence. Books were alphabetized by authors' names. "It makes life simpler if you know what you're doing," she explained.

Although she formed several lifelong friendships there with women "as close to my heart as my own family," she never warmed to Leonia. "I am more accustomed to the company of men because of my work, and I've never had time for science, my family *and* kaffeeklatsches. Frankly, I think large groups of women tend to get shrill." When the Mayers left Leonia, a woman complained to the local paper that Maria had not done enough volunteer work—as if her physics had not been volunteer work enough.

By early 1940, the Fermis and Mayers were afraid that American Fascists might take control of the United States. The pro-Nazi German-American Bund was active in New Jersey, and Maria and Joe resigned from a German-American social club near Leonia because its members were pro-Nazi. Despite her admiration for Prussian ways, "my mother was about as anti-Nazi as possible," Marianne recalled.

In many respects, World War II was a physicists' war. Physicists developed both the radar that protected Britain from German air attacks and the atomic bomb that ended the war. Physicists were in such short supply that even Maria Mayer got a paying job.

Her first break occurred that fall when the American Physical Society wrote her a letter addressed "Dear Sir" and named her a "Fellow." Then the day after Pearl Harbor, she received her first job offer, from Sarah Lawrence College, which she described as "a rather swell, but definitely not scientifically inclined, girls' school." When a Sarah Lawrence interviewer asked her if an interdisciplinary science course could be as important to women as learning how to regulate a furnace flue, Mayer was stunned. "I asked if the only reason students learned English was to read a cookbook," she remarked. "It was the right thing to say. Sarah Lawrence was trying to find out if I would be traditional and dull." She took the job, telling Sarah Lawrence, "You really ought to have a man to do this."

The college paid Mayer her first-ever salary—$2,800 a year part-time after twelve years in the profession. Mayer enjoyed teaching, although balancing two jobs and a family was sometimes hectic. She once dropped Joe off at the Leonia train station for his commute to Columbia and then forgot to drive home to get dressed herself. She drove the twenty miles to Sarah Lawrence in her bathrobe.

At Columbia after Pearl Harbor, Mayer took over Fermi's courses on twenty-four hours' notice and Urey assigned her top-secret atomic bomb research problems even before she got security clearance. She was learning a new field, atomic physics. By then, "Maria was absolutely frightened the Germans would get the bomb first," recalled Bigeleisen. "She had to do everything possible to see that the United States produced the bomb first."

Urey and Columbia were in charge of developing a supersecret method to enrich uranium for an atomic bomb. They needed to separate uranium U-235, which is easily fissionable, from the more common, natural uranium U-238, which is not. Columbia managed the Substitute Alloy Materials project, nicknamed SAM, for the federal government, and the government paid Maria Mayer. When she told Urey that she would not work Saturdays or when her children were sick, he assigned her to "side issues," like investigating the possibility of separating isotopes by photochemical reactions. "This was nice, clean physics although it did not help in the separation of isotopes," Maria Mayer said. Eventually she became the unofficial scientific leader of about fifteen people, mostly chemists.

During the war, she felt guilty—not about the bomb, but about leaving the children. Joe was away six days a week doing weapons research, a German maid proved to be physically abusive, and an expensive English nanny became psychologically abusive. Marianne decided that when she grew up she would stay home with her children.

SAM was good for Mayer's professional development, however. "Suddenly I was taken seriously, considered a good scientist.... It was the beginning of myself standing on my own two feet as a scientist, not leaning on Joe." She developed a reputation as an expert problem-solver. "Take her a problem and—zingo," marveled Bigeleisen. The day she returned to work from a gallbladder operation, she asked Bigeleisen what he was working on. He told her; she said it sounded interesting and asked if she could help. Her third sentence completed the problem's solution and revolutionized isotopic chemistry.

The war years were hard on Mayer's health. She had pneumonia and surgery for gallbladder and goiter problems. Yet nothing stopped her chain-smoking or heavy social drinking. She smoked constantly, often three or four cigarettes at once. She and Joe, who was also a chainsmoker, sat through seminars in a cloud of smoke. She favored vile-smelling, denicotinized Carl Henry cigarettes and nudged Joe for another before she finished the first. When Bigeleisen smuggled a bottle of rationed Ballantine scotch into her hospital room under his coat, she called him "a savior." During a tough, wartime session on the spectrum of Uranium-235, Bigeleisen provided Urey, Teller, and Mayer with a midmorning cocktail break that proved extremely popular. Marianne said her parents realized that smoking was bad for their health and tried to quit, although her mother never reached the same conclusion about alcohol. In the 1940s and 1950s, macho men and sophisticated women smoked and drank, and the drunker the party the more fun it was, Marianne observed.

When Joe went to the South Pacific to observe the Okinawa invasion, Mayer spent a month in Los Alamos working on Teller's hydrogen bomb. She analyzed the behavior of uranium compounds at the very high temperatures and pressures expected in a thermonuclear explosion. At her security clearance interview, an officer lectured her sternly, "There is one thing you must always keep in mind. Do not mention to anybody that there is a connection between the work going on at Los Alamos and at Chicago, Columbia, Hanford, Oak Ridge, etc., because that is Top Secret information." That was the first that Mayer had heard about any relationship between the installations. She was particularly irritated by his bungling because she had been forbidden to talk about her atomic bomb work with Joe. "Some friends say I am too dependent on Joe, but I've always told him everything," Maria said later. "Keeping from him that awful secret of the atom bomb research I did during World War II for four years was harder on me than all the prejudice of the years."

After the war, Mayer said she was glad that her particular project at Columbia had not contributed to the bomb. "We failed. We found nothing, and we were lucky, because we didn't contribute to the development of the bomb, and so we escaped the searing guilt felt to this day by those responsible for the bomb." Working on the bomb did not bother her at the time, though. And from private conversations with Urey as well as from her knowledge of physics and of Fermi's and Urey's research, she knew she was working on a fission bomb and what that meant, Bigeleisen emphasized.

When Japan surrendered, the Leonia crowd moved en masse to Chicago, the center of postwar scientific excitement. Her old Göttingen and Baltimore friend, James Franck, was already there. And when the University of Chicago formed an interdisciplinary institute to explore the nucleus, Fermi, Urey, Teller, Libby, and the Mayers joined the fun. Brilliant graduate students like future Nobel Prize winners T. D. Lee and C. N. Yang (chapter 11) flocked to the Windy City to study with Fermi.

In a new variation of the old volunteer theme, Maria became a "voluntary associate professor"—still unpaid. Later, she was promoted to "voluntary professor"—also unpaid. The Depression was over, but nepotism rules lived on. This time, Mayer did not care. Chicago was "*the* place to be," and Mayer was a major player in the "in" crowd. Chicago was the first place where she was not considered a nuisance, but was greeted with open arms, she said. Chicago was Göttingen revived.

Joe chaired Chicago's famous weekly meeting at the Institute for Nuclear Studies (now the Fermi Institute). Although he was a chemist, physicists respected him so much that they later elected him president of the American Physical Society. His only seminar rule was "Don't interrupt while someone else is interrupting." Attending the meetings was "like sitting in on a conversation of the angels," according to a participant.

Maria ran a physics-theory seminar as freewheeling as Joe's physics-chemistry session. She served on committees, helped hire faculty, advised graduate students, and helped set the tone for Chicago's notoriously difficult graduate examinations in physics. "We are interested only in the future Heisenbergs!" she announced loftily. The first year, four future Nobel Prize winners and thirteen future members of the National Academy of Sciences took the exam. One flunked.

"One of my favorite teachers at Chicago was Maria Mayer," reported Sam Treiman, a future Princeton University physics professor. "Professor Mayer taught solid, no-nonsense courses. She would have disdained even the hint of show biz. It happens, however, that she was a dedicated cigarette smoker, and in those days, it was quite acceptable for the professor to smoke in class. She often did. She would light up and lecture with a cigarette in one hand, a piece of chalk in the other. She puffed on the one and wrote on the blackboard with the other. They interchanged places in her hands frequently and in a seemingly random manner. Often, in the excitement of some physics development about to reach a watershed, she would come very, very close to writing with the cigarette or puffing on the chalk."

One morning Urey asked her for a calculation. It was based on a theory that she and Bigeleisen had developed. Urey wanted the answer in time for a four P.M. speech. "I suspected that he presented this as a challenge to Maria Mayer and me to see whether our method was as powerful as we claimed it to be," Bigeleisen recalled. Dividing up the work, they gave Urey his figures at two o'clock. By four P.M., the absent-minded Urey had already forgotten who had prepared the calculation; in his talk, he recommended attending Joe Mayer's class to get more information. Maria was so irritated that she called in a loud stage whisper from the back of the room, "What's wrong with *my* course?" She was competing even with Joe.

In the meantime, Bob Sachs, her former Hopkins student, had become head of the theory division of the new Argonne National

Laboratory outside Chicago. He later became director of the entire laboratory. Sachs asked Maria, "Wouldn't you like to earn some money?"

"That would be nice," she admitted.

"Why don't we arrange a half-time appointment for you at Argonne as a senior physicist?" Sachs suggested.

"But I don't know anything about nuclear physics [the major project at Argonne at the time]."

"You'll learn," promised Sachs.

So the student became the boss, Mayer entered a third field, and the federal government supported her financially while she worked on her Nobel-winning project. This time, though, it was her choice. She could have worked full time at Argonne, but she did not want to miss the excitement at Chicago.

She did not complain. "I had everything else I wanted—the biggest office, faculty status, and the Argonne National Laboratory paid me a nice consulting salary," Mayer said. "Most of our faculty friends were fighting for me." Her cardinal rule was, "*You don't rush your friends*." Like a politician, she realized she could not move too far ahead of public opinion.

Mayer liked Chicago life outside the university too. She and Joe bought an old three-story brick mansion in Kenwood, on the South Side of Chicago, at 4923 Greenwood Avenue. When Chicago's rich had abandoned the area, faculty families had moved in. The house had high ceilings, six fireplaces, and space for big vegetable and flower gardens. In a third-floor glass porch, Maria raised cymbidium orchids, and when Marianne was married, Maria coaxed them into bloom all at once: twenty plants with two thousand flowers.

With James Franck living nearby, Chicago's social life felt like old times in Göttingen. The Mayers' annual New Year's Eve party for a hundred or more senior scientists featured a twelve-foot-tall Christmas tree decorated Göttingen-style with tinsel and real candles. With orchids in every room and liquor flowing, there was a buffet supper downstairs, singing in the second-floor library, and dancing in the third-floor billiard room. The climax of the Mayers' social extravaganzas was an international conference held at the University of Chicago in 1951; Maria gave cocktail parties on four successive nights, with laboratory beakers for glasses. At the parties, Maria was animated and happy, the center of a crowd of scientists.

For a short time after the war, she was active in national politics, too. She, Urey, and Joe ardently supported the development of Teller's H-bomb in the early 1950s. In a postwar speech, she said that, without international control of nuclear energy, the United States should reduce its vulnerability to Soviet nuclear attack by building cities in long strips. On the other hand, she supported civilian control of nuclear energy and joined other scientists who lobbied in Washington against Pentagon control. Later, she and Urey changed their minds about the hydrogen bomb. During the 1960s, she opposed the Vietnam War and mediated painful arguments between Joe, who was a hawk, and her son, Peter, who was a dove.

Chicago was the scene of Maria Mayer's greatest scientific triumph, too. Wartime research had produced a wealth of data about isotopes, atoms of the same element that have differing numbers of neutrons. Why some isotopes are more abundant than others became a lively issue among scientists. Unstable nuclei tend to decay radioactively into other more stable elements; they change gradually from one element to another until they reach a stable form. Once they are stable, they do not change any more, so their numbers accumulate. Thus, the more stable the isotope, the more abundant it is in the universe. But why? Teller suggested that they find the answer. Teller soon lost interest, but Maria Mayer was fascinated by a series of odd clues.

- Why, for example, does an isotope with 126 neutrons hold onto its neutrons more strongly than an isotope with 127 or 128 neutrons?

- A few elements were much more abundant than contemporary theories could explain. All had nuclei with either 50 or 82 neutrons. Maria Mayer thought that "the excess stability must have played a part in the process of the creation of elements." But how?

And so it went, puzzle after puzzle, until she had uncovered a series of what she called "magic numbers"—2, 8, 20, 28, 50, 82, and 126. Nuclei with these numbers of protons *or* these numbers of neutrons are unusually stable. But why?

She began collecting data to support a nuclear shell theory. Such a theory had been considered during the 1930s and then set aside during the 1940s as Niels Bohr's liquid drop model of the atom solved one problem after another. An outsider to the postwar nuclear physics scene, however, Maria Mayer was not wedded to Bohr's model and was looking at the problem afresh.

Particles inside the nucleus orbit in shells, she suggested, "like the delicate shells of an onion with nothing in the center." After that comment, her old friend Wolfgang Pauli called her "The Madonna of the Onion." She looked at esoteric figures for energy levels, spins, angular momentum, potential wells, binding energies, radioactive-decay energies, isotopic abundances, and the like. If a vital fact was missing, she asked Argonne experimentalists to develop the information. No mere problem-solver, she dug deep into the fundamental processes of nature. "How the protons themselves are held together, how they interact with one another and with the uncharged neutrons also present in the nucleus—these are the great mysteries of nuclear physics," she realized.

She marshaled her evidence—without any theoretical explanation of its meaning—in a 1948 paper. She had "much better statistics of nuclei than were done ever before," according to Hans Bethe, dean of nuclear physicists. "The shells were established beyond doubt, but there was no theory."

One day she and Fermi were chewing over the problem in her office, where they met because Fermi did not like smokers in his room. As Fermi left to take a long-distance telephone call, he flung back a question, "Incidentally, is there any evidence of spin-orbit coupling?"

"When he said it," Maria recalled, "it all fell into place. In ten minutes, I knew." She experienced an almost physical reaction, an awesome process, as pieces fell together for her. She had the entire problem with its thousands of details worked out by the time he returned.

She floated home, excited, high, exalted. "I finished my computations that night. Fermi taught it to his class the next week."

"It was kind of a jigsaw puzzle," Maria Mayer said later, recreating the moment. "One had many of the pieces (not only the magic numbers), so that one saw a picture emerging. One felt that if one had only just one more piece, everything would fit. The piece was found, and everything cleared up.... Only if one had lived with the data as long as I, could one immediately answer: 'Yes, of course, and that will explain everything.'... In ten minutes, the magic numbers were explained." Winning the Nobel Prize would not be nearly as exhilarating as doing the work.

Her solution was totally unexpected. Spin-orbit coupling had never seemed important before. It has practically no effect on the atom's electrons, for example. Yet Maria Mayer had discovered that, inside the nucleus, it has a crucial effect.

To explain the theory to Marianne, she pictured a roomful of couples waltzing in circles, each circle enclosed inside another like the shells inside the nucleus. As the couples orbit the room, they also spin like tops, some clockwise and some counterclockwise. And as Maria pointed out triumphantly, anyone who has danced a fast waltz knows that it is easier to spin in one direction than in the other direction.

Thus, the couples spinning in the easier direction will need slightly less energy than the couples spinning in the more difficult direction. And that tiny energy difference was enough to explain the magic numbers.

Maria procrastinated about writing an article, just as she had her thesis. Two other physicists had produced a theory, and she decided to wait until their work was published. Joe was exasperated with her and told her she was carrying good sportsmanship too far. She finally submitted her paper in December 1949. By that time three Germans—Hans Jensen, Hans E. Suess, and Otto Haxel—had submitted their own paper outlining the same theory. Cut off from mainstream physics during World War II, they had not spent years working with Bohr's liquid drop model either. It was clear that they had conceived the same idea at the same time as Maria Mayer.

When she read their paper, she was "at first dismayed... for about five minutes." Then she realized that their work actually confirmed her theory. Together, they could convince more people faster. As Weisskopf said, "You know, now I believe it, now that Jensen has done it, too." And Jensen wrote her, "You have convinced Fermi, and I have convinced Heisenberg. What more do we want?"

For a theory that ran counter to decades of nuclear physics, the nuclear shell model was adopted remarkably quickly. The model has been so successful that physicists today find it difficult to imagine studying nuclear behavior without it. "The nuclear shell model is the central idea of nuclear structure.... The shell model lurks somewhere in every paper on nuclear structure," noted the nuclear theorist Elizabeth Urey Baranger, currently associate provost at the University of Pittsburgh.

Instead of competing with Jensen for priority over the theory, Mayer captivated him. He became the last in a long string of men charmed by Maria Goeppert Mayer. He called the shell model "*your* model," and she called it "*your* model." She said Jensen was "a dear, gentle man; we look at things in the same way, even our eyeglass prescriptions are identical." Jensen thought her "unbelievably modest."

They wrote a book together about the shell model. Actually, Mayer wrote most of it. In a letter, Jensen confessed to her that she had written 80 percent of the book—more than 95 percent of the important parts—and that he should not be listed as her coauthor. He complained that "I'm deeply ashamed that you've written so many chapters." He said he felt like "a parasite." Although Mayer did not drop his name from the book, she did list her name first. The book established both their reputations as the primary originators of the shell model theory, overshadowing Jensen's original collaborators, Haxel and Suess.

During the book-writing period, Jensen's letters mixed physics with affectionate remarks like "Precious Maria," "Always your little Hans," "Yours constantly," "Yours sincerely, The Scamp, the Spoiled Boy..." He confided, "This isn't much of a love letter—but physics is so very much less complicated." And when the book was finally finished, he was happy at being able to write "only love letters."

Once the book was complete, however, his infatuation cooled. Months went by between letters; he planned trips to the United States without knowing that Maria would be in Europe at the same time, and he started letters and finished them months later. He still depended on her for writing, however. Years later, when Jensen learned that Maria Mayer had had a stroke, he wrote to wish her a speedy recovery *and* to ask if she would write an article for a German journal: "maximum two typed pages, but the longer the better." He agreed to contribute a chapter to an anthology and then asked her to write it with him.

Maria Mayer was elected to the National Academy of Sciences in 1956, the same year that she suddenly lost the hearing in her left ear. By then, Fermi had died of cancer, the top graduate students had stopped coming to Chicago to work with him, and Teller and Libby had left. Urey moved soon after to a new University of California branch at La Jolla and, as always, invited the Mayers.

California offered Maria a full professorship—with pay. Overnight, Chicago's administrators discovered that they, too, could offer her a real professorship. She was mildly amused by Chicago's offer and delighted with California's. The Mayers moved in 1960. At age fifty-three, ten years after her revolutionary discovery, she finally had a regular, full-time, paid university job. She had fulfilled her father's dream: she was the seventh generation of professors in her family, and her son would be the eighth. It was too late.

Shortly after moving to California, while unpacking her books, Mayer suffered a devastating stroke. It paralyzed her left arm and blurred her speech. Although she tried to continue working, her health was never good again. Joe told her and all their friends that she suffered from a rare virus disease affecting her nerve endings; grave illnesses were less readily publicized then than today. Maria guessed she had had "a small stroke." She continued to smoke, though, and Marianne and her brother thought she drank more to numb the pain. Certainly, Marianne said, her health problems made her drinking more evident.

At four A.M. the morning of November 5, 1963, a Swedish

newsman telephoned their home. Joe answered, handed her the phone, and then raced to ice some champagne. Mayer and Jensen had won half the Nobel Prize, and Eugene Wigner, an old Göttingen friend who had moved to Princeton University, had won the other half. The local paper headlined the news, "La Jolla Mother Wins Nobel Prize." The Mayers celebrated at dawn with bacon, eggs, and champagne.

Once in Stockholm, Mayer relaxed and scrutinized the Swedish palace like a professional hostess. She gave it high marks: "So warm and alive, with roaring fires in every hearth, great Oriental rugs, and white lilacs in all the vases." She was the only living woman in the world with a Nobel Prize in science, and she and Marie Curie were the only women to have won a Nobel in physics. Sitting on a flower-banked dais before the king, she thought of all the people who had stood there before her, names she had heard as a child and friends from the past. Catching her eye, Joe realized that he had burst into tears that were streaming down his cheeks. Without Joe, she said, she would never have gotten to Stockholm.

In photographs, she looks small and frail, stepping gingerly down to King Gustav VI Adolf. Her arms are too weak to hold the gold medal or the heavily bound diploma, so a Swedish aide hovers nearby to carry them for her. After the ceremony, the king gave Mayer his arm and, as they swept through the reception room on their way to dinner, onlookers sank to their knees. "It was a fairy tale," she said.

Maria Mayer's last years were limited by her health. She had fulfilled her father's dream, and the Nobel Prize had made her a symbol of Superwoman: the brilliant professional with a happy marriage and successful children. She tried to continue working. As she said, "If you love science, all you really want is to keep *on* working. The Nobel Prize thrills you, but it changes nothing." As her health deteriorated, she acquired a pacemaker and published less. She died of a pulmonary embolism on February 20, 1972.

Joe gave her papers to the University of California at San Diego. They include personal letters and scientific notes; her daughter's report card from nursery school and travel plans to conferences; hand-copied notebooks of German poems and party menus—all mixed together in one woman's life.

* * *

9

Rita Levi-Montalcini

April 22, 1909–

NEUROEMBRYOLOGIST

Nobel Prize in Medicine or Physiology 1986

In a tiny bedroom laboratory, hidden away from the police, Rita Levi-Montalcini ground her sewing needles into micro-sized scalpels and spatulas. With miniature scissors from an ophthalmologist and tiny watchmakers' forceps, she had the tools for microscopic surgery. Her brother—fresh from the Fascists' most-wanted list—built an incubator. Her mother guarded the door with the lofty pronouncement, "She is operating and cannot be disturbed."

Levi-Montalcini assembled her secret, homemade lab after Italy's Fascist government forbade Jews from practicing medicine or science during World War II. Armed with a simple microscope, eggs, and burning enthusiasm, she studied the nervous system as it developed in embryonic chicks. When nightly bombing raids forced her into basement shelters, she slept clutching her microscope and slides. As anti-Semitism worsened, she moved her lab to the country. Bicycling over the hills, she begged farmers for eggs "for my babies." Casually, she asked, "Are there any roosters in the coops? Fertilized eggs are much more nutritious."

Levi-Montalcini's tiny homemade lab became her defense against the world's madness. As Germans marched across Europe, she found solace in observing the development of nerve cells. Her forbidden, bedroom studies laid the foundation for her discovery of growth factors, molecules that influence the development of immature cells. Today, scientists have identified numerous growth factors that help cells communicate with one another. Levi-Montalcini's nerve growth factor (NGF) may play a vital role in certain degenerative diseases of the central nervous system like Alzheimer's disease. Other growth factors help heal skin transplants on burn patients and repair damaged nerves in experimental animals. Someday they may help explain the development of cancer tumors.

Of her homegrown experiments, Levi-Montalcini says simply, "It was a pure miracle that I succeeded with such primitive instrumentation.... It cannot be repeated."

Levi-Montalcini never lost the zest for adventure and challenge that motivated her wartime exploits. She liked to learn new disciplines, leaping into the unknown, struggling with difficulties, and overcoming scientific challenges. She lived life like a roller coaster, alternating elation and despair, extravagant generosity and tempestuous competition, elegance and hard work. Ironically, she traced her tremendous will to survive as an intellect to a domineering father and a Victorian upbringing.

She was born Rita Levi in an intellectual Jewish family in Turin, a rich industrial city in northern Italy, on April 22, 1909. Her ancestors were Israelites who had come to Italy during the Roman Empire. Levis—both close and distant relatives—helped make Turin an intellectual center for Italian unification during the nineteenth century; during the 1920s and 1930s they helped turn the city into an anti-Fascist stronghold. Well-known Levis after World War II included writers Carlo Levi, Primo Levi, and Natalia Ginzburg. As an adult, Rita Levi added her mother's maiden name Montalcini to her father's surname to distinguish herself from the other Levis of Turin.

Her family, which had assimilated with local Roman Catholics, was not religious. When she played in the park, little Catholic girls asked, as they were instructed to do by their parents, "What is your name? What is your father's profession? What is your religion?" Puzzled, she asked her father what to reply. Adam Levi taught the three-year-old to repeat, "*Sono una libera pensatrice.*—I am a freethinker." Then he added, "You can decide when you are twenty-one whether you want to be Jewish or Catholic." Rita learned her lesson well because when her governess tried to convert her to Catholicism so that she could eventually enter the Kingdom of Heaven, she inquired, "And Mother and Father, will they come with us?"

Unfortunately not," the governess sighed. "Perhaps they'll only be able to join us when a dove that drinks once a year has dried up the sea."

"If that's how it is," Rita replied firmly, "I'm staying with them."

Although Rita's father wanted his children to question religious authority, he demanded instant obedience at home. An engineer, he owned an ice factory in southern Italy. But he was also an authoritarian, Victorian paterfamilias. His piercing gaze and imperious voice accentuated his energetic and domineering personality. His

Rita Levi-Montalcini in St. Louis.

short-lived temper was so explosive that his sisters turned the diminutive of Adam into his nickname: "Damino the Terrible." As a small child, Rita kept careful track of her father's nostrils; when they flared, he was about to lose his terrifying temper.

Rita Levi-Montalcini, eleven years old, 1920

Rita Levi-Montalcini on a boat to the United States, September 1947.

Rita Levi-Montalcini in her laboratory in Rome, 1985.

Despite the histrionics, Rita never questioned her father's love and concern for her welfare. He controlled all the minutiae of her daily life, but she never dreamed of disobeying his orders. When Rita and her fraternal twin Paola modeled new, entrancingly beribboned straw hats, he did not like them. The beautiful bonnets disappeared instantly and were never seen again. Rita's brother Gino wanted to be a sculptor, but his father ordered him to be an engineer. He obeyed and eventually became a famous architect, but he always wished he had been a sculptor.

Timid and submissive, little Rita lived with many fears: her father, monsters in the dark, long hallways, pogroms, and windup toys.

All the guardian angels who protected Rita's childhood were female: her Aunt Anna, her governess Giovanna, her beautiful mother Adele, and her twin Paola. A gifted painter, Adele Levi was a reserved and submissive wife. When Rita attended an opera with her parents, her father relished the part of the Foundry Owner, who threatened to break the young wife who did not love him. While the crowd roared its approval, Rita silently sympathized with the young wife. "It was not anger," Levi-Montalcini explained. "It was built-in. There was no anger whatsoever. I simply found the situation impossible.... Ever since my childhood, I had strongly resented the different roles played by my father and mother in all family decisions. I adored my mother and rebelled against this difference, which I also feared for myself as a future housewife." In second grade, she said that her fingers were "for sending kisses to mother," but she refused to kiss her father. While Paola shielded herself by communicating with no one, Rita clung to Paola.

When the twins completed fourth grade, Rita longed to continue her academic education. Ignoring her wishes, Adam Levi decreed that his daughters should attend a girls' finishing school and learn to be perfect wives and mothers. His aunts had earned doctoral degrees in literature and mathematics, and he blamed their unhappy marriages on their education.

Levi-Montalcini's years in finishing school were filled with confusion and despair. She learned none of the subjects required for entrance to a university—no mathematics, no exact sciences, no Greek or Latin. Her courses were mindless, and her classmates were interested only in marriage and motherhood. "I had no particular interest in children or in babies, and I never remotely accepted my role as a wife or mother," she recalled. When her beloved nurse Giovanna died of stomach cancer, Rita decided that she wanted to be a doctor. With no hope of attending medical school, she felt trapped and isolated in a dead end without escape. The position of married women was so debasing that she decided never to marry. She was twenty years old before she had the courage to tell her father the truth: She did not want to marry. Instead, she wanted to become a physician.

When Rita's mother pleaded her cause, Adam Levi reluctantly agreed to hire a tutor to prepare his daughter for university entrance examinations. "If this is really want you want, then I won't stand in your way, even if I'm very doubtful about your choice," he told her. Soon after consenting to Rita's plans, Adam Levi suffered a massive heart attack and died. Over the years, as the memory of her struggles with him faded, she began to worship his memory. By the time she wrote her autobiography in 1988, her beloved mother had become a nameless shadow and her father dominated the book as powerfully as he had ruled her childhood.

Enlisting her cousin Eugenia, Levi-Montalcini hired one tutor for mathematics and science and another tutor for Latin and Greek. She and Eugenia studied philosophy, literature, and history on their own. After only eight months of study, they took the examination. When their tutor called with the results, he announced joyfully, "Signorina Rita, you've both passed!" Levi-Montalcini led the list. She entered the University of Turin's medical school in 1930, determined to prove—to herself and probably to her father also—that she was as intelligent as any man.

In medical school, the three hundred male students spent an inordinate amount of time analyzing the physical charms of the seven female students. Whenever a particularly awkward young woman passed through the halls, the men raised their voices and loudly talked about "Greta Garbo in disguise." In such an atmosphere, Levi-Montalcini wore clothes that were as elegant and asexual as she could manage. As a friend observed, she behaved like a squid, ready to squirt at any young men who approached. She did not argue with them. "I just refused to accept poor treatment. It was like water off the back of a duck," Levi-Montalcini declared. "I wanted to spend all my time on research. I was not receptive to courtship. I dressed like a nun. I despised everything with a feminine flair. Women paid too much for it. I didn't want any sentimental contact with other students, only intellectual contacts. I didn't want any contact as a woman." After years of intellectual deprivation, she could finally be as fanatically devoted to learning as she pleased. She rejected the advances of several young men and told one named Guido, who became her wartime fiancé, that she would stroll with him in Valentino Park "on the condition that we talked only of cultural and musical subjects."

Ironically, no sooner had Levi-Montalcini escaped from her father's domination than she fell under the influence of another explosive and autocratic man, Professor Giuseppe Levi. Like Levi-Montalcini's father, the professor was famous for his rages. His tantrums—short-lived though they were—shattered those around him. Like Adam Levi, he controlled even the smallest detail of his family's life. He bellowed "stupid" at his children when they did something he disliked, like wearing city shoes in the mountains or talking with strangers in the train. One of his relatively mild remarks was "I beg your pardon, but you are a perfect imbecile." His sons inherited his anger, but tried to suppress it. At dinner once, one of the sons became so enraged that he grabbed his butter knife and, with silent fury, scraped the skin off the back of his hand. His daughter's description of her childhood, published in the 1950s, reminded Levi-Montalcini of her own upbringing.

Levi was larger than life as a teacher, too. Three of his students—Levi-Montalcini, Salvador Luria, and Renato Dulbecco—emigrated to the United States and won Nobel prizes in physiology. Levi's consuming passion to understand nature was contagious. Although his temper tantrums made Levi-Montalcini tremble with fear, he was different from her father in one vital respect: he liked his students to exercise their intellects. Furthermore, the same spontaneity that made him explode at a student filled him with enthusiasm at a piece of good work. Thanks also to Levi's spontaneity, he always spoke his mind—loudly and passionately—against Fascism. He thundered his disapproval in public buses and lecture halls, in public and in private. His courage was legendary, and his students loved him for it. Levi's passion for science well done, his disdain for shoddiness, and his tumultuous approach to life's dramas formed the backdrop for Levi-Montalcini's development as a scientist.

Above all, Levi was a magnificent histologist, skilled in the microscopic study of tissue structure. He made Levi-Montalcini expert in a new technique, staining embryonic chick neurons with chrome silver to make the nerve cells stand out in smallest detail. It was an elegant but simple method, and she used it later in her secret wartime lab. For her thesis, Levi-Montalcini studied collagen reticular fibers, which are the weft-like supporting fibers in different types of tissue. Soon she did not know whether she wanted to pursue research or to practice medicine.

After completing medical school in 1936, Levi-Montalcini worked with Levi for two more years, specializing in neurology and psychiatry. She was still torn between research and clinical practice. Then Benito Mussolini made the decision for her. In June, 1938, Il

Duce issued the "*Manifesto per la difesa della razza*"—"The Manifesto for the Defense of the Race." In it, the prime minister banned intermarriage between Jews and non-Jews and prohibited Jews from pursuing academic or professional careers, from studying or teaching at state schools, and from working for state companies or institutions.

Whatever Levi-Montalcini tried to do was either forbidden or dangerous. She practiced medicine secretly among the poor, but the racial laws prevented her from writing prescriptions. She could not study on her own because she could not use the university library. By March of 1939, she could not even visit the university for fear of endangering friends or being denounced. She worked in a research institute in Brussels, until just before the German invasion of Belgium. She was devastated. Her only alternatives seemed to be intellectual stagnation or emigration to the United States, which her family refused to consider.

Then an old friend from medical school asked about her current research. When she could not answer, he lectured her sternly, "One doesn't lose heart in the face of the first difficulties. Set up a small laboratory, and take up your interrupted research."

The idea had never occurred to Levi-Montalcini, but it appealed to her instantly. She felt like Robinson Crusoe setting off to explore a jungle. Her jungle, however, was composed of 100 billion cells of the human nervous system and the fibrous nerves that spread out from the cells like intersecting nets in all directions. Planning her workspace she realized that she did not need a large laboratory to stain nervous tissues with chrome silver. Chick embryos were ideal for home research because fertile eggs were cheap and readily available. In addition, the nervous systems of chick embryos are much simpler than those in the human brain.

Her brother built her an incubator, and she bought a binocular microscope. Then, when Giuseppe Levi was forced to leave the university, too, he joined the project. Levi-Montalcini called Levi her "first and only assistant," but it is difficult to imagine Levi meekly assisting a former student. Later, Levi-Montalcini downplayed his influence on her work. She told an interviewer in 1988, "I was on excellent terms with him, but I was always of the opposite opinion. He was a splendid man, scientifically and ethically.... I liked him as a man, but I didn't care too much about some of his ideas." During the war, however, they were coworkers. As the Italian press bannered Nazi slogans and her brother's name appeared on most-wanted posters for Resistance crimes, they buried themselves in problems of the developing nervous system.

First, she needed to find a bedroom-sized problem. She found her "Bible and inspiration" on an idyllic summer day in 1940. Italy's passenger trains were busy transporting troops, so civilians traveled in cattle cars. Sitting on the floor of an open-sided freight train, she dangled her legs over the side in the open air and enjoyed the scent of summer hay. Thus occupied, she idly read a scientific article by the eminent embryologist, Viktor Hamburger of St. Louis, Missouri.

Hamburger, a founder of developmental neurobiology, is one of the "supreme biologists of our time," according to John T. Edsall, editor of the *Journal of the History of Biology*. Hamburger was also a leader in the use of chick embryos for experimental research on nervous-system development. As a student in 1927, he had laid out a research plan of many years' work to elucidate the development of the nervous system. He hypothesized that it was influenced by signals derived from tissues like muscles and sense organs. Although he had no idea what the signals might be, his studies of the spinal cord in chick embryos suggested that the signals might influence the division and differentiation of neurons, the fundamental unit of the nervous system. Reading Hamburger's article, Levi-Montalcini decided that she could reproduce his experiments in her bedroom.

Reaching her arms into her glass incubator, she operated on three-day-old chick embryos under a low-powered dissecting microscope. Using her micro-sized scalpel, she amputated a tiny limb-bud from each embryo. Then, over the next seventeen days, she sacrificed a few embryos each day. Using her needle-sized spatula and the eye surgeon's scissors, she extricated each embryo from its egg and dissected it. Then, to her brother's horror, she scrambled the remains of the eggs for dinner.

After dissecting the embryos, Levi-Montalcini sliced the spinal cord into thin sections and stained them to be visible through her microscope. Clustered together within the embryonic spinal cord are the particular neurons that she wanted to study. A neuron is composed of a bulbous cell body with thin fibers extending from it. Neurons are similar in all vertebrates, from chicks to humans. One type of neuron, the motor neuron, has its cell body in the spinal cord and extends its fibers out to the embryonic limbs. When this neuron is activated, it makes the muscles contract and the limbs move. It was these neurons that Levi-Montalcini focused on. When a limb is amputated, the motor neurons in the spinal cord all but disappear. Hamburger thought they had failed to proliferate; Levi-Montalcini and Levi concluded that they had proliferated, started to grow, and then died. In Levi-Montalcini's tiny bedroom lab, she and Levi had laid the foundation for the modern concept of nerve cell death as a part of normal development.

Far from discouraging her, Levi-Montalcini's dangerous and

cumbersome working conditions excited her. And the Fascists' persecution of the Jews actually intensified her pride in her heritage. When Italian journals rejected their articles because her name and Levi's name were Jewish, what seemed like misfortune turned out to be luck. The articles were eventually published in Belgian and Swiss publications that could still be read in the United States.

"Many years later, I often asked myself how we could have dedicated ourselves with such enthusiasm to solving this small neuroembryological problem while German armies were advancing throughout Europe, spreading destruction and death wherever they went, and threatening the very survival of Western civilization. The answer," she concluded, "lies in the desperate and partially unconscious desire of human beings to ignore what is happening in situations where full awareness might lead one to self-destruction." Whatever her motivation, she had found her life's passion—the central nervous system.

As the Allied bombing of Italy's industrial cities intensified, Levi-Montalcini and her family fled to a country village for safety. Her lab moved from the bedroom to a dining room corner. Despite power failures every few days and a scarcity of eggs, she continued to work until the Italians overthrew Mussolini and German troops poured into Italy. At that point, the Germans began rounding up Jews in earnest for the extermination camps. Hastily, Levi-Montalcini closed her lab, and she and her family fled south to Florence. Rita and Paola forged the family's identity papers, but they made a potentially fatal error; the papers were dated over a year but numbered consecutively, as if issued at one time. Levi-Montalcini worried constantly that they would be discovered, particularly after Giuseppe Levi arrived at their rooming house and announced himself to their landlady as, "Professor Giuseppe Levi—Oh, no, I keep forgetting: Professor Giuseppe Lovisato." The landlady had already suspected that Levi-Montalcini's family was Jewish, but she said nothing to the authorities.

After the war and a brief period treating sick refugees, Levi-Montalcini returned to Turin as Levi's assistant. She was depressed after the long struggle and no longer enjoyed research. A year later, though, Hamburger wrote her. He had read her articles in the Belgian and Swiss journals and wanted her to visit Washington University in St. Louis for a few months. Specifically, he wanted to know whose theory was correct—his or hers.

Elated again, Levi-Montalcini ended her engagement to Guido and prepared for her trip. She had decided definitely never to marry; she could not adjust her own life and work to meet another person's needs. She did not want to repeat her mother's marital experience. Furthermore, between World War II and her father's opposition to her education, her career had already been delayed by a decade. At thirty-eight, she was ready to concentrate on her life's work. She had concluded that "in scientific research, neither the degree of one's intelligence nor the ability to carry out one's tasks with thoroughness and precision are factors essential to personal success and fulfillment. More important...are total dedication and a tendency to underestimate difficulties, which cause one to tackle problems that other, more critical and acute persons instead opt to avoid." She planned to take advantage of every opportunity in the United States. She sailed in 1946, intent on joining the world's tiny band of neurobiologists.

After the sinister atmosphere of Fascist Italy, the campus of Washington University felt like a Garden of Eden without the snake. It occupied the grounds of the 1904 World's Fair, and instructors held classes on its lawns. In Turin, students were allowed in the library for only a few minutes at a time. In St. Louis, students put stockinged feet on the library tables, chewed gum, and slept on their books. The only jarring note involved young women, who knitted socks for their boyfriends during lectures.

After only an hour's talk with Hamburger, Levi-Montalcini was sure that she had come to the right place. Well over six feet, he towered over Levi-Montalcini's five feet, three inches. But Hamburger was kindly and—unlike Giuseppe Levi and her father—gentle and soft-spoken. The notion of competition was foreign to him, and he would never hurt another person's feelings. Levi-Montalcini decided that working with him would be delightful.

"Conceptually, it was perhaps crucial for the discovery of the nerve growth factor that we came from entirely different scientific backgrounds," Hamburger explained. "I came from experimental and analytical embryology, of which Rita hadn't the foggiest idea....Rita was a neurologist from medical school and knew the nervous system, of which I had only the foggiest idea. And she brought to St. Louis a most important tool, the silver staining method for staining nerves."

Their styles were different too. Hamburger was analytical and historically oriented, interested in where a problem had been as well as where it was going. Hamburger was rooted in the embryology of his teacher, Hans Spemann, a German Nobel Prize winner. "Viktor's whole career was taken in the direction of something like the nerve growth factor, but his style was taking small steps forward. He was more restrained and cautious," according to Robert Provine of the University of Maryland. "To use a baseball metaphor, Viktor was a singles hitter with a very high batting average. Rita was more willing to take the big swing, with the chance of making a home run."

Despite their opposite approaches, Levi-Montalcini and Hamburger became close personal friends and colleagues. When Hamburger's wife suffered a mental breakdown, she was confined to institutions and never returned home. "Rita helped me a great deal in this personal crisis, which coincided with her most important discoveries around 1950," Hamburger recalled. "So we moved very close, personally and scientifically."

Their perceptions of each other may have been different, however. Hamburger saw before him a simply dressed refugee from war-torn Europe: "Her whole manner was very humble and modest. She didn't speak the language, and the first two years she felt very awkward. She spent half her salary to send CARE packages to her family." From Levi-Montalcini's point of view, she had come to St. Louis to settle a dispute and be recognized as a scientist. Despite their friendship, that sense of competition never disappeared.

Levi-Montalcini's twenty-six years at Washington University were the "happiest and most productive years of my life.... I felt at home the day I landed. There is great cordiality, generosity. And America is a society in which merit is genuinely rewarded—you cannot say the same for Italy." In St. Louis, for the first time, her work could be her life. She was in her laboratory from early morning until midnight. On Sundays she enjoyed occasional Mississippi riverboat excursions, lab trips to the Ozarks, and holidays with Italian-American friends. But most of her friends were other scientists. She spent her summer vacation in Italy with her family; St. Louis was for work.

Late one autumn afternoon in 1947, Levi-Montalcini was peering through her microscope at her latest series of slides, mentally reconstructing the hour-by-hour development of neurons in chick embryos. She had been studying the problem day and night for months, counting the nerve cells as they came into being with and without the presence of a limb.

Suddenly, that afternoon, the pieces fell together in her mind. The cells acquired personalities. Migrating from one portion of the central nervous system to another, they looked like armies maneuvering on a battlefield. In later slides, the neurons began to die. The membranes of their nuclei became indistinct, faded away, and their bodies shriveled. Then hosts of macrophages—amoebalike cells that swallow dead cells—moved on the scene, like crews removing corpses from a battlefield. Within hours, all evidence of the nerve-cell soldiers had disappeared.

Levi-Montalcini was stunned and intoxicated with excitement. Racing to Hamburger's office, she led him back to her lab to show him the slides. She had found direct evidence that large numbers of neurons die during the course of their normal development and that amputating an embryonic limb makes even more of them die. The life of a developing neuron depends on some kind of feedback signal or hormone from the limbs. And without it, they die. When Hamburger left, she put a Bach cantata on her office record player. Elated, she thought she smelled Italian truffles in the St. Louis breeze.

Experiences like that convinced Levi-Montalcini that her strong suit was a highly developed intuition. "I have no particular intelligence, just average intelligence," she insisted. But intuition is "something that comes to my mind, and I know it's true. It is a particular gift, in the subconscious. In the night it happens, particularly to me and Paola. It's not rational." While Levi-Montalcini regarded her discoveries as virtual epiphanies, others complained that she was forgetting her own hard work and dedication—and the contributions of other scientists like Giuseppe Levi and Viktor Hamburger. But for Levi-Montalcini, life is a series of dramatic crises, from the deathbed scenes of her friends to the thrilling moments when her mind synthesized masses of data into a coherent pattern. And when her intuition told her something, she declared, "I know. I'm sure it's right."

By 1953, Levi-Montalcini and Hamburger had published two papers together. "All the observations and the experiments were done by her," Hamburger said. "I was the chairman of the department and was very busy and was not involved in the actual doing of the laboratory work. But we talked every day about it. We were in constant communication about it and she showed me her slides and told me what she had discovered, and I was very enthusiastic and encouraged her....She has a fantastic eye for seeing those things in microscope sections....and she's an extremely ingenious woman."

On a blizzardy day in 1950, Hamburger told Levi-Montalcini about an experiment that one of his students had published. Elmer Bueker was a part-time Ph.D. student who was supporting his family by teaching high school biology. He had wondered whether fast-growing tumor tissues affect the nervous system's growth in the same way that the fast-growing, complex tissues of a developing limb-bud do. So he grafted a mouse tumor onto a chick embryo. He concluded that the tumor produced the same effect as a developing limb on the nervous system and published his results.

When Levi-Montalcini reproduced Bueker's experiment, she was so astonished she thought she might be hallucinating. The bundles of nerve fibers had thrust eagerly and luxuriantly into the tumor, surrounding the tumor cells, choking off their blood supply, and then moving on to surround the embryo's growing organs. The fibers looked as if they were searching for something. Later, she thought that perhaps her mouse tumors had been more potent than Bueker's.

In any event, her next step was a bold one. She hypothesized that the tumor must have released some potent growth factor to make the nerve cells grow so spectacularly. At first she thought that the nerve fibers must have absorbed the growth factor when they touched the tumor. But then she transplanted the tumors to a membrane that is outside the body of the embryo and that communicates with it only via the circulatory system. The tumor still made the nerve fibers grow extravagantly. So she concluded—long before she had the actual evidence—that something mysterious was coming out of the tumor to make the nerve fibers grow so extravagantly.

Euphoria and excitement took over. She was desperate to speed up her experiments, to work more efficiently, to find answers faster. Working constantly, she was pushing the field to its limits, learning new disciplines and working with experts in other fields, noted her student Ruth Hogue Angeletti, now a professor at Albert Einstein College of Medicine. Soon Levi-Montalcini would even form interdisciplinary partnerships with biochemists.

A medical school friend, Hertha Meyer, had become an expert in a new technique: growing tissues in vitro (literally, in a glass dish). If Levi-Montalcini could study the process in vitro, her experiments would take hours instead of weeks. So she got a grant to visit Rio de Janeiro, where Meyer had emigrated to escape the Nazis. Carrying two tumor-infected mice onto the plane in either her handbag or her pocket—all good stories vary a little in the telling—Levi-Montalcini smuggled them through customs into Brazil.

There, amidst the exoticism of carnival, she struggled to make the nerve cells grow in a dish. It was "one of the most intense periods of my life, in which moments of enthusiasm and despair alternated with the regularity of a biological clock," she recalled. Despairing, she made one last try. She put a piece of tumor next to—but not touching—a bit of nerve tissue in a drop of clotted blood. Then she watched. Within hours, the tiny piece of tumor had made the nerve fibers grow out fantastically, miraculously, in a halo, like rays from the sun. The effect was so spectacular that she has never tired of repeating the experiment. The tumor was exuding something that made the nerve fibers reach out in all directions. That glorious halo remains the characteristic test for the presence of Levi-Montalcini's nerve growth factor, the NGF. Levi-Montalcini returned to St. Louis in 1953 thrilled and excited, expecting to spend a few quick months isolating and identifying the factor.

Luckily, Hamburger had found a young biochemist to help her. The first person Hamburger tried to hire had turned him down flat, saying the job was too difficult. But Stanley Cohen, then a young postdoctoral fellow at Washington University, signed on. "I might as

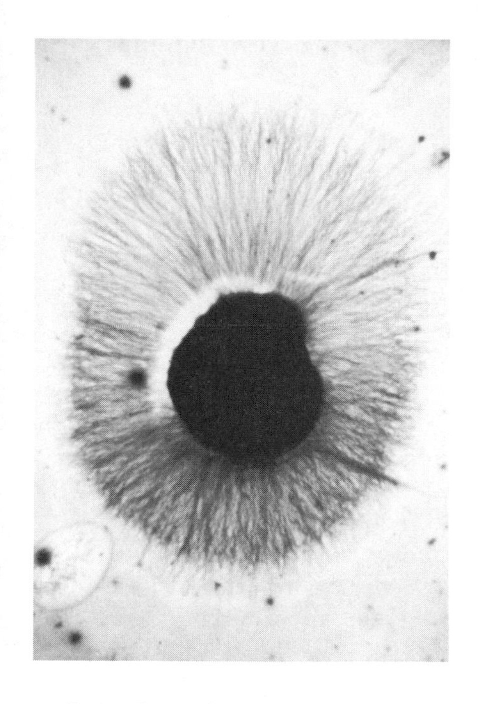

The halo of nerve fibers growing from chick embryo tissue after incubation in the nerve growth factor discovered by Rita Levi-Montalcini. This photograph, magnified greatly, was taken through a microscope.

well go down in flames on some important problem as on a trivial one," Cohen told himself. Hamburger bowed out of the project, which was now a biochemical problem. And instead of a month, Levi-Montalcini spent the next six years—"the most intense and productive years of my life"—trying with Cohen to identify the nerve growth factor.

Cohen was a clarinet-playing, pipe-smoking biochemist from Brooklyn. He shared his office with a scruffy but cheerful mongrel named Smoky. The son of an immigrant tailor and a housewife, Cohen received a college education only because Brooklyn College had a free-tuition policy and his grades were high enough to get him in. For his Ph.D. thesis, he dug five thousand earthworms from the University of Michigan campus and studied their metabolism.

Cohen was modest, steady, and unassuming, but he liked watching Levi-Montalcini's fireworks. "She worked like a fiend. She had a great drive to succeed," he realized. As he told her, "Rita, you and I are good, but together we're wonderful." He knew biochemistry and she knew neuroembryology, and their skills complemented each other. To get Cohen enough growth factor to analyze, Levi-Montalcini spent a year in grindingly dull work, growing mouse tumors. She became so adept that, when she turned her head to talk to a colleague, her hands continued dissecting under the microscope. Eventually, she and Cohen identified their compound in a solution of proteins and nucleic acids. But which was the active ingredient: the protein or the nucleic acid?

Cohen went to see Arthur Kornberg, an enzyme biochemist who was working at Washington University at the time and who later also won a Nobel Prize. Kornberg had a suggestion. Snake venom contains enzymes that break down nucleic acids but leave protein untouched. Try snake venom, he said, referring him to a colleague, Osamu Hayaishi, who was purifying enzymes from snake venom.

When Levi-Montalcini tested the snake venom on a bit of nerve tissue, she was amazed. The venom produced a stupendous halo—a much bigger halo of nerve fibers than the mouse tumor had. Compared to mouse tumors, the snake venom proved to have three thousand times more growth factor. After purifying the factor from commercially available venoms, Cohen figured that mouse tumors and snake venom could not be the only two natural sources of the growth factor; mouse tumors and snake venom had too little in common. Then he made an inspired guess. The venom-producing snake gland has a mammalian equivalent: the salivary gland. Testing various animals, he discovered that the salivary gland of a male mouse is phenomenally rich in nerve growth factor. One drop of a male mouse gland in fifty liters of solution produced a spectacular halo. With a cheap and ready source of the nerve growth factor, Cohen was able to purify it.

The team of Cohen and Levi-Montalcini was fabulously successful, but it broke up in 1959. Calling Levi-Montalcini into his office, Hamburger told her that Cohen had to go. As chair of the university's zoology department, Hamburger had pushed Levi-Montalcini's promotion to full professor through the university the year before. But he could no longer justify keeping a biochemist in a zoology department. Today science is interdisciplinary, and zoology departments routinely hire biochemists. But the practice was uncommon in 1959, and Hamburger said he could not pay a professor who could not teach zoology courses.

Cohen moved to Vanderbilt University in Nashville and continued working on another substance, the epidermal growth factor, which he had discovered in St. Louis as part of the NGF project. EGF stimulates the growth of cells in the skin, cornea, liver, and other organs.

Once more, Levi-Montalcini plunged into despair. Her six years with Cohen had been her most creative period, and his departure felt like the tolling of a funeral bell. Unsettled and dissatisfied, she was unsure of what to do next. She knew that many neurologists still did not believe in NGF. Both NGF and EGF were completely novel biological phenomena and hard to incorporate into existing knowledge. For a time, she took up a different research problem, the nervous system of cockroaches. Then she realized that she could not give up or abandon NGF.

Some colleagues contend that Levi-Montalcini overdramatized the scientific community's lack of interest in NGF. After all, they note, she was elected to the prestigious National Academy of Sciences in 1968. Nevertheless, she began to promote her discovery persistently and doggedly.

To illustrate the vital importance of NGF for the development of the nervous system, Levi-Montalcini developed a series of elegant proofs with the help of Cohen and graduate student Barbara Booker. When nerve growth factor was injected into newborn rodents, they developed an excessive number of neurons. Then, injecting antibodies to the nerve growth factor into newborn rodents, she showed that their developing nerves virtually disappear.

Further research on antibodies against NGF, conducted with a young Italian biochemist, Piero Angeletti, resulted in a classic, frequently cited review article. And in 1972, Levi-Montalcini's postdoctoral fellow Ruth Hogue Angeletti and a young Washington University biochemist, Ralph Bradshaw, identified the precise sequence of NGF's amino acids. So NGF was definitely real and important.

Armed with her evidence, Levi-Montalcini was unstoppable. Her carefully developed talks were famous for extending far beyond the usual fifty minutes. Deep into one talk, a renowned neuroscientist started grabbing her slides out of the projector, hoping to cut her off; Levi-Montalcini just filled in the blanks as if nothing had happened.

Traveling to a French scientific conference, she found herself without a reservation for the flight she wanted; every seat was filled. A colleague inside the plane watched as the pilot was summoned back into the terminal; then, suddenly, the scientist saw two figures crossing the tarmac toward the plane. They were Levi-Montalcini and the pilot—carrying her luggage. Levi-Montalcini climbed regally into the copilot's seat, and the plane took off. Her colleague never figured out whether she had used her charm or her temper—or a combination of both—to hitch a ride.

She acquired an entirely new image to sell NGF. Old-fashioned elegance gave way to an aristocratic, womanly chic. The transformation made her feel good and was good salesmanship. Her bun became a dramatic swoop, and an elegant uniform satisfied her own finishing-school standards and her busy schedule. Still slim as a fashion model and straight-backed, she designed a basic high-necked, sleeveless dress with a matching jacket. She had them made in Italy in silk and brocade. And she wore them every day summer and winter with four-inch-heeled shoes, her mother's single strand of pearls, a magnificent gold bracelet, and an antique brooch. Arriving at work each morning, she put a lab coat over the silk for dissecting mice. Before teaching her large and popular lecture course, she stopped by her office to dab a bit of perfume behind each ear. At the end of a day, she was still spotless. According to a particularly persistent story, an airline lost her luggage on her way to speak at Harvard University. Refusing to appear in her wrinkled traveling outfit, she lectured in an evening gown—the only fresh dress she had with her.

She enjoyed entertaining and became famous for her elegant dinner parties. Alone, she ate yogurt with plain rice and steamed vegetables. But for guests, she jetted in fresh truffles and developed a basic banquet, with fabulous variations. For appetizers, chicken liver pâté with raw pistachios and Marsala wine; cheese curls; and Belgian endive leaves stuffed with caviar and sour cream. For the main dish, a beef *filet en chemise* with cheese puffs and vegetables. Then a salad and frozen zabaglione for dessert. Levi-Montalcini always makes an impressive show. In Italy, she has a full-time cook, and friends cannot believe that she can prepare a meal. She does not dissuade them. Cooking, she emphasizes, was only a hobby.

As always, she lived intensely, with style and verve. Her St. Louis secretary served afternoon espresso on a tray to lab workers each day as they reported on their work. A paper she gave her postdoctoral fellow Robert Provine to edit was filled with flamboyant, sweeping statements. When Provine pruned out the flourishes, she complained mournfully, "You took my beautiful prose, and you turned it into boiled spinach." Her dry humor has a touch of aristocratic understatement. She was discussing a complicated point with Provine when two streakers raced by—a naked bicyclist with an equally nude girl perched on his handlebars. Levi-Montalcini turned regally to Provine and deadpanned in her rich Italian accent, "Bob, do they do this often here?" A friend calls her "La Regina"—in Italian, "The Queen."

She moves mountains to help people in need. She gave a technician's family a new refrigerator and arranged a job for a poor youth. She brought the daughter of a tyrannical Italian professor and her fiancé to the United States so that they could marry. She was kind and supportive to the undergraduates in her lecture class. Her secretaries remained her friends for decades, remembering how she helped them with mailings, French lessons, and the like.

With her peers and competitors, on the other hand, she could conjure up tempestuous visions of her father and Giuseppe Levi. "She places a great importance on intelligence, and she couldn't tolerate stupidity very well," Cohen learned. "If somebody said something that she thought was stupid, she'd tell him it was stupid." Levi-Montalcini speaks her mind, whether she is talking to an eminent professor or a street cleaner, a friend said.

"She had a lot of fights with a lot of people, myself included," admitted Ralph Bradshaw, who became chair of the biochemistry department at the University of California at Irvine. "Rita was extremely possessive of NGF. She viewed it as her private property. It became her child.... There's almost no one in NGF at one time or another who hasn't been at odds with her."

Referring to Levi-Montalcini's tumultuous emotions, a colleague asked Provine how he liked the "Levi-Montalcini Roller Coaster." Provine replied, "Overall, it was a good ride. But sometimes it felt like working for Maria Callas and Marie Curie." But then he added, "Great ideas are a dime a dozen. The great scientist is one who delivers. And Rita delivered."

When a student inadvertently hurt Levi-Montalcini's feelings, she screamed at the student—at length and in front of colleagues. Then the mood passed, and it was over. "Her temper doesn't last. She doesn't hold a grudge," according to her former secretary and longtime friend, Martha Fuermann. "Two things are very important to her: her work and her twin, almost in that order.... Her research has to go right. She's so wrapped up in her work, it's her life."

"Now she accepts her position as the originator of a large field, but in the [beginning], you had to wear your asbestos suit if she got your paper or grant to review," explained Ruth Hogue Angeletti. "It's a classic problem. You find something really exciting. You're the mother of the field."

Levi-Montalcini was homesick for her family, too, especially her twin Paola. Securing a National Science Foundation grant in 1961, she started a small research unit in Rome. Soon she was spending six months of the year in Rome and six in St. Louis. Italian university positions are doled out by seniority, not merit, and Levi-Montalcini had lost her place in line when she went to St. Louis and became a United States citizen. The Italian biochemist Piero Angeletti, who alternated places with her during her traveling years, helped get funding from the Italian government for an independent research institute for Levi-Montalcini to direct. The institute became the vehicle for her return home. She and Angeletti planned to run the institute together; then, at the last minute, he pulled out to accept a more lucrative position with a pharmaceutical company. Their friendship did not survive the blow.

Running the institute was not easy, given Italy's bureaucracy. Her researchers sometimes worked for months without pay because of governmental snags. But they stayed, out of loyalty and devotion. By 1979, when Levi-Montalcini retired as its full-time director, the Laboratory of Cell Biology was one of the largest biological research centers in Italy. She still works there as a guest scientist.

Growth factors finally came into their own during the 1980s. Nerve growth factors were the first clear example of a class of molecules that provide a regulatory link between targets in the body and the nerve cells that innervate them. Then it was discovered that many oncogenes—cancer-causing genes—are mutated counterparts of growth factor genes. For cancer to develop, one of these growth factors or receptors must go awry at a critical point during cell division. The oncogene connection transformed growth factors into a hot new research field.

It is clear now that the nerve growth factor keeps cells from dying in their early embryonic stages. Without nerve growth factors, half of some kinds of cells would die; with NGF, they survive. NGF affects particular kinds of cells, whether they are in the central nervous system, the peripheral nervous system, or the brain. It also appears to link the immune and the nervous systems of the body. Scientists think there must be many other growth factors besides NGF for other kinds of nerve cells, although only a few have been discovered so far.

As the years passed and NGF's importance became more obvious, Levi-Montalcini's rough edges softened and her roller coaster smoothed its ride. "Then, around 1981–82," Bradshaw recounted, "she buried the hatchet with literally everybody."

In 1986, the partnership of Levi-Montalcini and Cohen shared the $290,000 Nobel Prize for Medicine and Physiology. The recognition was especially gratifying after the years of struggle and doubt. But the prize bore bittersweet fruit too. Some scientists felt that the Nobel Committee should have honored Hamburger along with Levi-Montalcini and Cohen. After almost a decade of peace with her colleagues, Levi-Montalcini had become embroiled in controversy again.

"Many neuroscientists are puzzled by the omission of Viktor Hamburger from the prize," wrote Dale Purves and Joshua Sanes in *Trends in Neurosciences* in 1987. No one disputed that Levi-Montalcini had discovered the nerve growth factor, that she deserved the prize, and that the Nobel is given for discoveries, not a lifetime of accomplishment. However, many agreed with Purves that Hamburger had established the research model, the paradigm, that led to the discovery of NGF. "Hamburger set the stage in the 1920s, 1930s, and 1940s.... His exclusion tends to obscure a line of research that now spans more than fifty years."

Alerted to the complaints, a member of the Nobel Committee

called Levi-Montalcini and asked if the group had made a mistake in omitting Hamburger. "No," she answered, "He was in Boston and I was in Rio."

As the dispute lingered on, Levi-Montalcini began to feel that she had to defend the Nobel Committee for its decision to omit Hamburger. Interviewed by *Omni* magazine, she declared, "I've been on excellent terms with Viktor Hamburger, too. He has been an excellent chairman of the department. He was very gentle with me—never any disaffection, despite the fact that the Nobel came to me and not to him. But I believe this is correct. Viktor is a very learned person who's always done excellent work. But he never discovered NGF."

In 1988, her autobiography *In Praise of Imperfection* was published. The Alfred P. Sloan Foundation has funded a series of autobiographies written by famous scientists for nonscientists. As of 1992, Levi-Montalcini was the only woman in the series. When Purves reviewed the book for *Science* magazine, he complained that Levi-Montalcini had presented a superficial and fairy-tale view of science and how it is accomplished. Further, Purves complained, she had slighted the contributions of Viktor Hamburger and Giuseppe Levi to her career. The book does not credit Piero Angeletti, either, for his help in getting Levi-Montalcini back to Italy.

Hamburger, however, was awarded the National Medal of Science in 1989, the nation's highest scientific award. He said he was hurt, not by the Nobel Prize Committee, but by his old friend's remarks in her *Omni* interview and her autobiography. They have "a very ambivalent relationship, to put it mildly," he conceded. "On the surface, we get along well. I resent very much what she did to me. She never had great respect for my science." She, in turn, was hurt when she visited St. Louis in October 1991 and the ninety-one-year-old Hamburger did not make time to have dinner with her.

In Italy, the Nobel turned Levi-Montalcini into a national heroine. In her eighties, she still has the magnificent carriage of a model and works full time in her four-inch heels and her reed-slim suits. She uses her fame to promote the cause of Italian science and scientists. According to a local joke, the pope is instantly recognized—provided he appears with Rita Levi-Montalcini. When she became the first woman admitted to the Pontifical Academy of Sciences, the pope extended his hand for her to kiss his ring; Levi-Montalcini shook his hand instead. When an Italian-American politician visits Rome, she must sit on the banquet dais. If Parliament proposes a new tax plan, she must give the policy speech explaining its effect on science. Where abortion is debated, television crews call on Rita Levi-Montalcini to represent women scientists.

Meanwhile, the importance of the nerve growth factor continues to increase. Levi-Montalcini's discovery has become an important clue to one of biology's central mysteries: how life starts as a single embryonic cell and then marvelously differentiates, eventually producing a complex organism of many different cell types, each in harmonious relationship to one another. Before the nerve growth factor, little was known about how organs signal developing nerve cells to link up with them or how messenger chemicals tell nerve cells when to grow and when to stop growing. The nerve growth factor was the first of several hundred signals now known to affect cells and organs.

When Levi-Montalcini discovered NGF in 1952, she could never have imagined all its potential applications to medicine. Growth factors already speed burn healing and diminish the side effects of chemotherapy and radiation therapy. The nerve growth factor itself is now known to belong to a family of factors called neurotrophins, which may help slow the degeneration of the nervous system in diseases like Alzheimer's and Parkinson's. Someday the nerve growth factor may help heal peripheral nervous system damage in diabetics and prevent such damage to chemotherapy patients. Meanwhile, biotechnology companies are searching for other growth factors that may stimulate the growth of motor neurons in patients suffering from spinal-cord damage. A bone growth factor may regulate the formation of the embryo's shapeless tissue into skeleton and then heal broken bones. In 1991, researchers at Washington University used growth factors to transform limp muscle into hard, well-formed bone inside experimental rats.

Above all, growth factors have helped change our views of the nervous system. Today, scientists think of the nervous system not just as an organ that monitors and controls the body, but also as an organ that is controlled by the body. The body's influence on the nervous system is considered as important today as the influence of the nerves on the body.

Levi-Montalcini still directs wide-ranging research, but she concentrates her volunteer energies on degenerative diseases of the nervous system. She is president of the Italian Association for Multiple Sclerosis, because NGF exacerbates the inflammation of multiple sclerosis while the factor's antiserum helps reduce the inflammation.

Levi-Montalcini and her beloved sister Paola share a double apartment that includes Paola's art studio. They are as close as ever. When they are apart, Levi-Montalcini telephones home several times daily. But mixing the life of a celebrity and a research scientist at eighty-plus years is not easy. Levi-Montalcini lives in a welter of appointments made and unmade. She has a chauffeur for her Alfa

Romeo Lotus and a car telephone to communicate with the institute. She keeps two secretaries busy—one in English, the other in Italian. She promotes Italian science and women scientists wherever she goes. Her only concession to age is to hide a heavy sweater under her silk jacket.

As for retiring, she declares, "The moment you stop working, you are dead.... For me, it would be unhappiness beyond anything else. ...I don't work for the sake of mankind. I work for my own sake." For, as she quotes Dante,

> Take thought of the seed from which you spring.
> You were not born to live as brutes.

* * *

Zabaglione Coffee Ice Cream
à la Rita Levi-Montalcini

1. **Beat 10 egg yolks** with **4–5 tablespoons sugar** until thick and light-colored.
   Add slowly:
   3 demitasse cups strong coffee
   3 demitasse cups liqueur, such as rum, brandy, or coffee liqueur.
2. In the bottom of a double boiler or large pot, bring a small amount of **water** to a simmer. Pour **the egg mixture** into the top of the double boiler or into a metal bowl placed on top of the pot. Do not let the water boil hard.
3. Stir with a whisk until the mixture thickens. (If the mixture is too runny or if it curdles or forms lumps, put it in a food processor to "homogenize" it. *If it is still too runny,* mix it with another **6 egg yolks beaten with a small amount of sugar and liqueur** and reheat gently.)
4. Pour the mixture into a bowl. **Add 4 to 5 teaspoons of coffee powder,** stirring to mix and cool.
5. Add **16 ounces of cream,** freshly whipped with a small amount of sugar.
6. Add ½ **wine glass of strong liqueur: brandy, rum, coffee, or raspberry liqueur.**
7. Pour the mixture into a ring mold, cover, and freeze overnight.

8. Turn the ice cream onto a plate and dust the top with a **mixture of coffee and sugar.** This gives the surface a translucent appearance, with stripes, if you like. Return the ice cream to the freezer.

9. Remove the ice cream from the freezer and keep it in the refrigerator for an hour before serving. In the hole in the center and around the rim, put **candied fruit, marrons glacés, kiwi,** slices of **sugared oranges, mango or other fruits,** sprinkled lightly with **liqueur.**

According to Levi-Montalcini, the ice cream keeps for months (or years) in the freezer.

10

Dorothy Crowfoot Hodgkin

May 12, 1910–July 29, 1994

PHYSICAL CHEMIST

Nobel Prize in Chemistry 1964

WHEN DOROTHY CROWFOOT HODGKIN was a small child in 1914, she was left in England with her younger sisters and a nursemaid while her parents returned to the Middle East. Trapped there by World War I, Dorothy's mother saw her children only once in the next four years.

Far from being devastated by the experience, Dorothy Hodgkin believes that it made her independent. Certainly, the self-reliance, gentle compassion, and zest for challenge that helped her as a child enabled her to face an adulthood of crippling disease and political controversy.

Dorothy Hodgkin liked doing the impossible. She enjoyed life at the edge of the scientific frontier. As a result, she made not one brilliant breakthrough, but a series of them, deciphering the atomic structure of one medically important substance after another, each larger and more complicated than the last. Using X-ray crystallography, she uncovered the structure of penicillin during World War II. Later, she solved the structure of vitamin B_{12}, the cure for pernicious anemia, and the structure of insulin, the lifeline for diabetics.

Even more important, she reinvented crystallography with every major molecule she solved. She chose projects that everyone else considered impossible. Then, with relentless determination and sparkling imagination, she devised new methods that expanded crystallography's technical capabilities far beyond what other chemists considered possible at the time. More than any other scientist, she personified the transformation of crystallography from a black art into an indispensable scientific tool.

She helped establish one of the characteristic features of contemporary science: the use of molecular structure to explain biological function. She helped transform organic chemistry, liberating it from the chore of structural determination. A confidante of former Prime Minister Margaret Thatcher, she worked all her life to improve relations between East and West. She has been called "the cleverest woman in England" and "a gentle genius."

Dorothy Crowfoot was born in Cairo, Egypt, then a British colony, on May 12, 1910. Her father, John Winter Crowfoot, supervised Egyptian schools and ancient monuments for the British government. Her mother, Molly Hood Crowfoot, had attended a French finishing school, and though she had no academic training, she was a self-taught expert in botany and ancient weaving. She drew the illustrations for the official study *Flora of the Sudan.*

During a family vacation in England, Dorothy's childhood was shattered by the outbreak of World War I. Fearing a Turkish attack on British colonies in the Middle East, her father installed four-year-old Dorothy, two younger sisters, and a nursemaid named Katie in a house near his mother's home outside Brighton. Then, in the grand tradition of the British Empire, Dorothy's mother left her children in England and returned to the colonies to care for her husband. There she was trapped by submarine warfare for the rest of World War I. From the time Dorothy was four years old until she was eight, her mother managed to make only one visit home. Hodgkin understood. "It was so damned dangerous, you see," she explained years later.

When her mother returned to England after the war to spend a year with her daughters, she discovered a gentle eight-year-old with enormous violet eyes and a wide, enthusiastic smile. Dorothy was easy to underestimate, unassuming and quiet-spoken all her life. She never pushed her way into a conversation, but when others argued, she spoke up shyly, going right to the heart of the issue. Eventually, the Crowfoots settled their children near relatives in Geldeston, close to England's easternmost point. The girls lived there year-round, while their parents returned each summer from the Middle East, Dorothy's father for three months and her mother for six.

Dorothy's early education was sketchy at best. She moved from one small private school to another. When she was ten, she attended a small class organized to improve the education provided by governesses. Its little chemistry book started off with experiments to grow copper sulfate and alum crystals. Entranced, Dorothy repeated the experiments at home.

Chemistry and particularly crystals were a traditional hobby for British women of leisure throughout the nineteenth century. Latin

Dorothy Crowfoot Hodgkin with atomic models of (from left to right) penicillin, insulin, and vitamin B_{12}.

Dorothy Crowfoot Hodgkin with daughter Elizabeth, in the early 1940s.

Dorothy Crowfoot Hodgkin at the time of the Nobel Prize.

Drawing of Dorothy Crowfoot Hodgkin's hands by Henry Moore, 1978.

and Greek were considered suitable only for gentlemen; ladies and working-class men were relegated to practical subjects like chemistry. Ladies owned chemistry sets marketed specifically for them. They grew crystals, conducted experiments, published research in women's chemistry journals, and attended scientific lectures. Michael Faraday, the giant of nineteenth-century physics, learned his chemistry from *Mrs. Marcet's Conversations in Chemistry*.

As Dorothy discovered, a crystal is a solid composed of atoms arranged in a regular and repeated pattern. The size is immaterial; some mineral crystals weigh thousands of pounds. Snowflakes and diamonds are crystals, but so are most metals, including iron and copper. The most important feature of a crystal is the regularity of its atomic structure and the repeat of its pattern. Crystalline patterns are like three-dimensional wallpaper.

When Dorothy enrolled in a nearby government-supported secondary school at age eleven, she boarded during the academic year with a town family. After school, she and her sisters volunteered for the League of Nations and various peace campaigns. Their mother was vehemently opposed to war—World War I had killed her four brothers—and when she attended a League meeting in Geneva, she took Dorothy along. Dorothy also began her lifelong membership in the Labour party during her girlhood. When her school held mock elections during national campaigns, it was easy to find Conservative and Liberal party candidates, but Labour attracted few students. Finally Dorothy's mother asked, "Why don't you run for Labour?" Did she win? "No, no, good gracious no," Hodgkin said later. "I got six votes."

In addition to world peace, Dorothy's main interest was chemistry. Her mother was delighted and encouraged her at every turn. When Dorothy was thirteen, she and her sister Joan spent six months with their parents in the Sudan. Molly Crowfoot took the girls to visit a family friend, soil chemist A. F. Joseph, at the Cairo branch of the Wellcome Laboratory. Geologists there showed them how to prospect for gold. When Dorothy practiced panning in their parents' yard, she found a shiny black mineral and thought it might be manganese dioxide.

"Please, can I analyze this mineral and find out what it is?" she asked Joseph. With his help, Dorothy discovered that her sample was ilmenite, an iron and titanium mix. Impressed, Joseph gave her a real surveyors' box full of reagents and minerals. He also advised her to buy a standard textbook of analytical chemistry. Back in England, Dorothy set up a small, attic laboratory at home. For her sixteenth birthday, in 1926, Molly Crowfoot bought her a children's book by the Nobel Prize–winning physicist William Henry Bragg. In it, he described how he could shine X rays through a crystal to discover the arrangement of its atoms. Bragg and his twenty-two-year-old son William Lawrence won the Nobel Prize for the technique, which had revolutionized physics and chemistry and would do the same for biology. Dorothy had found her lifelong work.

She was having school problems, though. When she enrolled in secondary school, she was behind the other students and "terribly behind in arithmetic." She caught up only to discover that girls were not allowed to study chemistry in her school. Luckily, the instructor was a woman, and Miss Deeley made a special exception for Dorothy and a friend to join the class. By the time Dorothy graduated from high school in 1928, she was determined to study chemistry at Oxford University. At that point, she learned that she was totally unprepared to pass the Oxford University entrance examination. She had not studied Latin or a second science, and both were required of all entering students. Luckily, her mother could tutor her in botany. Her mother's sister, Aunt Dorothy Hood, volunteered to pay Oxford's tuition.

After Dorothy passed Oxford's entrance examinations, she took a six-month vacation with her parents in Jerusalem. Her father had become director of the British School of Archaeology there. While the Crowfoots excavated Byzantine churches, Dorothy recorded the patterns of their mosaic floors. She had such a good time that she considered switching from chemistry to archaeology. She liked both subjects for the same reason: "You're finding what's there and then trying to make sense of what you find."

While Hodgkin was vacationing with her parents, the women of Oxford University were waging a critical battle between coeducation and "virility." In 1927, Oxford University had one woman student for every five men. The men panicked. Cambridge University, with one woman for every eight or nine men, was obviously more manly, they argued. Oxford's "virility" was at stake. The women never had a chance. The university passed regulations limiting the number of women students to fewer than eight hundred so that Oxford would remain an institution for gentlemen's sons and a few gentlemen's daughters. It would not become truly coeducational until the 1970s.

Such attitudes kept Hodgkin closely chaperoned at Oxford. Men were not permitted to attend Sunday tea parties at women's colleges. Women students were not allowed to have lunch or tea in a man's room without a chaperone and prior permission from the dean. Women medical students dissected cadavers in private rooms separated from men. Women were not permitted to join Oxford's debating club, and only a few sang with the University Opera Society.

As far as the Hodgkin family was concerned, however, Dorothy was a responsible adult. For the first time, she was allowed to take charge of her younger sisters during their Christmas vacation. Until then, they had spent holidays with relatives. When one of her sisters became seriously ill, Hodgkin nursed the child day and night. She never even thought of notifying the college that she would return late. The choice between studies and children seemed obvious.

The choice between chemistry and archaeology was becoming clearer, too. She visited a German laboratory in Heidelberg one summer, and an Oxford professor loaned her a key to his laboratory so she could work through her other holidays. Soon she knew a great deal about inorganic chemistry. She still had "lurking questions," though. Chemists had erected captivating edifices to describe the physical world, but she kept wondering if they were accurate. "Would it not be better if one could really 'see' whether molecules...were just as experiment suggested?" she thought.

Longing to "see" molecules, she decided to specialize in a new field, X-ray crystallography. Max von Laue, a German physicist, had

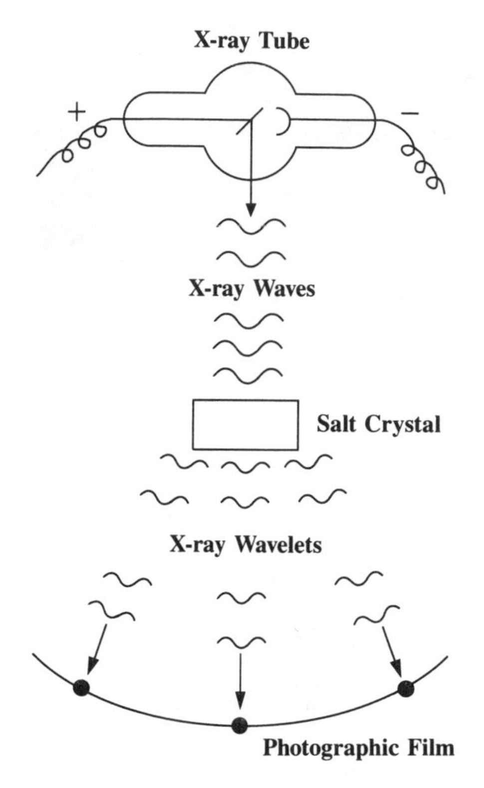

Figure 10.1. X-Ray Crystallography.
X rays pass between the atoms of the crystal and are reflected at angles onto photographic film.

won a Nobel Prize in 1914 for his discovery that X rays can be scattered by the atoms of a crystal. The Braggs won their Nobel Prize a year later for showing that, when the crystal scatters the X rays onto a photographic plate, they make spots reflecting the arrangement of the atoms within the crystal.

In principle, crystallography is simple. The Braggs beamed X-ray waves through a crystal and studied the waves that emerged and struck a photographic plate.

When visible light strikes a group of evenly spaced parallel lines, the light is separated into different wavelengths. The effect is seen on a compact disk when it reflects a rainbow of colors. A crystal works the same way, on an atomic scale. X rays and the spaces between the atoms in a crystal are roughly four thousand times smaller than light waves and CD grooves. As the X rays pass between the atoms of the crystal, they are scattered by the electrons surrounding the atoms and

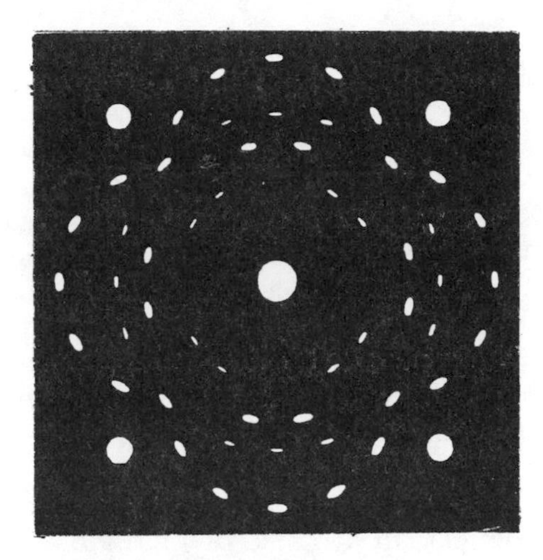

Figure 10.2. Diffraction photograph.
By studying the intensity of the spots on her photographs, Dorothy Hodgkin could calculate the position of the atoms in the crystal.

are reflected at an angle. Continuing on their separate ways, they strike a photographic plate and form spots. By analyzing the relative darkness and lightness of the spots and their positions on the photographic plates, the Braggs could deduce the locations and sizes of the atoms in the crystal.

Hodgkin soon discovered that crystallography, in practice, is more difficult than Bragg's book suggested. It was more like analyzing a jungle gym from its shadows. She took dozens of photographs of each crystal from different angles. She gauged by eye the intensity of thousands of spots on photographic plates. The most intense spots reflected patterns in the crystal where the X rays had struck large numbers of electrons. The centers of the atoms are most likely to be found in the middle of the electrons. She made lengthy and tedious mathematical calculations.

Despite her skills, Hodgkin could not find a job after her university graduation in 1932. It was her childhood friend from Egypt, A. F. Joseph, who came to her rescue. Meeting an eminent chemistry professor on a train, Joseph asked for advice about Hodgkin's prospects. The professor recommended a job with John Desmond "J. D." Bernal at Cambridge University. Bernal was pioneering the use of X rays to study biological crystals, especially proteins, the most diverse and important chemical constituents of cells. By decoding their molecular structure, he hoped to understand their properties, reactions with other chemicals, and their synthesis from simpler compounds. Hodgkin scraped together enough money to work with Bernal for one year by combining a £75 research grant from Cambridge and a £200 gift from her devoted Aunt Dorothy.

Bernal lived life with gusto. His friends called him "Sage" because he was knowledgeable about so many topics, from science to women, politics, and art. Bernal married one woman, maintained households with two others, and entertained innumerable lovers on the side, according to his biographer Maurice Goldsmith. Bernal was raised a Roman Catholic in Ireland but became a Communist during the 1930s because he regarded Marxism as more humane and rational than capitalism. He was interested in left-wing artists: Paul Robeson sang in his apartment, and Picasso drew on his wall. As a friend said, Bernal was "so very pale in face and so red in outlook." He himself recommended printing his biography on variously colored paper. As Hodgkin dryly noted, "Oral tradition differs as to the colour of the pages: black and white, certainly, for science; red for politics; blue for arts; purple or yellow for his personal life."

Bernal was a visionary scientist, a fabulous conversationalist, and a firm believer in equal opportunities for women. Bernal and both Braggs hired women scientists; altogether, they turned crystallography into one of the few physical sciences employing significant numbers of women. Bernal included everyone in the informal life of his lab. At lunchtime, someone would buy fresh bread, fruit, and cheese from a market, and someone else would make coffee on a gas ring in the lab. Bernal would chat about anything from anaerobic bacteria in the bottom of Lake Baikal to the origin of life, Romanesque architecture in French villages, Leonardo da Vinci's military inventions, poetry, or painting. Hodgkin thought of those lunchtime talks as sparkling visits to enchanted lands.

Even the atmosphere in Bernal's lab was electric. It made Hodgkin's hair stand on end—literally, that is, from static electricity. Electric wires hung from the ceiling in casual disregard for safety. "It was a miracle nothing happened," Hodgkin thought. Actually, there were several accidents. A physicist who bumped over some apparatus got a massive radiation dose. Bernal and another colleague nearly killed themselves setting up an X-ray tube. After fitting it together with string and sealing wax, they connected it to an old transformer. While Bernal held the ground wire, his co-worker snipped through it. The electric shock threw them both across the room.

As Bernal's assistant, Hodgkin revealed a unique combination of extraordinary gentleness, shining brilliance, and an iron determination to understand and solve scientific problems. Her friend, the

Nobel Prize–winner Max Perutz, thought of her as "the gentle genius."

Hodgkin certainly had plenty of raw material to work on. Bernal's desk was covered with crystals that friends and colleagues had sent him to study. As a result, Hodgkin made some of the earliest X-ray studies of vitamin B_1, vitamin D, sex hormones, and protein crystals. She became famous for "clearing Bernal's desk."

Two momentous events—one tragic and the other bright with hope—coincided one winter day in 1934. The joints in Hodgkin's hands had become inflamed, tender, and painful, so her parents took her to consult a London specialist. He diagnosed an extremely severe case of rheumatoid arthritis. At the time there was no effective treatment for the disease, now known to be caused by the immune system's attack on its own tissues. Although Hodgkin rarely mentioned the pain she endured, her hands and feet gradually became badly crippled. Through it all, she never permitted her handwriting to change, and she continued to work nimbly on minute details with exquisite accuracy. She liked to "think with her hands."

The same day that Hodgkin learned about her rheumatoid arthritis, Bernal took the first X-ray photograph of a protein crystal. A friend had brought him an enormous crystal of pepsin, an enzyme that aids digestion in the stomach. The regularity of the spots on the X-ray photograph proved unmistakably that proteins can be crystallized and that the arrangement of atoms in crystalline proteins would some day be "seen." Bernal was so thrilled that he could not sleep that night; walking the streets of Cambridge, he dreamed about the future when X-ray crystallographers would explain the structure and function of complex proteins. The day that began for Hodgkin with so much pain ended with a vision: some day she would be able to "see" molecules of great biological and medical importance.

Partway through Hodgkin's year with Bernal, Somerville College offered her a job teaching chemistry. Somerville was the women's college at Oxford University where she had been a student. She did not want to leave Bernal's lab, so Somerville sweetened its offer: She could remain with Bernal for another year and then spend a second year at Somerville finishing her doctoral research. Only then would she have to teach. Jobs were scarce and Hodgkin needed money, so she reluctantly agreed to leave Cambridge in 1934.

Hodgkin's early years in Oxford were lonely. As a woman, she was excluded from important parts of Oxford's scientific life. "Prewar Oxford was a masculine stronghold and the science faculties even more so," explained one of her first students, Dennis Parker Riley.

The biggest problem—and one she minded a great deal—was that the chemistry club uniting Oxford's chemists did not permit women to belong or attend meetings. Women could attend general sessions, but not the small weekly talks about current research. Hodgkin was never invited to address the club.

Fortunately, Oxford's chemistry students had a comparable organization, and the students did invite her to speak. A crowd of chemistry professors came and listened with close attention and obvious respect. Riley was so impressed with their reaction that he asked Hodgkin to be his research adviser. His request raised eyebrows throughout Oxford. "Here I was," said Riley, "a member of a prestigious [men's] college choosing to do my fourth year's research in a new borderline subject with a young female who held no university appointment but only a fellowship in a women's college." Male faculty members held dual positions from their colleges and the university at large. Hodgkin had only her Somerville College post and pay. Hodgkin was delighted, however. She handed Riley a book about crystallography and a tube of crystals and told him to get started. As Riley put it, she led him to the deep end and invited him to dive in. She did not inquire whether he could swim.

Besides scientific companionship, Hodgkin missed Cambridge's laboratory facilities. Hodgkin's lab was located in the basement of the Oxford University Museum among medieval architectural motifs, dinosaur skeletons, dried beetles, and anthropological exhibits. Some of the kinder terms used to describe her facilities were "ghoulish," "primitive," and "an eighteenth-century engraving." Her room was a basement–ground floor affair with a fretted Gothic window so high above the floor that no one could see out of it. Just under the window was a gallery, reachable only by a rickety circular staircase. There, by the window's light, Hodgkin kept her polarizing microscope. To study one of her precious crystals, she climbed the ladder, clutching the crystal on a thin glass fiber in one arthritic hand and holding onto the ladder with the other. Legend has it that she never lost a crystal.

Next door was an X-ray room like a 1930s science-fiction thriller, festooned with electric cables suspended by string from the ceiling. Blithely ignoring health hazards, Hodgkin and her students analyzed their data at a large, paper-strewn table in the middle of the X-ray room. The only safety precaution was a sign: "Danger—60,000 Volts." During World War II, an overzealous Home Guardsman who brandished his bayonet at a late-working student touched an electric wire and almost electrocuted himself.

Her lab equipment was antiquated or nonexistent. England's economy had not recovered from World War I, and Hodgkin's department budgeted only £50 a year for apparatus. The laboratory did not even have a refrigerator for storing crystals, and when Hodgkin's student bought a cheap one, the administration severely reprimanded him.

Hodgkin decided that holiday visits to Bernal's lab would give her enough scientific companionship, and she could manage among the dinosaurs. But she realized that she *had* to have modern equipment. No one at Oxford, however, knew how to apply for a research grant. So she went to a senior professor. Would he please ask a British chemical firm for £600 for equipment? she asked. Despite an otherworldly air, Hodgkin was direct, practical, and down-to-earth. She always talked facts, not generalities. And she knew that one of her biggest weapons was being "a lone girl" among Oxford's male professors. Faced with such a formidably prepared applicant, the professor agreed to find Hodgkin money for new cameras and X-ray tubes.

She wanted to settle down and focus in detail on one class of biological compounds. She chose the sterols, especially cholesterol. A number of chemists had proposed formulas for these crystalline fatty alcohols, and one even won a Nobel Prize for his work. All their formulas turned out to be wrong. With a crucial observation from Bernal, a London chemist was finally able to derive the correct chemical formula for cholesterol. He could not explain, however, how the carbon, hydrogen, and oxygen atoms were arranged three-dimensionally within the molecule. Hodgkin decided to find out. She was choosing—as she would throughout her career—an important biochemical problem just beyond the limits of what everyone else considered feasible. Determining its structure would be the most complicated X-ray investigation of an organic molecule yet undertaken. Yet Hodgkin and her research student Harry Carlisle succeeded in determining its three-dimensional structure. For the first time, X rays had revealed the structure of a molecule that synthetic and organic chemists could not decipher. Hodgkin got an almost childish pleasure out of triumphs, and she jigged for joy around the lab.

Hodgkin was exploiting several new techniques developed during the early 1930s, and she soon developed an unbeatable repertoire of skills. Linus Pauling had produced a set of commonsense rules about how atoms form crystals. The sizes and electrical charges of atoms determine their arrangements as molecules; their electrical charges must be neutralized, not just on average over the entire molecule, but also locally over each small region. Finally, the atoms must fit snugly together.

A physicist, A. L. Patterson, demonstrated that a mind-boggling series of mathematical computations would relate the spots on the photographic plates to the distribution of the atoms within crystals. For even a simple crystal, Hodgkin had to do tens of thousands of calculations. They produced an electron density map that looked like a topographical map of a mountainous country. The mountains corresponded to the electron-packed regions of the crystals. The peaks of the mountains indicated the most likely places to find atoms. Hodgkin became a legend at interpreting electron density maps, sensing the relationship between the peaks and valleys on the charts and the atoms and empty spaces within a crystal.

The mathematical calculations took months and years with hand adding machines. In 1936, C. A. Beevers and H. Lipson wrote Hodgkin, offering to sell her for £5 two sturdy wooden boxes filled with eighty-four hundred paper strips. These Beevers and Lipson strips sped up the calculations, and Hodgkin was so delighted with her purchase that she kept their letter as a souvenir.

Only an early crystallographer could love a Beevers and Lipson strip. Each paper is approximately a half-inch wide and seven inches long. All eighty-four hundred of them are covered with typewritten numbers—values for cosines and sines of waves with different heights and frequencies. The strips are arranged in order in the boxes and must be kept in sequence or they became almost impossible to locate. Dropping a box was a disaster. Nevertheless, Beevers and Lipson strips were ten thousand times easier than a slide rule. They saved looking up the sines and cosines for every X-ray angle and saved multiplying them by various factors. They also reduced all calculations to simple addition. In short, Beevers and Lipson strips were a primitive but ingenious computer system.

Unfortunately, Pauling's rules, Bragg's and Patterson's maps, and Beevers and Lipson strips did not give Hodgkin all the information she needed. Half was still missing. The spots on the photographic film did not reveal enough. So she made another crystal and compared the two: the new crystal was just like the first except that it contained an additional heavy atom like mercury. Comparing the two closely related crystals helped her get the missing half of the information and locate all the atoms. She was ready to solve extremely complex crystalline structures.

In the meantime, word was getting around that Hodgkin, like Bernal, had remarkable insight about X-ray photographs. Soon her desk was also cluttered with crystals. An Oxford professor whetted her appetite with a newly prepared insulin crystal. "Though insulin was much too large a molecule for me to study in detail then, I could not resist growing the crystals, making measurements and thinking about them," Hodgkin admitted. A few years later, while walking along Oxford's main street, she met penicillin pioneer Ernst Chain. He was thrilled; penicillin had just cured four mice infected with streptococcus. "Some day we will have crystals for you," Chain promised Hodgkin excitedly. Soon she was dreaming about insulin *and* penicillin.

In the meantime, at age twenty-seven, Dorothy had fallen in love. Her future husband, Thomas L. Hodgkin, came from a long line of historians, scientists, and other academic stars. He was a great raconteur, a bit of a bohemian, and a bon vivant. He loved good food and wine and was a magnificent cook who kept his dinner guests in gales of laughter. He was warmhearted and good-humored. While in Palestine in 1936, Thomas had become so sympathetic with the Arab rebellion against British colonial rule that he resigned his job, returned to England, and joined the Communist party. There he sank briefly into what the *London Times* called "gloom and inactivity," only to be rescued by Dorothy.

The Hodgkins were married in December 1937, and two years later, the Friends Service Council hired him to organize adult education classes among the unemployed in Stoke-on-Trent. Thomas's ancestors had been Quakers, and he and Dorothy always sided with the underdog. He loved the job and, for almost ten years, taught in the north of England and spent weekends with Dorothy in Oxford.

"They really adored each other. Thomas really kept her amused," Max Perutz reported. Thomas once told Dorothy's student Barbara Rogers Low that "in any marriage there's always someone more devoted to the marriage than the other person"; Thomas seemed to be talking about himself. He was certainly protective of frail-looking Dorothy. Once when he left on a trip, he asked crystallographer Keith Prout, "Will you look after dear little Dossie?" Snorted Prout later, "She was the last person who needed looking after."

"It was a remarkable marriage because early on Thomas decided she was the more creative of the two and that she was going to have a chance," observed Anne Sayre, author of *Rosalind Franklin and DNA*. Dorothy and Thomas had three children: Luke, born in 1938, Elizabeth, in 1941, and Toby, in 1946. After World War II, when Thomas began teaching at Oxford University, he was the one who took them to the dentist and the zoo and who stayed home evenings so Dorothy could return to the lab. "There was great sweetness in this, no martyrdom," Sayre noted.

Thomas was involved in British politics and loved to argue. Dorothy, on the other hand, was interested in world peace movements and disliked arguments. Politics for Dorothy meant personalities and people, not issues and trends. "Dorothy was always at the personal level," realized her student and longtime friend Thomas Blundell of Birkbeck College, London. She sympathized with the people of the Soviet Union, just as she did with individuals from other places. With Bernal and Thomas in and out of the Communist party over the years, Dorothy thought about joining too from time to time. "But joining the party was regarded as a serious commitment. A friend gave me a big list of reading to do in preparation—and I never quite finished it," she admitted.

Combining two careers and three children proved "reasonably easy," Dorothy reported. Her arthritis improved dramatically during each pregnancy, her hands straightening out somewhat and the pain receding. Eventually it was discovered that the hormone cortisone reduced the inflammation of rheumatoid arthritis. She listed her hobbies in *Who's Who* as "archaeology, walking, and children." She could switch easily from deep concentration on a calculation to conversation with a child. When Luke started violin lessons, he wanted his mother along for the first session, and Dorothy cut short a grant meeting with a Rockefeller Foundation official, who was charmed.

"We had good help at home," Hodgkin explained. "Some time in the war, Edith and Alice came, elderly refugees from the blitz, and stayed helping us for the rest of their working lives, and Thomas's mother was marvelously helpful—she took the children off for one whole day a week to Queen's College where her husband was then provost. During the war, the children were very young but I had no sense of guilt in continuing with scientific work—it seemed the natural thing to do at that period." She walked home to lunch with the children each day and was usually home by five P.M. for dinner with them. She rarely worked late.

By 1940, Britain was at war and Hodgkin was finishing her cholesterol studies. She dropped by Chain's lab to ask if she could look at his penicillin crystal. Penicillin, a small protein, looked "just the right size for a beginner," she joked. She thought that knowledge of its molecular structure would make its mass manufacture from related compounds easier. Before penicillin, even a boil or a scratch from a rose thorn could be lethal. With millions of war casualties anticipated, a drug that prevented bacterial infection was a vital weapon. British and American authorities declared penicillin a "high security secret" and only those working in the field could get research reports. Some of the first penicillin crystals made in the United States were flown to Hodgkin.

At first, penicillin looked hopeless. It was far more difficult to analyze than a sterol, full of false starts and blind alleys. Unbeknown to anyone, penicillin crystals come in several shapes and the Americans and British were working with four different types. Furthermore, even mild changes in conditions change the way the molecules pack within a penicillin crystal. Hodgkin had only one graduate student, Barbara Rogers Low, to help her on the project. Low, studying the photographic films, could see that molecules sat on top of one another, overlapping and hard to separate one from another. "It was a bit of a mess; not a total mess, but a bit of a mess," Low thought.

The biggest problem was that no one knew the chemical groups that make up penicillin. Hodgkin and Low were starting more or less from scratch. And much of what the chemists thought they knew turned out to be incorrect. When Hodgkin discovered that part of the molecule consists of a beta-lactam ring, a knighted chemist assured her that she was dead wrong. Another chemist promised, "If penicillin turns out to have the beta-lactam structure, I shall give up

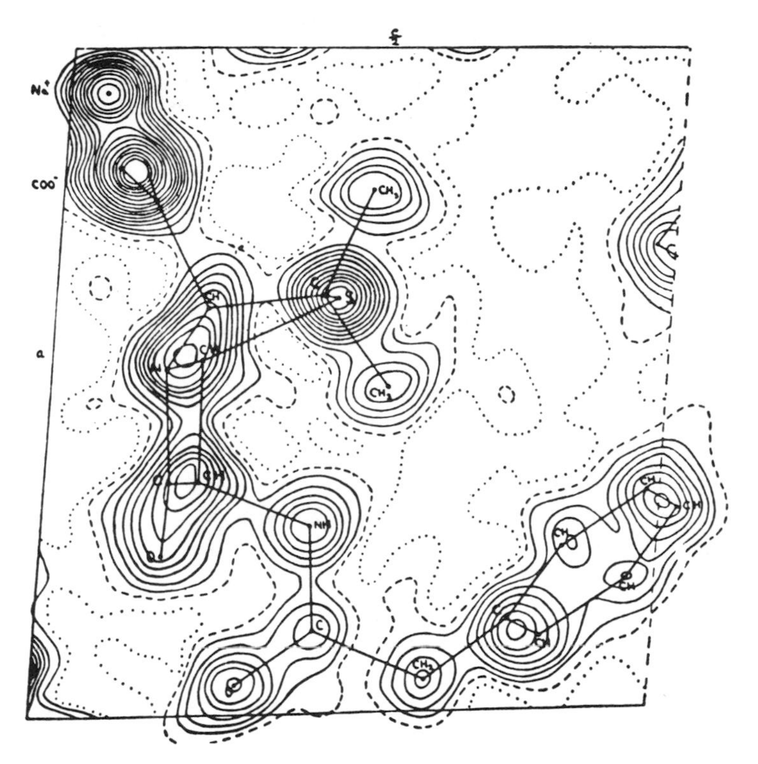

Figure 10.3. Electron density map of a penicillin molecule. The map, which resembles a topographical map of a mountainous region, shows the peaks where electrons are concentrated. Dorothy Hodgkin obtained this map by analyzing several hundred spots on X-ray diffraction photographs.

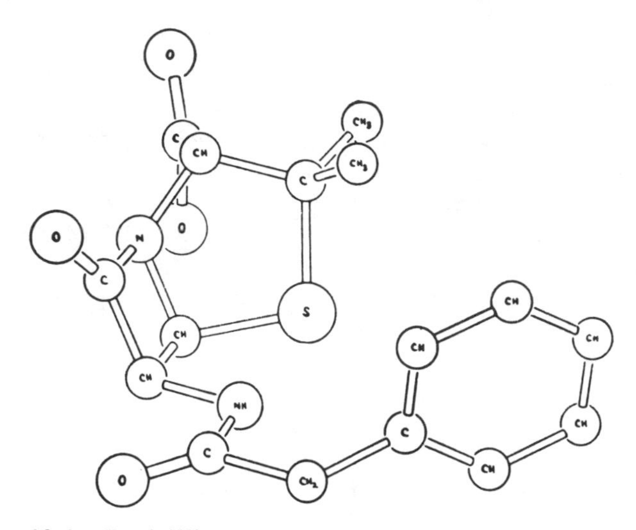

Figure 10.4. Penicillin.
The arrangement of atoms in a penicillin molecule, as Dorothy Hodgkin deduced it using electron maps like those in figure 10.3. The five-sided thiazolidine ring is at left, attached to the four-sided beta-lactam ring.

chemistry and grow mushrooms." Hodgkin was right, however; all penicillins are made of a five-sided thiazolidine ring attached to a beta-lactam ring. No one had ever seen components joined in such an unusual way. And no one became a mushroom farmer, either; the gentleman stayed in chemistry and eventually won a Nobel Prize.

Hodgkin's laboratory conditions were still primitive. The Rockefeller Foundation's officer knew of only two other labs as poorly equipped as hers, and one of them consisted of two tables piled on top of each other. The museum was unheated, and the floor felt cold and damp. Equipment budgets were still paltry. When the wire that held her thermometer broke, Low asked the department administrator for two more inches. He looked in his account books and bellowed, "I gave you two inches of wire three months ago. What have you done with it?" Finally, the sheer magnitude of the data collection and mathematical calculations seemed endless. When Low put a number in her electrical calculating machine, it ground away interminably with an "*Mmmmmm, kerplunk.* I think I spent years of my life sitting in front of that thing with the Beevers-Lipson strips."

Conditions gradually improved, though. When Bernal went into war research, he gave Hodgkin his equipment. The Rockefeller Foundation awarded her the first of many modest grants. And most important, halfway through the war, she got help from one of the world's first computers. An early IBM analog computer was more like a supermechanical calculating machine than a modern electronic computer. It tracked ship cargoes during the day, however, and penicillin atoms at night. Low helped write the first three-dimensional computer program and punched data onto cards, carefully avoiding any cards manufactured in the north of England because they were so damp that the machine chewed them up. Her formulas so confused the keypunchers that she began to disguise the penicillin atoms as ship cargoes.

"It was a nice day when we could set up the model first precisely in three dimensions and rang up our friends to come and see what penicillin actually was," Hodgkin wrote. It was 1946, four years after she had started work in earnest on penicillin. By the war's end, hundreds of thousands of servicemen had been treated with penicillin. After the war, drug makers used Hodgkin's model to develop scores of semisynthetic penicillins. They grew the two basic rings with bacteria and synthetically added different sidechains to attack particular bacterial infections.

Penicillin proved that Hodgkin was a master crystallographer. The Nobel Committee later called it "a magnificent start to a new era of crystallography." Financially, however, she was broke. She and Thomas each earned approximately £450 a year, not enough for two homes. "I realized that most of my male colleagues had parallel university appointments (and salaries), and it struck me that I might also have one," Hodgkin recalled.

Once again, Hodgkin approached a friendly male colleague for help. And again, she was persuasive. In 1946, she became a university lecturer and demonstrator, and the family finances stabilized. She did not become a university reader—the equivalent of an American college professor—until 1957. Her lab did not get modern quarters until 1958. And in 1960, when she got an endowed chair, her Wolfson Research Professorship came from the Royal Society, not Oxford. Other honors came first from outside Oxford too. In 1947, at the age of only thirty-seven, she was admitted to the Royal Society of London, the nation's most prestigious scientific organization. She was only the third woman elected to the 287-year-old society. Sometimes it paid to be a woman, Hodgkin thought; England in the early postwar years was liberal enough to want to appoint women to the society.

Hodgkin's life after World War II became increasingly complicated. Thomas was interested in African history and traveled extensively. Thomas did much to stimulate the study of African history in Britain and to help abolish the myth that Africa had no history before its contact with the West. A great supporter of African nationalist Kwame Nkrumah, Thomas spent extended periods at the University of Ghana. The Hodgkins were commuter spouses again.

Thomas's travels were not the only complications in Hodgkin's life, however. During the mid-1950s, they rented a large house at 94 Woodstock Road in Oxford so that Dorothy's sister Joan and four of her five children could join the household. Dorothy's two younger children were still at home, and six cousins made a large and boisterous household. The Hodgkins' cluttered house almost always had an overnight guest, whether a Third World leader, a scientist, or a child's school friend. Guests got the impression of almost total disorder. When Thomas was asked if so-and-so was staying with them, he replied cheerfully, "I don't really know. But maybe." A green bulletin board, like those posted in hotels and bars, hung by the front door for visitors' mail. When Linus Pauling came to dinner, Dorothy announced, "I've got to get Thomas at Heathrow Airport." Telling the guests to get their dinner from the oven, she left. After Hodgkin solved the structure of insulin, an enormous Carlsberg Beer truck rolled up to her lab each Christmas and delivered six crates of Danish lager for Dorothy. Hodgkin claimed she never knew why it appeared, though she thought the Carlsberg Foundation might own an insulin company. In any event, she threw a big Christmas party each year and invited her students to drink up.

Her lab was bigger and busier too after World War II. Its quarters spilled over into an enormous, barnlike library, used for the famous 1860 evolution debate between biologist Thomas Huxley and Bishop Samuel Wilberforce. Her research group was multinational and multidisciplinary and included many women. She attracted to chemistry several women who would not ordinarily have been interested in the subject; future prime minister Margaret Thatcher was Hodgkin's student. Right-wing Thatcher and left-wing Thomas often argued politics at dinner. "I liked her personally," declared Dorothy, who kept politics out of her laboratory.

Hodgkin ran the lab like a firm but gentle parent. To avoid being a remote administrator, she limited her research group to approximately ten workers. "She dealt on a level of tact unknown in our generation," observed crystallographer David Sayre. Commenting on a crazy paper, she breathed concern, "I'm not sure, but I think you might find..." Former student Tom Blundell recollected, "We always felt we were members of her family." Behind her back, friends called her "Mother Cat"; her students were known as "Dorothy's cats." Perutz thought she radiated motherly warmth.

The Rockefeller Foundation investigator certainly thought so. Reporting home, he wrote that the lab is "kept under good strong scientific discipline by their gentle lady boss who can outthink and outguess them on any score. A lovely small show reflecting clearly the quality of its director....She conducts the affairs of her small laboratory on a most modest, almost self-deprecatory scale. Never willing to ask for help unless she is absolutely certain that it is needed, she pitched her last request to the foundation at a level which has now turned out to be clearly inadequate." In academia, this ranks as almost purple prose.

Hodgkin's lab was as cluttered and informal as her house. With her crippled feet zipped into ankle-height slippers, she hummed church hymns as she worked. She always emerged for afternoon tea, though. Lab members alternated bringing the snacks each week, and anyone with a birthday, a new baby, or a comparable celebration brought a large, iced cake.

Hodgkin enjoyed mothering her students. "You never felt that if you walked into her office you'd feel small. You weren't made to feel she was anything special, although you knew she was—very," remembered Judith Howard, who became the only woman professor of chemistry in an English university when she was hired by the University of Durham in 1991. When Jenny Pickworth Glusker asked for a letter of recommendation for a job in American industry, Dorothy arranged for a much better research position at the Fox Chase Institute for Cancer Research in Philadelphia, where Glusker still works. Princeton University still banned women students, but Dorothy bribed its chemistry department into hiring a woman post-doctoral fellow by pointing out that the young woman could show them how to computerize their crystallographic data. Decades later, Hodgkin beamed at the memory.

Hodgkin's enthusiasm was not limited to her own group, however. "Lots of scientists are enthusiastic about their own results, but it's much rarer to be excited about other people's, and she's one of those," Perutz stressed. Hodgkin came one day to compliment him on his studies of hemoglobin, and then, "in her gentle way, told me what was the next step I should take, to solve the structure in three dimensions, drawing my attention to a paper which I never had read and never would have read."

Thanks to Hodgkin's penicillin work, she now knew researchers in the pharmaceutical industry. One day in 1948, Dr. Lester Smith of the Glaxo drug company gave Hodgkin some deep-red crystals that he had just made. They were tiny, needle-like prisms of vitamin B_{12}, the so-called anti–pernicious anemia factor. People who do not get enough vitamin B_{12} to make red blood cells die of pernicious anemia. The structure of vitamin B_{12} is so complicated, however, that chemists understood only about half of it. They needed to know the entire structure before they could hope to manufacture the life-saving vitamin in bulk.

Hodgkin knew nothing about the molecule, other than that it was intermediate in size, with roughly one hundred non-hydrogen atoms. Penicillin's sixteen or seventeen atoms had absorbed four years of her life. A structure as complicated as vitamin B_{12} had never been solved by X-ray analysis. That night, she took two beautiful X-ray photographs of Smith's crystals. The photos convinced most crystallographers that B_{12}'s molecular structure could not be deciphered with the techniques then available. Hodgkin thought differently. She thought B_{12} was solvable, and she started work.

For six years, Hodgkin and her group collected data on vitamin B_{12}. They grew more and larger crystals and took twenty-five hundred X-ray photographs of them. At one point, a colleague became so discouraged that he dumped an enormous mess of solvents into a container—everything from water to ether and acetone—and left in disgust for a bike trip on the Continent. When he returned, a tarry black mass had formed. At the bottom were a few tiny, rocklike crystals of a vitamin B_{12} variant. No one has ever managed to duplicate his "experiment." Chipping away at the clump, Hodgkin uncovered a single crystal less than one millimeter long. "The wizardry with those fingers of hers was astonishing," her student John H. Robertson marveled.

Hodgkin was cooperating with the Glaxo corporation, while John G. White of Princeton University was working with Glaxo's American competitor, Merck & Company. "We were formally supposed to be rivals," Hodgkin conceded. "But after a time, when we were dreadfully bogged down, we went into collusion. I think we were regarded as wholly unreliable by the firms to which we were attached, and we ended up publishing jointly."

Knowing that vitamin B_{12} has one cobalt atom and a cyanide group, she replaced the cyanide group with a group containing selenium and studied the differences between the two crystals. By now, Hodgkin was a master at interpreting the topographical maps of atomic positions. With mischievous glee, she loved spotting false clues. Glancing casually at students' photographs of brand-new crystals, she gaily identified them, leaving the students humiliated and wondering what was left to do. She discovered an entirely new feature in organic chemistry, the corrin ring, considered one of X-ray crystallography's major contributions to organic chemistry.

Colleagues began to refer to her uncanny "woman's intuition." Sometimes her confidence seemed almost illegitimate and embarrassing. But her instinct was "the product of her phenomenal knowledge of the relevant chemistry and physics, her long experience, and her marvelous memory for detail," Robertson realized. She learned to trust her hunches, because they were almost always right. Spotting some dubious calculations, she worried, "I don't somehow feel in my bones that they are right."

Spotting some dubious calculations, she worried, "I don't somehow feel in my bones that they are right."

No matter what she did with vitamin B_{12}, though, she kept coming back to the same problem: B_{12} looked like a molecule of porphyrin, a compound closely related to chlorophyll. When Hodgkin finished chipping out her new crystal, she gave it to Jenny Glusker to decipher. She kept Glusker in the dark, though, about her own thoughts on the problem. Many chemists still regarded X-ray crystallography as black magic. "Any error in chemical formulation, even if only a detail in the structure as a whole, would have aroused doubts about the reliability of crystallographic methods," White explained. Early in the investigation, for example, when she made an interim report on B_{12} at the Stockholm conference in 1951, the audience was obviously skeptical. Knowing their attitude toward crystallography, she refused to publish anything until she was confident of every detail. Above all, she did not want to bias Glusker's investigation. At the end of the year, Glusker discovered that vitamin B_{12} and porphyrin are exactly alike, except for a small detail.

They finally had all the data they needed. Now the problem was analyzing it. Beevers-Lipson strips and adding machines would never do the job. At that point, Kenneth Trueblood, an American visiting Oxford during the summer of 1953, came to see Hodgkin. He and three graduate students at the University of California at Los Angeles had programmed one of the world's first high-speed, electric computers for crystallographic calculations. By great good fortune, the SWAC—the National Bureau of Standards Western Automatic Computer—was on UCLA's campus and available to Trueblood free of charge. Casually, he offered to help.

Finally, Hodgkin had a computer sophisticated enough to handle nature's complexities and six years of data. She and Trueblood worked "by post and cable." Hodgkin mailed him her data, and he ran it on his computer and mailed back beautifully accurate results. When they sent telegrams, they wrote in code to save money. Hodgkin was overjoyed at their progress. When a typographical error made the size of an atom wrong by a factor of ten, she splurged and telegraphed, "Cheer up. Send everything airmail." Through it all, Trueblood never saw her visibly upset. She was always gentle, but firm. Crystallographers soon became some of the most enthusiastic users of early computers. Thanks to Trueblood's help, Hodgkin announced the complete structural formula of vitamin B_{12} in 1956—eight years after she had started. Its atoms were $C_{63}H_{88}N_{14}O_{14}PCo$.

"What the penicillin structure was for the 1940s, the vitamin B_{12} structure was for the following decade: the most important achievement of X-ray analysis in the field of natural-product chemistry," according to Jack Dunitz of the Laboratorium für Organische Chemie in Zurich. "Nothing short of magnificent—absolutely thrilling!" W. L. Bragg cheered, adding that penicillin "broke the sound barrier!"

Dorothy published her penicillin studies under her maiden name "Crowfoot" and announced vitamin B_{12} as "Hodgkin." Years later some scientists still did not know that the Crowfoot of penicillin fame was the Hodgkin of B_{12} fame.

So far, Hodgkin's life had progressed from one shining triumph to another, with her family, her career, and her political beliefs forming one seamless entity. As Cold War tensions escalated, her scientific drive and her political convictions clashed head on. Hodgkin had long been concerned about the splintered nature of the crystallographic community. Her friend, the eminent crystallographer Dame Kathleen Lonsdale, used to say that crystallographers roost, like cuckoos, in the nests of many academic departments and industries. True interdisciplinarians, they work with chemists, biochemists, physicists, geologists, engineers, mathematicians, and others.

"When the war ended, we decided the first thing that crystallographers had to have was an international union of crystallographers to encourage everyone to meet and exchange information. It seemed a perfectly harmless thing," Hodgkin explained. So she helped form the International Union of Crystallography.

From the beginning, the union tried to include everyone in the field, regardless of nationality. Germans were invited as soon as the war ended, and when their country was partitioned, East Germans were admitted too. The organization agreed never to meet in a country that did not allow every member to attend. The United States Department of State and the British Foreign Office were appalled. The United States refused to let Eastern Europeans and Soviets enter the country. For the first time, Hodgkin's scientific and humanitarian interests clashed. And as the Cold War worsened, her problems began to mount.

"The next thing that happened, I was invited to Pasadena in 1953," Hodgkin explained years later. She was winding up her B_{12} studies, and Linus Pauling had organized a conference to discuss the helical structure of protein amino acids that he had discovered by model building. Hodgkin wanted very much to go and applied routinely to the United States State Department for a visa. The visa application asked for the name of every organization she had ever joined. Without thinking, she wrote in "Science for Peace," a small group that included several Communists. Although she had visited the United States several times before, the State Department refused to grant her another visa. Unofficially, she was told it was because of Science for Peace. Without a visa, she could not enter the United States. Pauling's 1953 conference was a pivotal meeting in the development of molecular biology, and Hodgkin was extremely disappointed.

Hodgkin's friends thought she was singled out because of Thomas's and Bernal's membership in the Communist party. Ironically, however, Thomas had no problems getting visas to the United States. Then Bernal suggested that she take her first trip to Moscow with him to discuss ways to improve scientific relations with the Royal Society. The net effect of the episode was to deprive Hodgkin and the United States of the chance to exchange scientific information and to send Hodgkin to the U.S.S.R. Hodgkin liked Russia very much. As usual, she responded to people with affection and pleasure. "People were so nice and just so glad to see you," she reported.

In the future, whenever Hodgkin wanted to attend a scientific conference in the United States, she had to secure a waiver from the U.S. attorney general stating that "yes, yes, you had these awful connections with peace organizations but it's in the best interests of the United States for you to come and talk about vitamin B_{12}," Hodgkin recalls. "So Pauling organized a second invitation, and all of my American friends wrote to the Attorney General supporting my visit. It was organized under what I always thought was a rather bogus meeting to celebrate the fortieth anniversary of X-ray diffraction in the U.S. I was the main speaker, about B_{12}, which we'd just worked out. We were just so happy to see each other that it was totally enjoyable." In 1990, when the Soviet Union was disintegrating and Hodgkin was eighty years old and wheelchair-bound, the State Department finally relented and approved her passport—"for indefinite re-entry."

Aware of her controversial position, Dorothy never discussed politics with her students. Too many planned to study or work in the United States, and associating with left-wingers could hurt them. Even the Rockefeller Foundation checked whether grant applicants like Rosalind Franklin (chapter 13) were likely to be refused visas by the United States government. With Glusker, for example, Hodgkin discussed politics only once, when she said that the Girl Guides scouting organization had formed her political values, particularly her interest in international cooperation. It was in the Girl Guide spirit that Hodgkin helped Chinese crystallographers set up an insulin-research project years before the United States government recognized their country.

Dorothy was visiting Thomas at the University of Ghana during the fall of 1964 when she learned that she had won the Nobel Prize in Chemistry that year. She shared the prize with no one. At fifty-four, she was the fifth woman and the first British woman to win a Nobel in science. The *Daily Mail* headline put her triumph in perspective: "Nobel Prize for British Wife."

She and Thomas celebrated that evening at the Institute of African Studies, watching ceremonial court and hunter dances. The Hodgkin family was "somewhat dispersed," as the Nobel Committee noted with understatement. Luke was teaching mathematics at the University of Algiers and Elizabeth was teaching at a girls' school in Zambia while Toby was volunteering in plant genetics in India. Dorothy's sister Diane, the wife of a Canadian geographer, heard the news near the North Pole and cabled "From Pole to Pole"; her husband was near the South Pole. The year before, Thomas's cousin Alan Hodgkin had won the Nobel Prize for physiology and medicine; news accounts pointed out that the Hodgkin children have Nobel winners on both sides of the family.

Many scientists stop doing research once they win a Nobel Prize. But Hodgkin was working on the most complicated structure of her career—insulin's 777 atoms. Emotionally attached to insulin for thirty years, she had tried to decode it off and on as techniques improved. But as she observed later, "I seem to have spent much more of my life not solving structures than solving them." Then, when Fred Sanger identified the sequence of amino acids in insulin's molecule, she was overwhelmed with excitement. She desperately wanted to discover insulin's three-dimensional arrangement.

Insulin is a small protein and computers were now easily available, but the project was enormous, nonetheless. Its structure is "so complicated and irregular," she observed, "it is not surprising that for a brief time we thought it was not there at all." It was late one night in 1969 when her team put the finishing touches on a six-part molecule, roughly triangular in shape, with three pairs of molecules enclosing two zinc atoms at the core. When Perutz heard the news, he posted a sign in Cambridge: "Late Night News from Dorothy Hodgkin: Insulin Is Solved." She had analyzed seventy thousand X-ray spots. In the 1969 *Nature* article announcing insulin, the names of her team members appear alphabetically; Hodgkin is third from last. She did not give the first major lecture on insulin either. It was delivered by Tom Blundell, her young postdoctoral fellow at the time. Hodgkin introduced his talk by pointing out that she had started working on insulin before Blundell was born!

Thanks to crystallography, many organic chemists were now suffering from an identity crisis. At the beginning of Hodgkin's career in the 1930s, chemists had loftily told crystallographers, "Of

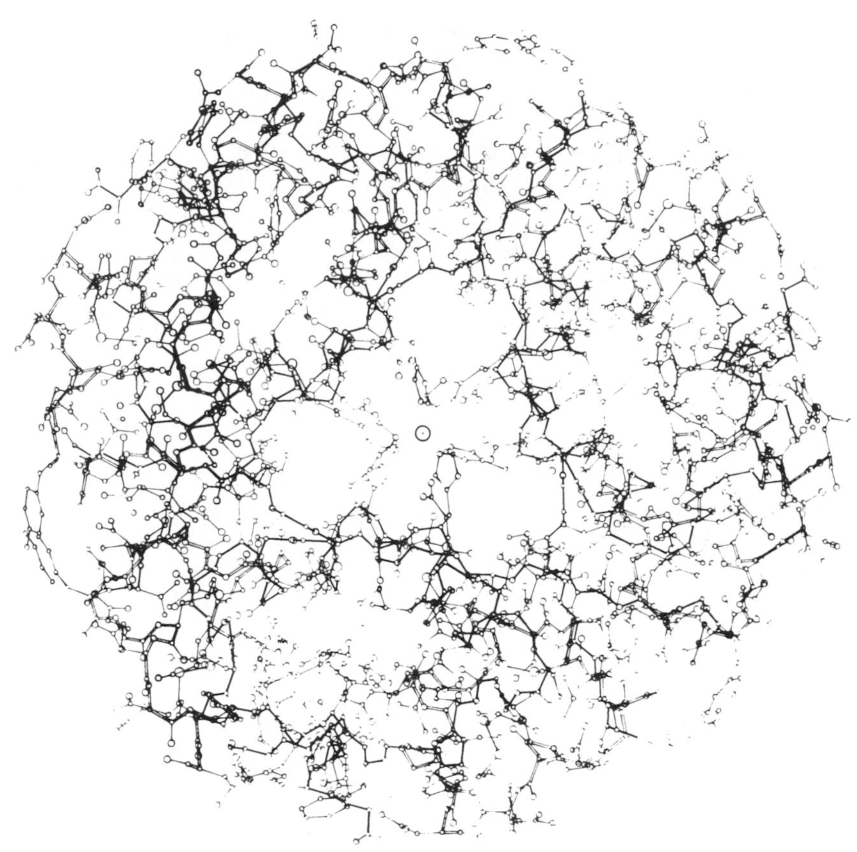

Figure 10.5. Insulin.
The atomic structure of a cluster of six insulin molecules. Dorothy Hodgkin deduced this structure by analyzing approximately 10,000 spots from X-ray diffraction photographs.

course, you are only telling us what we already know." By the early 1960s, X-ray crystallography was not only necessary, sometimes it was the *only* method for discovering the three-dimensional shape of an organic molecule. X-ray analysis—personified by Dorothy Crowfoot Hodgkin—was faster, more accurate, and more illuminating than chemical degradation. One of the traditional missions of organic chemistry—elucidating the molecular constitution of natural substances—had become obsolete, and its leaders muttered about "the crisis in organic chemistry." Later, they realized that Hodgkin's work had actually freed them from the chores of structure determination to do other, more exciting research.

Hodgkin received a letter from Buckingham Palace in 1965, the year after the Nobel Prize. Far from being thrilled, she dreaded opening it. "I left it sealed, fearing that they wanted to make me *Dame* Dorothy," she confessed. When she finally read it, she was relieved and pleased that Queen Elizabeth wanted to give her the Order of Merit—a much greater honor, which carries no title. The only other woman who had ever received the Order of Merit was Florence Nightingale in 1907. Hodgkin received the decoration in a private audience with Queen Elizabeth. At an Order of Merit banquet, the artist Henry Moore sat next to her and was so moved by her gnarled and twisted fingers that he asked if he could sketch them. His drawing hangs in the Royal Society alongside her official portrait. Of the two works, she tells friends, she prefers the drawing of her hands.

Throughout the 1960s and 1970s, Hodgkin continued working for peace organizations. She campaigned against the Vietnam War and visited Hanoi and China. When she returned from Hanoi, she spoke to a student group at Oxford. The students came prepared with arguments against her having visited North Vietnam. But Hodgkin did not talk about politics at all. Instead, she talked about people she had met, friends she knew from previous visits or from her daughter's year-long stay in North Vietnam during the war. Some, like Perutz, had long thought that Hodgkin "shut her eyes to the evil things that went on" in Communist countries. But watching Hodgkin talk with the Oxford students, Blundell felt that "Dorothy saw politics as individual personalities, in terms of the people, not in terms of dogma or political convictions the way that most people do." Hodgkin operated on a personal level in politics, just as she did in her lab and home.

Physicist Rudolf E. Peierls asked Hodgkin in 1975 to be president of the Pugwash Conferences on Science and World Affairs, an organization that campaigns for world peace and disarmament. As Hodgkin tartly explains, "Pugwash is a respectable government-oriented body." Peierls told her, "It isn't much of a burden—just attending the yearly conference and making a speech, though it would be okay to omit the speech if you want." Then he felt guilty, because Hodgkin came to every meeting and worked hard. The same phenomenon occurred when she was elected chancellor of Bristol University. Instead of being a ceremonial figurehead, she visited the university frequently, even helping to solve administrative disagreements. As a friend observed, "She's a tough old stick."

Hodgkin explained her international philosophy in an article for the *Bulletin of the Atomic Scientists* in 1981. "How to abolish arms and achieve a peaceful world is necessarily our first objective," she explained. "The conflicts that have occurred in the last twenty years have taken place in some of the poorest countries of the world, driving them still deeper into poverty and sharpening the contrast between their needs and the cost of arms. If some—and preferably all—of the million dollars spent every minute on arms were turned to the abolition of poverty from the world, many causes of conflict would vanish."

Recalling the genocide in Cambodia and Uganda, she confessed that "what bothers my mind is that we all knew about the events and seemed powerless to do anything, so that eventually only armed intervention by some nation, frowned on by almost everyone, could save the oppressed." Instead, the United Nations could have been strengthened, fact-finding and conciliation missions dispatched rapidly, and the compulsory jurisdiction of the International Court of Justice extended.

Maintaining her personal style of diplomacy, Hodgkin wrote her former student Margaret Thatcher in the late 1980s, suggesting that the prime minister should not oppose the Soviet Union so vehemently without having visited it. Thatcher invited Hodgkin to her weekend home and listened for hours. After that, Thatcher visited the Soviet Union, established close rapport with Soviet President Mikhail Gorbachev, and, in part because of her highly televised trip, won her 1987 reelection campaign.

"Thatcher invited her for lunch once a year or so," Kenneth Trueblood explained. "Once Dorothy had just come back from meeting with Gorbachev, and the next day she had a lunch meeting with the queen and then a week later she was lunching with Margaret Thatcher. You see, when she was in the Soviet Union and in China, everybody talked with her because she was a Nobel Prize winner. And she was known not to be judgmental. She took people at their word," Trueblood said. The *London Times* reported that she "may be or may not be, as has been said, the cleverest woman in England."

By then, Hodgkin felt that the romantic era of her life was over. Thomas had retired in 1970 and died of emphysema in 1982. Typically, rumor attributed his death to an exotic lung mold from Africa. "Too much smoking," practical Dorothy retorted. In crystallography, modern computers were trivializing the difficulties of X-ray analysis. Determining the molecular structure of crystalline substances is now so much easier that several thousand are analyzed yearly.

Dorothy retired in 1977 to a golden stone house in the Cotswolds north of Oxford. Originally three row houses, it stretches along a lane with a steep flowered hillside behind. A village lies in front in a hollow. Wheelchair-bound from arthritis and a broken pelvis, Hodgkin continues to travel to scientific and peace conferences. And her face still lights up as she reports, "I'm fortunate to have nine grandchildren and three great-grandchildren." She has looked frail for years, though. When a friend put her in a taxi, the driver was afraid to take her to the train station.

"Will she be all right?" he asked anxiously.

"Sure," the friend replied. "She's only going to Moscow."

* * *

Dorothy Crowfoot Hodgkin died at home at the age of 84 on July 29, 1994.

11

Chien-Shiung Wu

May 31, 1912–February 16, 1997

EXPERIMENTAL NUCLEAR PHYSICIST

WAVING A CHINESE FLAG, Chien-Shiung Wu marched a band of several hundred student agitators through the back streets of Nanjing to avoid being seen by government officials or newspaper reporters. Turning in at the presidential mansion, they occupied the courtyard.

Hours passed. Patriotic demonstrations—even songs and slogans—were illegal in China just before the outbreak of World War II. With Japan armed and threatening to invade the north, the Chinese government did not want to provoke any trouble. Toward midnight, as snow began to fall, the President of China finally emerged from the mansion and approached Chien-Shiung Wu.

General Chiang Kai-shek listened as Wu and the other student leaders urged him to stand firm against the Japanese. The underground student movement was the voice of Chinese nationalism, and Wu's group had already organized an effective boycott of Japanese goods among Nanjing families. Chiang listened courteously to their demands and agreed to do what he could.

Although Wu had been a leader in the student underground since high school, she had been a "Courageous Hero" since birth. Because she signed her physics articles "C. S. Wu," no one knew her as "Courageous Hero." Yet that is what her name means in Chinese and what her father hoped she would become. A fervent believer in equal rights for women, he provided her with the finest education that China could offer and gave her a piece of advice. "Ignore the obstacles," he told her. "Just put your head down and keep walking forward."

Thanks to her father, Wu's childhood was wonderfully happy. She was born in 1912 in Liuhe, a small town about thirty miles from Shanghai, the business and industrial capital of China at the mouth of the Yangtze River. Her father, Wu Zhongyi, had quit engineering school to participate in the Chinese Revolution of 1911, which toppled the Manchu dynasty. He had read widely about Western democracies and women's emancipation so, after the revolution, he returned home to Liuhe and opened the region's first school for girls. "I want every girl to have a school to go to," he told his daughter. "I want everyone who has suffered to have a place to go to air his sufferings."

Wu and her father were like two very good friends. He encouraged her and her two brothers to ask questions and to solve problems. The family lived simply in a house filled with newspapers, magazines, and books. In the evenings, the family read together. Her father was principal of his school, and her mother Fan Fuhua helped him by visiting local families. She urged the parents to give their daughters an education and to stop binding their feet.

When Wu was nine years old, she graduated from her father's school. It offered only four elementary grades at the time, although it expanded later to include a high school. Her graduation precipitated a family crisis. She wanted to continue her education, but the only possibility was boarding school. Her father consulted his great-grandmother, the family matriarch. She had never attended school, but she knew how to read and write and was seldom seen without a book. She declared that her great-great-granddaughter should attend the finest school possible, no matter how far away. An excellent girls' school was located in Suzhou, about fifty miles inland from Shanghai on the way to Nanjing. One of her father's friends taught there, and he promised to look after Wu.

The Soochow Girls School offered a complete western curriculum. Professors from leading American universities often lectured there, including several from Columbia University where Wu eventually became a faculty member. Except for winter and summer vacations when she returned home, she stayed in Suzhou from the time she was ten years old in 1922 until she graduated in 1930 at the age of seventeen.

Its high school was divided into two parts: an academic school and a normal school to train teachers. Because the teachers' training program was free and its graduates were assured jobs, Wu enrolled there. One day, listening to her friends' conversation at meals, she realized that her friends in the academic school were learning far more about science and foreign languages than she was in the teachers' training program.

Luckily, all the students shared one dormitory, so she was able to organize a simple, yet practical solution to the problem. Each evening as her friends finished their homework, they turned their science books over to Wu for the night. Working late, she taught herself

Chien-Shiung Wu at the time of the parity experiment.

Chien-Shiung Wu's passport picture from China, 1935.

Chien-Shiung Wu and her uncle, who helped pay for her trip to the United States, before she left China in 1935.

Chien-Shiung Wu with Wolfgang Pauli, discoverer of the neutrino, about 1940.

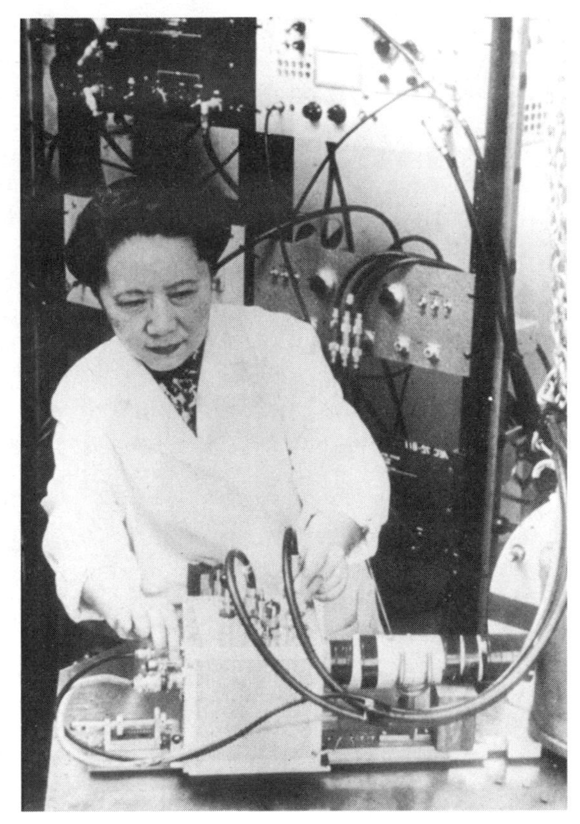

Chien-Shiung Wu at Columbia University, 1957.

mathematics, physics, and chemistry. She developed the habit of what she called "self-learning." Gradually, she came to the realization that she enjoyed studying physics more than anything else.

When Wu was in high school, friends recruited her to lead their chapter of China's underground student movement. They argued, with some justification, that Wu's academic record was so stellar that the school could never expel her for political activities, particularly since her father, a former revolutionary himself, could be counted on to support her activities. So Wu began representing the students at meetings and peaceful strikes, which they were clever enough to schedule just before summer vacations when everyone wanted to go home early anyway.

Wu graduated from Soochow in 1930 with the highest grades in her class. That summer, she received a letter announcing that she had been selected to attend China's elite National Central University in Nanjing. When her parents asked what she wanted to study, she told them, "Physics." But, she told them, she did not know enough math or science, so she would probably have to study education.

Her father strongly disagreed. "There's ample time to prepare yourself," he announced. That evening when he returned home, he asked his chauffeur to bring in a package from the car. Inside were three books on advanced mathematics, chemistry, and physics. Wu was elated. After studying all summer, she enrolled at Nanjing as a mathematics student. As she gained confidence, she transferred into physics. "If it hadn't been for my father's encouragement, I would be teaching grade school somewhere in China now," she said years later.

Despite Wu's activities with the student underground in Nanjing, she was the university's top student. When a group of professors from different departments met, they bragged to each other about their best student. Later, they learned that each professor had been bragging about one and the same person—Chien-Shiung Wu.

After her graduation in 1934, she spent a year teaching in a provincial university and a year doing research in X-ray crystallography at the National Academy of Sciences in Shanghai. Her instructor there was a Chinese woman physicist who had earned her Ph.D. from the University of Michigan. Because China had no graduate instruction in physics, the woman urged Wu to go to the United States for more training. Wu's uncle, who was extremely fond of his niece, offered to help pay for the trip. He had traveled to France during World War I on a work-study program, started China's first long-distance bus company, and made a fortune.

Wu sailed from Shanghai in 1936, intending to earn a Ph.D. at the University of Michigan as quickly as possible and then return to her family in China to help modernize her country. She never saw her family again.

Within days of Wu's arrival in San Francisco, she changed all her plans. Through family friends, she was introduced to a young Chinese physics student at the University of California at Berkeley. His name was Yuan Chia-liu, which he soon Americanized to Luke Yuan. He was the grandson of Yuan Shikai, a famous general who had served the Empress Dowager at the end of the Manchu dynasty. He had been elected the first actual president of the Republic of China over the head of Sun Yat-sen, but in his last years he had attempted unsuccessfully to become the first emperor of his own dynasty. Yuan's father, who had advocated democracy in China, opposed the attempt. Because an obedient son was not supposed to criticize his father, he wrote the general a poem:

> Rain and wind are always on the mountain peaks.
> You're already high enough.
> Don't climb any higher.

Then, having defied his father, he went into exile, barely able to support himself by writing articles and selling his calligraphy. Yuan remained with his mother and relatives. When he arrived in Berkeley to study physics, he had twenty-four dollars in his pocket.

Professors later teased Wu, accusing her of having stayed in California in order to be near Luke Yuan. But she actually had three other reasons. First, she had heard—correctly—that women were not permitted to use the University of Michigan's student union building. She had attended a coeducational university in China and considered herself first and foremost a scientist, not a woman. She did not want to settle for second-class treatment. Second, Michigan had more than six hundred Chinese students, and she did not want to spend all her time in the United States with other Chinese.

Third and most important, Berkeley physics was at the height of its fame. Ernest Lawrence was building a cyclotron to smash atoms together and would soon win a Nobel Prize for his work. Robert Oppenheimer, who later ran the Manhattan Project to build the atomic bomb, was teaching the new European quantum theory about the behavior of atoms and subatomic particles. They, in turn, attracted other eminent scientists to Berkeley. The thought of competing with some of the most brilliant minds in physics did not faze Wu. As she explained later, "I didn't see a man there who could compare with my father." So, she stayed in Berkeley.

Her introduction to Berkeley's International House, a dormitory for foreign and American graduate students, was a disaster. Entering the cafeteria for breakfast, she expected to find her usual bowl of rice. Instead, there were long lines of strange and unidentifiable foods that she was expected to order by name and pay for with paper money called scrip. After studying the situation, she left without eating. That afternoon, she wandered off campus in search of recognizable food. In a local bakery, she found rolls, tea, and a new friend, Ursula Schaefer, who later married the future Nobel Prize–winning physicist Willis Lamb.

Ursula Lamb had also arrived at International House the night before. She was German, loved everything Chinese, and was also appalled by Berkeley's student cafeteria. Within a few days of arriving in the United States, Wu had found both her future husband and a lifelong friend. She also found a practical solution to the food problem. Through a friend, she located a kindly Chinese caterer who allowed students to eat his banquet leftovers for twenty-five cents a dinner. Wu, Lamb, and Yuan often ate there. The two women also taught each other about their respective cultures. Lamb took Wu to see Wagner's opera "Parzival," which Wu found excruciatingly funny. In return, Wu introduced Lamb to Chinese theaters in San Francisco's Chinatown. Wu recommended books for

Lamb to read about China and borrowed her copy of *The Bible as Literature.*

Wu became the belle of Berkeley. She was what the men called "a smasher—a gorgeous, willowy girl with a smile that would melt anybody's heart." Lamb thought she had a "smile that brought the sunlight right into your lap." Berkeley men rhymed "Wu" with "woo" and years later reminisced about her exotic Chinese dresses with their high collars and slit skirts. She ordered them in different fabrics made to her measurements in China or Taiwan. Even late in life, she rarely wore Western clothes.

Despite her popularity, most students called her "Miss Wu." "Chien-Shiung" was hard to pronounce properly, and Wu was a dignified, controlled, and rather formal person. Much to everyone's amusement, for example, she called Lamb's fiancé "Mr. Lamb," even after they married. Only Robert Oppenheimer and her closest friends called her "Jiejie," an affectionate term for "elder sister" in Chinese.

Lamb soon realized that her friend's charming appearance and modest behavior hid a formidably rational and honest personality. Wu was, above all, truthful. She never learned to tell "white lies." So she normally hid her opinions behind a smile, and her smile behind her hand, in traditional Chinese fashion. Not even Lamb ever saw her teeth.

Occasionally, if Wu disapproved of something Lamb had done, she lectured her friend with utter frankness. Lamb took Wu's criticism as an indication of how much she cared for a friend's welfare. "She was the only one who gave me real hell," Lamb recalled. "I've had some dressings down from her, but blue. Nobody else would dare to say those things, and I went with my tail between my legs for days. But I always realized that it was meant in the spirit of 'I'll tell you what nobody else will tell you.' It was salutary." Wu's tirades actually solidified their friendship. Even late in life, Lamb considered Wu "an absolutely, totally reliable friend. If anything happened, she would be right there. She's very reserved, but very, very human and warm."

During the 1930s, when Wu was at Berkeley, the most exciting field of physics was nuclear physics. One of Ernest Lawrence's assistants was Emilio Segrè, who shared a Nobel Prize in 1959. Segrè was studying the nucleus, and Wu joined his group. Segrè had a reputation among the students for being grouchy and difficult, however. Sure enough, when Wu left an uncapped mercury bottle in the lab, he left her a gruff note: "The vapor is poisonous. Do you want to see your grandchildren?"

Despite his brusqueness, Segrè treated Wu like his own daughter.

"Wu's will power and devotion to work are reminiscent of Marie Curie, but she is more worldly, elegant, and witty," he said. Many times, he told her not to work so long in the lab: "Miss Wu, you have to spend some time reading physics. You have to stand back from things and see the whole picture." Years later, she repeated his advice to her son, Vincent, when he also became a physicist.

At the end of her first year at Berkeley, the physics department recommended Wu and Yuan for fellowships. The university administration was prejudiced against Asians, however, and no Chinese physicist had ever been granted a fellowship. So the physics chairman decided instead to give them two hundred dollars in return for grading papers. Even with free rent, Yuan did not have enough money to live on, so when the Nobel Prize–winning physicist Robert A. Millikan of the California Institute of Technology telegraphed him about a six-hundred-dollar fellowship, Yuan moved immediately to Southern California. Wu and Yuan did not know it, but they would spend the next forty-five years commuting in order to be together.

Coming down to breakfast at International House in July 1937, Wu was shocked to see huge newspaper headlines: "Japan Invades China." Stranded in the United States, she would be cut off from her family, from any news of their safety, from their financial support, and from any possibility of returning home until the war was over. She was still absorbing the news when a secretary arrived and asked her to report immediately to the chairman of the physics department. He assured her, "We are all very, very sorry. But you don't have to worry about yourself. We will take care of you now."

When she returned to her room, she found two bouquets on her bed. Attached to the flowers were notes expressing the sympathy of the two young Japanese women who lived in International House.

By December, the Japanese had captured Shanghai, Suzhou, and Nanjing. In Nanjing, Japanese troops went on a rampage, raping or murdering an estimated 42,000 civilians. Japanese soldiers took souvenir snapshots of the massacre and left their film to be developed with Chinese shop clerks, who smuggled copies to the newspapers. Pictures of the atrocities appeared all over the world, including San Francisco.

Wu was unable to return to China for thirty-six years. For eight of them, China was fighting Japan, at first alone and then allied with the United States, Great Britain, and Russia. Until the Japanese surrender in 1945, Wu heard nothing from her parents or brothers.

Wu worked hard to forget the war. She adopted her father's advice: "Just put your head down and keep walking forward." As she explained, "I have always felt that in physics, and probably in other endeavors, too, you must have total commitment. It is not just a job. It is a way of life." Marie Curie became her idol; she too had left her native land and through hard work and determination become a physicist. When university officials worried about Wu's working late nights in the lab and walking home alone after dark, another graduate student, Robert R. Wilson, who had a jalopy and also worked late, was recruited to drive her back to her dorm. Every morning around three or four A.M., Wilson dropped by her lab to say, "Miss Wu, it's time for you to go home." Wilson later directed the Fermi National Accelerator Laboratory in Batavia, Illinois.

With nowhere to go if she failed, the pressure to succeed was enormous. "When she took an exam, there was always a crisis," Wilson recalled. "She'd become very willowy, and you'd expect her to faint away. Then she'd pass the exam, and there'd be a triumphant celebration in a Chinese restaurant."

She never felt completely at home in English. "He's" and "she's" became confused, and articles and verbs tended to disappear at exciting moments. Her pronunciation was often unclear. As a result, she began a lifelong practice of writing her talks out before presenting them orally. Once she became so involved in a speech that she wrote her equations on the blackboard Chinese-style from right to left.

The harder she worked, however, the happier she felt. She did not try to conquer the world of physics. Instead, she decided, "You come in order to work and to find your way. You must work very hard at the beginning. It is hard to push the door open and to get inside a subject. But once you understand it, it is *very* interesting."

She began work on a two-part thesis. First, she studied the electromagnetic energy given off when a particle going through matter slows down. In 1939, the German chemists Otto Hahn and Fritz Strassmann had split the atomic nucleus in a fission process that Lise Meitner explained. Wu began to study the phenomenon. For the second half of her thesis, she focused on the radioactive inert gases emitted when the uranium nucleus splits.

After Wu received her Ph.D. degree in 1940, she stayed on at Berkeley for two years as a research assistant. She soon became known as a local fission expert. The great Danish physicist Niels Bohr suggested that she give a colloquium talk explaining the process. Oppenheimer called her "the authority." By 1941, she was well known outside the university. A local newspaper reporter thought she looked "as though she might be an actress or an artist or a daughter of wealth in search of Occidental culture." She made a lecture tour through the United States in 1941.

In 1942, when scientists were trying to start a self-sustaining nuclear fission reaction at Hanford, Washington, the chain reaction started up for a few hours and then mysteriously quit. Enrico Fermi, suspecting that one of the substances produced by fission was poisoning the reaction, was told, "Ask Miss Wu." Xenon, one of the gases that she had studied for her Ph.D thesis, had caused the problem.

Despite her growing prominence, Berkeley's physics department refused to hire her. Of the nation's top twenty research universities at the time, not one had a woman physics professor. Furthermore, West Coast hysteria about Asians had reached a peak, and Japanese-Americans had been contained in internment camps. Any Asian was suspect. Discrimination against the Chinese was particularly misconceived because, by 1942, China had been fighting Japan for five years, longer than any other nation. Segrè never forgave Berkeley for not hiring Wu permanently. "They would have had a star," he told her.

By 1942, however, Wu was ready to move on. Most of Berkeley's physicists had left to do war research; Oppenheimer was already in charge of the atom bomb project being developed at Los Alamos, New Mexico. Although Wu was a recognized authority on fission, she was not asked to join the group. Instead, she and Yuan were married in the garden of Robert Millikan, who had supervised Yuan's thesis. Then the young couple moved East.

Yuan had a job designing radar devices at RCA laboratories in Princeton, New Jersey, and Wu got a job teaching at Smith College in Northampton, Massachusetts. They met for weekends in New York City. Wu went to Smith because a dean from the women's college had visited Berkeley before the war and suggested to Wu that "when you get your degree and you want a job in the States before you go back, call me." Wu was an assistant professor and enjoyed teaching, but Smith, like other women's colleges at the time, did not have the financial resources to give its faculty members time off from teaching to do research. Wu was primarily interested in research.

At a conference in Boston, Wu ran into Ernest Lawrence from Berkeley. By that time, the United States had been at war for a year and there was a severe shortage of physicists. Scientifically speaking, World War II was a physicists' war. The development of radar helped Britain hold out against the German air force, while the atomic bomb would end the war against the Japanese. Physicists developed both technologies and were in great demand.

"Are you happy not doing any experiments?" Lawrence inquired.

"I feel sort of out of the way," she admitted.

Lawrence promptly recommended her to several universities that were looking desperately for physicists—male or female—to temporarily replace professors on defense leave. As a measure of the shortage, Wu, who two years before could get a job only in a woman's college, now got offers from eight universities. Among them were Princeton, Columbia, and the Massachusetts Institute of Technology, all of which refused to admit women as students at the time. She chose Princeton. For the first time in several years, she and Yuan could be together.

At age thirty-one, Wu was Princeton's first woman instructor. Most of her students were navy officers sent to Princeton for accelerated engineering training. "They were good students," she said, "but they were afraid of physics, and first you had to get them over the fear."

After a few months, Wu was called for an interview by the Division of War Research at Columbia University in New York City. Two physicists questioned her about physics all day, but carefully revealed nothing about the lab's secret projects. They conducted the interviews, however, in their offices.

At the end of the day, they asked, "Now Miss Wu, do you know anything about what we're doing here?"

"I'm sorry," she smiled, "but if you wanted me not to know what you're doing, you should have cleaned the blackboards."

They burst out laughing and suggested, "Since you already know what's going on, can you start tomorrow morning?"

In March 1944, after two years of teaching at Smith and Princeton, Wu returned to research. Working at Columbia in a converted Nash automobile warehouse, she helped to develop sensitive radiation detectors for the atomic bomb project.

After the defeat of the Japanese in 1945, Wu finally got word from her family in China. They were well, and her father had directed one of the most famous exploits of the Chinese theater of war. He had engineered the construction of the Burma Road, a single-lane, one-thousand-kilometer highway built by hand labor through the Himalayan Mountains between Burma and China. For several years, the Burma road was the only route for transporting Allied supplies to the Chinese army.

Equally good news came from Columbia. Wu was one of the few Manhattan Project physicists asked to remain at the university after the war. Columbia had one of the finest physics departments in the United States, and she became a senior investigator on one of the largest federal research grants there. In 1947, her son, Vincent Wei-chen Yuan, was born, and Wu hired a kindly woman to care for him. Wu, Yuan, and Vincent moved into a Columbia apartment two blocks from her laboratory so that she could move quickly between home and lab. Later, she listed "a nice husband," a home close to work, and good child care as three requisites for the successful married woman in science.

Yuan became an eminent authority designing accelerators at Brookhaven National Laboratory on Long Island, about two hours from Columbia. Each Monday morning, he commuted by train to Brookhaven, returning home on Fridays for weekends with the family. Yuan was extremely skilled at experimental techniques, and by six o'clock on Friday he was usually in Wu's lab, helping her and her students with their apparatus.

Just as Wu and Yuan were settling into American life, they were given an opportunity to return home to China. National Central University offered both Wu and Yuan professorships, suggesting that they spend one more year in the United States gathering laboratory equipment before returning home. Chiang Kai-shek and the Chinese Communists, however, were fighting a civil war. As the Communists solidified their control of China, Wu's father advised her not to come back. The move home would have been irrevocable because, to keep Chinese scientists from traveling to Communist countries after the war, the U.S. State Department routinely refused them reentry visas. In any event, Wu and Yuan had decided that they did not want to raise Vincent in a communist country. They began the lengthy naturalization process and became United States citizens in 1954.

After the Chinese Communists' victory in 1949, all official communication between the United States and China ceased. Although Wu received letters from her family three or four times a year via Hong Kong or Saigon, politics could not be discussed and a visit home was impossible. By 1973, when Wu was finally allowed to go to China, her parents and brothers were dead. Her elder brother had been killed during the Cultural Revolution, and her younger brother had been confined to one room and interrogated repeatedly until he committed suicide. Several friends had also killed themselves, and Yuan's sister had suffered greatly.

From then on, Columbia was home, although she retained her Chinese dress, her preference for Chinese food, and her Chinese name. She helped keep track of the best Chinese chefs in New York as they moved from restaurant to restaurant, and Friday lunches with Columbia's Chinese physicists became a gourmet tradition. Her refusal to adopt an English first name created difficulties. The United States Immigration Service, not realizing that she was a woman, confused her for several years with a Philadelphia bank-fraud artist of the same name. By the time Columbia's administration sorted out the mixup, Wu's immigration file was several inches thick.

Getting organized at Columbia, she told herself, "Now I must find my field." One of the most important choices any researcher makes is picking a significant topic to study. "If you choose the right problem, you get important results that transform our perception of the underlying structure of the universe," explained the Nobel Prize–winning theorist Chen Ning Yang. "If you don't choose the right problem, you may work very hard but only get an interesting result." Wu had an uncanny ability to chose important problems.

After thinking a while, Wu decided, "Beta decay has lots of problems. After you clean up the problems, then you try to see if the evidence agrees with Fermi's theory." Fermi had proposed a theory in 1933 to predict how the nucleus behaves during beta decay. Beta decay, the field in which Lise Meitner made important contributions, is one of three types of radioactivity. Fermi's theory—if correct—would have been extremely useful, but experiments conducted in the United States, England, and Russia contradicted his hypothesis.

Wu always stressed that, to do important work, "You must know the purpose of the research exactly, what you want to get out of it, and what point you want to show." So first she tried to understand beta decay. The process occurs when the nucleus of a large atom ejects a superfast electron and a neutrino and, in the process, changes into another element. Normally, electrons surround the outside of each atom and do not exist inside the nucleus. In beta decay, however, a neutron inside the nucleus breaks apart, forming a proton, an electron, and a neutrino. Bursting out of the nucleus at enormous speeds, the electron and neutrino rid the nucleus of excess energy. The proton remains inside the new, more stable nucleus.

Experimentalists were confused because Fermi's theory predicted unequivocally the number of electrons that would come out at particular speeds. According to Fermi, most of the electrons would burst out of the nucleus at very high speeds. Yet all the experiments produced enormous numbers of slow electrons.

In a series of careful experiments, Wu discovered that previous researchers had used radioactive materials of uneven thickness. Electrons traveling through thick sections had simply ricocheted off more atoms and lost more energy. When they emerged into the open air, they were traveling at slower speeds than electrons from thin sections containing fewer atoms. Wu used a uniformly thin radioactive material and got exactly the electron speeds predicted by Fermi.

"Her beta decay work was important for its incredible precision," observed another Nobel Prize winner, William A. Fowler, at Caltech. "Our lab was working in the same field. She did rather better than we did.... She established a brilliant reputation. Those who tried to repeat her experiments and those who were competing with her found that she was always right. She always chose to do the significant and important experiments—no matter how difficult they were. And they were very difficult to do."

"She became known as someone whose work you could believe," recalled her former graduate student Leon Lidofsky, now a professor at Columbia. "One of the things I learned from her was that, if you got a result that didn't agree with someone else's, you had to be able to show what they'd done wrong as well as what you'd done right. Otherwise, no one would know whose data to trust. She had a very, very strong sense that things had to be done right, that when you had finished, you had to believe that what you had done was right, so that you could go on from there and use the data. If it was done sloppily, it wasn't worth doing because the results weren't reliable."

Many physicists believed that Wu's beta decay work was good enough to win a Nobel Prize. Unfortunately, it did not fit the rules of the award. The physics prize is given for a discovery or invention. "She had straightened up a big mess in physics quite elegantly, but it wasn't quite a discovery," explained Noemie Koller, one of Wu's graduate students, now a professor and dean at Rutgers University.

From 1946 to 1952, Wu was "completely submerged" in beta decay. She began to revere Lise Meitner for her work on beta decay. As one experiment followed another, Wu's normally objective articles began to sparkle with highly subjective words like "satisfaction," "happy," "fruitful," "pleasant," and "great prosperity."

She worked herself as hard as her students. Lidofsky thought Wu was one of the most beautiful women he had ever seen. "She was exquisite. But in addition, she was a very strong personality," Lidofsky declared. "She was very, very single-minded."

Emilio Segrè recalled fondly, "She is a slave driver.... She is the image of the militant woman so well known in Chinese literature as either empress or mother." When she had a special article or talk to finish, she went to bed early and got up at four A.M. to work. She often reminded her son that the inventor Thomas Edison had defined genius as "one percent inspiration and ninety-nine percent perspiration." And she reminded herself of her father's advice to "ignore the obstacles and keep walking ahead."

"It was very exciting," Koller remembered. "But she was rough—very demanding. She pushed the students until they did it right. Everything had to be explained to the last decimal. She was never satisfied. She wanted people to work late at night, early in the morning, all day Saturday, all day Sunday, to do things faster, to never take time off." Students claimed that Wu was disappointed with a Jewish student who observed the Sabbath and could work only six days a week.

"She wanted the best result, the best measurement, the most

precise understanding, or the best explanation. She wanted us to understand what we were doing," recalled one of her students. "She'd give praise only at select times—usually only when other people were present."

Late on a Saturday night on the way home from a business trip, Wu asked the taxi driver to pass by her lab. All its windows were dark. Early Sunday morning, she telephoned Koller excitedly, "Equipment all alone. Nobody working. Equipment all alone." She thought the students should be as excited about their work as she was, Koller realized.

Fiercely competitive, Wu told her students not to show their data to visitors until it was published because it might be stolen. When guests pried, she switched into a particularly convoluted form of Chinese-English. She could talk and talk in a charmingly soft voice and yet somehow never answer the question. "I was a woman when it paid to be and wasn't one when it paid not to be," she confided to a friend. As Koller recalled, "Maybe she felt that she was in enemy territory. We thought she was a great professor. We didn't realize her tenuous position, how she had to fight."

Nothing could keep her away from the lab. She loved to tinker and, when she thought that her students had adjusted their apparatus improperly, she did not hesitate to "fix" it. Hoping to get more uninterrupted time with their experiments, students gave her two tickets to take Vincent to a special showing of a children's movie. When the film started, the students settled down to work, confident of having several peaceful hours ahead of them. In walked Wu, beaming happily. "I sent him to the movie with the nursemaid," she explained.

Some of Wu's students during the 1940s and 1950s thought that Vincent did not get enough attention from his mother. "She worked late at night," a former student said. "Her son would call and say he was hungry. He'd call and call. Then the next day she'd say how great her son was, he was so hungry that he'd opened a can of spaghetti and eaten it."

According to Vincent, however, "It was an okay way to grow up." He attended a boarding school on Long Island for the first through fourth grades; the school was near Brookhaven where his father worked and the two returned home together for weekends. About boarding school, Vincent commented later, "I didn't dislike it; I don't think I'd want to stay in them all through my school career. But it was very similar to being at camp."

In fifth grade, he switched to the Collegiate School in Manhattan and then the highly competitive Bronx High School of Science. During the eighth grade, he stayed in a French boarding school to learn French while his father was on sabbatical in Europe. Vincent became a physicist like his parents, but remains grateful that neither of them pushed him into science. "They were very careful not to stand over me and say I should be doing this or that," Vincent remarked.

During these years, Wu acquired a famous nickname, "The Dragon Lady," from the "Terry and the Pirates" comic strip. The Dragon Lady was a glamorous but dangerous Chinese beauty modeled after the imperious wife of General Chiang Kai-shek. Outsiders used the nickname for Wu more than the students who actually worked with her, according to Leon Lidofsky and others. She was no more aggressive or demanding than Columbia's male professors, emphasized William Havens, director of Columbia's Nuclear Physics Laboratories at the time. As Koller recalled, "We never called her Dragon Lady, at least not with any conviction. She was the most human of the professors at Columbia during the 1940s and 1950s. None of them had any regard for students. It was a very, very self-centered bunch. It's not surprising she was fiercely competitive."

Her students regarded her as a strong and dominating mother-figure. Evelyn Hu, now a microfabrication expert at the University of California at Santa Barbara, sensed in Wu "a motherliness, a sense of concern...and vulnerability." Treating her students like family had advantages and disadvantages though. Like a worried parent, Wu could castigate her students in no uncertain terms, telling them precisely what they were doing wrong.

She pushed her Chinese students especially hard, believing that mediocre Chinese could not survive the widespread prejudice against Asians. "I think her slave-driving in part came because she was aware of what it took to succeed," explained Ursula Lamb, Wu's graduate-school friend who became a history professor at Barnard College, the women's division of Columbia.

"She was a slave driver," Lamb contended. "She was abrupt and rough, and maybe more rough than an American would be. If something annoyed her, if she felt that a student was performing below par, she gave him hell, because she was anxious that he wasn't going to make it. But maybe her students wouldn't allow for that. They'd take the words but not the spirit. Her manner of being abrupt is being personal and caring in a way that is simply different. For her to break through her politeness and to tell somebody something honestly is a measure of compassion and caring—not aggression, as it is in the United States."

Despite the success of Wu's beta decay research, she did not become a Columbia faculty member until 1952. As a full-time researcher with no teaching responsibilities, she was easily overlooked by department administrators, William Havens realized. At the time, every other permanent professor at Columbia taught classes. So Havens arranged for Wu to teach part time, and, in 1952, Wu was promoted several steps at once. Once she was an associate professor with tenure, she had a permanent job and her paycheck was secure. She was ecstatic.

There would always be "a place in my heart especially reserved for beta decay," Wu declared. Nevertheless, she spent the years 1952 through 1956 looking for a new field to explore. One early spring day in 1956, Tsung Dao Lee, a young Chinese-American physicist at Columbia, came to her office for advice. His questions quickly revived her old flame. Lee and Chen Ning Yang of the Institute for Advanced Study in Princeton, New Jersey, were working together on a mysterious puzzle created by a newly discovered particle, the K-meson.

Thanks to new and larger atom-smashing accelerators built during the 1950s, physicists were discovering one new subatomic particle after another. Eventually Fermi complained, "If I could remember the names of these particles, I would have been a botanist." When the mu-meson was found, physicist I. I. Rabi snapped, "Who ordered that?" While protons, neutrons, and electrons explain all of ordinary matter, almost two hundred other subatomic particles have been discovered. Most are artificial exotics that survive in giant accelerators for only a fraction of a second. They include the K-meson.

The riddle of the K-meson arose from the fact that, when it decays radioactively, it sometimes produces two particles and sometimes three. At the time, physicists assumed that it was actually two different particles. When Lee and Yang learned that every K-meson has the same weight and properties, they guessed—correctly—that the K-meson might be one particle decaying in two different ways.

If Lee and Yang were right, particles inside the atom sometimes violated basic laws of physics. The laws of parity and symmetry say that molecules, atoms, and nuclei behave symmetrically. That is, nature does not care whether observers look at it directly or in a mirror. An experiment conducted in a mirror world should be the same as one conducted in our world. As a result of Emmy Noether's mathematical breakthrough during World War I, the laws of parity and symmetry had already explained much about molecules, atoms, and nuclei. So why not inside the nucleus too?

Lee and Yang, however, suspected that the particles inside a nucleus might sometimes favor one direction or another. In short, they might be sometimes right- or left-handed. They might sometimes violate the time-honored law of parity.

Physicists are guided by experimental evidence and mathematics. If the two disagree, it is easier to believe a simple, mathematical theory than a mass of confusing experiments. So most physicists wanted to believe that all interactions between the particles inside a nucleus were symmetrical. They assumed that when a nucleus decays and ejects electrons, roughly the same number of electrons would come out one end of the nucleus as the other end. Lee and Yang guessed the opposite: that an electron might prefer to come out one end rather than the other. If so, nature inside the nucleus—for some mysterious reason that is still not understood—would not always behave symmetrically.

Sitting in Wu's office, Lee asked her if anyone had ever proved experimentally that parity is always valid inside the nucleus. Wu told him to check the literature. "The literature" turned out to be a one-thousand-page book of graphs and tiny-type tables summarizing forty years of data compiled by hundreds of physicists. Lee and Yang actually plowed through the book. When they finished, they realized that no one had ever proved that the particles inside an atomic nucleus always obey the law of parity. They wrote an article pointing out the lack of experimental proof and suggesting several experiments that could be done to settle the question.

"I know of nobody at that time, in the summer of 1956—and that includes Lee and Yang—who believed that it would not be symmetrical," Yang explained later. "We wrote our paper only because we thought it should be tested." But nobody would do an experiment; it was too difficult. As an experimentalist told Yang, he would do it just as soon as he found a very bright graduate student who was willing to become a slave. "Nobody believed it would happen and, because it was so difficult, they wouldn't tackle it. But Wu had the perception that right-left symmetry was so basic and fundamental that it should be tested. Even if the experiment had showed it was symmetrical, it would still have been a most important experiment," Yang asserted.

In the meantime, Wu figured that the chance of parity's being violated inside the nucleus was only one in a million. Nevertheless, she said to herself, "This was a golden opportunity for a beta decay physicist to perform a crucial test, and how can I let it pass?" Not everyone has a chance to prove or disprove a basic law of nature. And thinking over the experiments she might do, she was tempted to try some extremely difficult techniques that no one had ever attempted before.

That spring, Wu and Yuan had booked passage on the luxury ocean liner, the *Queen Elizabeth*, to celebrate the twentieth anniversary of their departure from China. They planned to sail to Europe, attend a physics conference in Geneva, and continue on to the Far

East for a lecture tour. Suddenly, Wu realized that she had to do the experiment immediately, before anyone else realized its importance and did it first. Yuan agreed to take their sentimental anniversary trip alone, and Wu went to work.

Her experiment was so complex that it eventually required months of preparation just to design and test the equipment. Atomic nuclei, like all forms of matter, move constantly in every direction with heat energy. She needed to get rid of as much of this random heat energy as possible in order to see the direction in which the electrons were ejected. She planned to make the nuclei of radioactive cobalt so cold that they almost stopped moving. Then a powerful magnet tens of thousands of times stronger than Earth's magnetic force would make the slow-moving nuclei align themselves, like tiny magnets, all in one direction parallel with the magnetic field. Since temperature and magnetic fields do not affect radioactivity, the cobalt nuclei—chilled and lined up like toy soldiers—would keep right on disintegrating and emitting electrons. With luck, the nuclei would stay aligned for perhaps fifteen minutes—long enough for Wu to detect whether the radioactive nuclei were ejecting most of their electrons one way, as Lee and Yang had suggested. Even today, the experiment would be challenging. With 1950s technology, it was extraordinarily difficult.

Moving fast, Wu lined up her collaborators before Lee and Yang had even finished their article. The National Bureau of Standards, a federal agency in Washington, D.C., had one of the few laboratories in the United States that could cool material almost to absolute zero. At absolute zero, all motion stops, but in practice no one on Earth has ever achieved such a low temperature. Ernest Ambler, a pioneer in orienting radioactive nuclei under cold temperature conditions, had moved from Oxford University to the bureau a few years before. He and several other physicists there—Raymond W. Hayward, R. P. Hudson, and D. D. Hoppes—agreed to work with Wu.

For the next six months, Wu averaged four hours of sleep a night as she raced back and forth between classes and students at Columbia and the experiment in Washington.

According to the plan, a two-stage refrigeration system cooled the radioactive cobalt nuclei. In the first stage, helium gas, cooled to liquid form, would lower the temperature of the nuclei a few degrees Kelvin above absolute zero. Then a tiny box made of cerium magnesium nitrate crystals (CMN) would cool the cobalt nuclei to within a few thousandths of a degree of absolute zero.

Much of the equipment was homemade. Crystallographers assured Wu that growing ten CMN crystals one inch in diameter would take experts months. But she had neither the time nor the money. Instead, her team consulted a thick, fifty-year-old German chemistry book found on the top of a dusty library shelf. In it was a recipe for making CMN crystals. Wu's graduate student, Marion Biavati, made the first crystals in a beaker on her kitchen stove while she was cooking dinner. When Wu took the finished crystals to the bureau in Washington, she said, "I was the happiest and proudest person in the world." A dentist's drill was used to bore holes in the crystals because it exerted its pressure inward and would not shatter the crystals. Duco cement stuck the crystals together in a tiny box, but Duco lost its stickiness at the temperature of liquid helium, so soap was used instead. Palmolive was good, Ivory was better, but nylon string was best of all.

University faculty members sometimes regard national laboratory employees as nine-to-five clock-watchers. But the bureau team was working day and night. No one had ever done such an experiment with beta rays, and many of the techniques had to be redesigned. Hoppes slept in a sleeping bag on the floor by the equipment; each time the CMN refrigeration system kicked in, he phoned the team to race to the lab, no matter what time of night.

Nevertheless, Wu worried that the Washington crew was not working hard enough. She was always soft-spoken, quiet, and polite, but very intense, Ambler recalled. She could visit the Washington team for only a day or so every two weeks, so she did not waste a minute. At one point, the bureau team redesigned part of the apparatus while Wu was at Columbia. When she returned, Hayward reassured her, "It's okay, Miss Wu." From deep down inside her, Wu sighed, "It's not okay." When she disapproved of a team member's technique, she sighed mournfully to another team member, "He's not careful." To assert their independence, several members of the bureau's team continued their regular lunchtime bridge game; Wu could not imagine lunches lasting more than fifteen minutes. Behind her back, some of the Washington team called her "The Dragon Lady." Hayward thought that Wu was "very, very, very competitive—because of some deep-down insecurity perhaps. It was as though she was afraid she wasn't going to get credit."

As Wu and the bureau team checked and rechecked their results, word of the experiment leaked out. The results were unmistakable: more electrons came out one end of the nuclei than the other. Particles inside the nucleus do not always behave symmetrically. The law of parity—once considered sacred—was sometimes violated. Finally, around two o'clock the morning of January 9, 1957, everyone was satisfied. Hudson opened his desk drawer and pulled out a bottle of French champagne and little paper cups for a toast. The next morning, when lab workers found the champagne bottle in the trash, they knew the experiment had succeeded.

Wu still remembers the rush, with "everything turning out so good, lots of people, lots of pressure, you understand the theory to make comparisons—it was very, very good," she sighed contentedly. "These are moments of exaltation and ecstasy. A glimpse of this wonder can be the reward of a lifetime."

Soon competing physicists started producing experiments showing that other particles inside the nucleus can also violate the law of parity. Once the news was out, Wu and her teammates had to race to avoid being scooped. In one afternoon, they wrote an article summarizing their nine months of work. When they finished, Ambler raised a delicate question. How should the authors be listed? If they appeared alphabetically, Ambler would be first and Wu last. With a sigh, Wu indicated that that would not be the correct approach. So, "like the perfect bloody Englishman," Ambler asked, "Would you like to go first, Miss Wu?"

"I don't regret that," Ambler explained later. "She was senior, and she'd brought the idea to us. And that was courtesy if nothing else."

That Monday, Columbia called a press conference to announce the news. Ambler came up from Washington to attend, but some bureau employees thought their role had been downplayed. "It wasn't Wu. But many of the other people in the field just assumed that we were a bunch of stumblebums and that she was the outstanding physicist," Ambler said. When he went back to Washington, he learned that bureaucrats there were also irritated with him. He had not filled out the proper forms to make Wu a guest worker at the lab.

The experiment stunned the physics community. Robert Oppenheimer was so surprised that he telegraphed a physicist friend: "Walked through door." Wolfgang Pauli, Wu's old friend who had discovered the neutrino, had offered to bet "a very large sum" that her experiment would fail. When he learned her results, he joked, "I am glad that I did not conclude our bet. I can afford to lose some of my reputation but not some of my capital." The legendary physicist Richard Feynman, returning from South America, paid an unannounced visit to her lab to ask questions.

"Wu's experiment pointed to a whole new way of looking at things that produced a lot of new developments," according to Fowler. It was a large and essential step toward one unified theory explaining both electromagnetic forces and the weak forces that are responsible for several forms of radioactivity.

Surprisingly, Wu and the parity experiment captured the world's imagination. As the *New York Post* wrote, "This small modest woman was powerful enough to do what armies can never accomplish: she helped destroy a law of nature. And laws of nature, by their very definition, should be constant, continuous, immutable, indestructible." Lee, Yang, and Wu appeared on the front page of the *New York Times* and in *Time* and *Newsweek* magazines. Wu was feted, honored, and quoted all over the world. As she joked, it was the first time anyone had won a prize, "not for establishing a law, but for overthrowing it." In Israel, the prime minister wrote Wu about a book he had read on Indian Yoga. "Does it make sense from the point of view of physics?" he wrote her. Wu even made the cocktail party circuit. Referring to men and women in science, Clare Boothe Luce, a former United States ambassador to Italy, quipped that in the case of C. S. Wu, parity was most definitely preserved.

Ten months later, Lee and Yang won the 1957 Nobel Prize for physics. Wu was extremely disappointed that she had not won it, too, and many other physicists believe that she should have shared the prize. In the beginning, no one else had been willing to tackle the experiment, although anyone with an accelerator that produced muons could have settled the issue far more easily than Wu had with beta decay. However, Lee and Yang had originated the theory; the National Bureau of Standards team had provided the cooling temperature technology, and another experimentalist at the University of Chicago had also started work on an experiment as a result of Lee's and Yang's article. Wu's priority was further clouded by the entry of other physicists into the competition at the last minute, once news of her results leaked out.

The Nobel Committee has made similar rulings before. In 1914, Max von Laue suggested the principle of X-ray diffraction in a conversation that he never published; two university colleagues did an extremely difficult experiment proving that the technique worked and published their results. Yet von Laue won the prize for originating the idea; the experimentalists who made it a reality did not.

Although Wu lost the Nobel, she won every other available prize and a string of firsts: the first Wolf Prize from the state of Israel; the first woman to receive the Research Corporation Award; the first woman to win a Comstock Award from the National Academy of Sciences, an honor given only once every five years; the first woman president of the American Physical Society, and so on. She became a full professor at Columbia and was given an endowed professorship in 1972. She became the seventh woman in the National Academy of Science and collected honorary degrees from more than a dozen universities, including Yale and Harvard. She was the first woman to earn an honorary doctorate of science from Princeton University. She won the National Medal of Science, the nation's highest science award, from President Gerald Ford.

Although Wu mellowed and relaxed a bit, nothing slowed her zest for physics. When she and her old friend Pauli flew together to a conference in Israel, they talked about beta decay the entire trip. The German was amazed and wrote his sister, "Frau Wu is as obsessed with physics as I was in my youth. I doubt whether she ever even noticed the light of the full moon outside."

In another famous experiment, Wu used her knowledge of beta decay to confirm—not topple—another law of physics. Richard Feynman and Murray Gell-Mann of Caltech had hypothesized a new law of nature: the conservation of vector current in beta decay. According to their theory, the force involved in beta decay is more analogous to the electromagnetic force than had been suspected. Experiments conducted in Berkeley, Russia, and Geneva failed to confirm their hypothesis, though. At a 1959 physics meeting at MIT, Gell-Mann pleaded with Wu to test his theory. "How long did Yang and Lee pursue you to follow up their work?" he asked. Wu told him she was too busy. Finally, in December 1963, she did an experiment that confirmed the law. She had contributed another important step toward the present-day unified theory of fundamental forces.

After years of focusing on beta decay, Wu branched out into other fields. She studied sickle cell anemia and searched for new kinds of exotic atoms that may live only a billionth of a second. She traveled half a mile underground in a salt mine near Cleveland, Ohio, to detect the super-slow radioactivity of selenium 82, which has a half-life of 100 billion billion years. (The mine filtered out background radiation from cosmic rays.)

Since retiring in 1981, Wu has traveled extensively, lecturing and teaching, advising scientists in China and Taiwan, and encouraging American women to become scientists. She is the first living scientist with an asteroid named after her, and she enjoys inviting friends to visit the "Wu Chien-Shiung Asteroid" as soon as interplanetary travel becomes possible. She maintains ties with her former students. Sitting next to a scientist on an airplane, she noticed that he was reading a chemistry article by a Lidofsky. "Is that Leon Lidofsky?" she asked the passenger. "No, it's his son, Steven Lidofsky," came the reply. Wu, who had known Leon Lidofsky's children since birth, later told Leon proudly, "You know, I felt just like a grandmother." Wu also managed to retain the friendship of both Lee and Yang—who ended their ten-year collaboration bitterly and do not speak. In a rare joint appearance, they both came to her retirement party and stayed at opposite ends of the room.

Wu often spoke about the need for the United States to invest more in education and research in order to stay economically competitive. And as a former student revolutionary, she was appalled by the Chinese government's crackdown on students in Tiananmen Square in 1989. The government could have defused the atmosphere by meeting and talking with the students, just as Chiang Kai-shek had done with her group years before, she complained.

She remained concerned about the position of women in science. At a conference, she commented dryly, "Men have always dominated the fields of science and technology. Look what an environmental mess we are in."

At another conference, she commented frankly, "Bringing a womanly point of view may be advantageous in some areas of education and social science, but not in physical and mathematical sciences, where we strive always for objectivity. I wonder whether the tiny atoms and nuclei, or the mathematical symbols, or the DNA molecules have any preference for either masculine or feminine treatment."

So why, she asked, has the problem of women in science not been solved? "I sincerely doubt that any open-minded person really believes in the faulty notion that women have no intellectual capacity for science and technology. Nor do I believe that social and economic factors are the actual obstacles that prevent women's participation in the scientific and technical field."

Then, sounding much like her father, the revolutionary feminist of 1911, she answered her own question: "The main stumbling block in the way of any progress is and always has been unimpeachable tradition."

* * *

Chien-Shiung Wu died in New York City on February 16, 1997.

12

Gertrude B. Elion

January 23, 1918–

BIOCHEMIST

Nobel Prize in Medicine or Physiology 1988

CAREFULLY FILED AWAY in Gertrude Belle Elion's small and cluttered office are several folders filled with loving stories, including these:

Dear Ms. Elion:

I opened my newspaper this morning and through many tears read of your great honor, the Nobel Prize. My daughter Tiffany was stricken with herpes encephalitis in September, 1987. A neurologist said the only hope for her was possibly the drug acyclovir.

I have thanked the Lord so many times that he blessed you with the determination, stamina, love, and patience to work all of the long hours, days, months, and years it takes to invent a new drug. Tiffany is a senior in high school this year and doing great. May the Lord bless you beyond your wildest dreams.

—Tiffany's mother

* * *

Dear Ms. Elion:

I am one of the grateful individuals who has directly benefited from your research that resulted in the discovery of Imuran. I received a kidney transplant from my brother almost seven years ago, and my quality of life is superb.

Sincerely, Sharyn

* * *

Gertrude Belle Elion.

Gertrude Belle Elion at five and a half years of age.

Gertrude Belle Elion, May 1933, when she was fifteen years old.

Dear Ms. Elion:

While reading the article about your Nobel Prize, I was overcome with a sense of trembling and amazement. I have a little boy who was diagnosed two years ago with acute lymphocytic leukemia. Since that time, he takes every night two pills of 6-mercaptopurine, better known to us in the family as 6-MP. My son and I long wondered who was responsible for this wonderful gift. We now know. And so it is with inexpressible gratitude for having contributed to the saving of one human life so very dear to me and so many other human lives that I write to say to you in the simplest, and hence, the most profound and sincere of terms, thank you!

Rabbi P.

* * *

Dear Doctor Elion:

Thank you! Your hard work and relentless dedication were involved in the cure of my son's reticulum cell sarcoma when he was fifteen years old. His prognosis was terminal after an exploratory operation revealed a massive tumor in and

Gertrude Belle Elion with collie, Lollipop, and George Hitchings in 1960.

Gertrude Belle Elion and George Hitchings with Elvira Falco in right rear, Tuckahoe, New York, 1948.

outside the stomach and into the gall bladder, also many smaller tumors in the abdominal cavity. He was inoperable. He received 6-mercaptopurine and prednisone therapy with massive radiation. His tumors were never surgically removed. Today, seventeen years later, he is a happily married man and a chemist. I have always asked the Almighty to guide and inspire the many researchers in their work. Now I finally know who I am praying for.

Sincerely, Jim's mother

* * *

Dear Dr. Elion:

After a very severe case of shingles, Zovirax saved my eyesight. If you ever feel unappreciated for any reason, please take out this letter and reread it.

Sincerely, A.M.

* * *

Gertrude Elion—Trudy to her friends—keeps these letters and others like them nearby because they bring her so much joy. For Trudy Elion, biochemistry is not an abstract science. Her quest to cure diseases has always been inspired by people. Personal tragedies and the patients who take her compounds have kept the challenge fresh. Discovering drugs is not just a career for Trudy Elion—it is her mission and her life.

Elion is a unique figure in drug research. She is one of the few scientists in the industry and one of the few Nobel Prize winners in science without a doctorate. She spent years teaching school, taking secretarial courses, and working in marginal laboratory jobs—once without pay—before she could get a job in chemical research. It was not until late in World War II, when the nation's supply of male chemists finally ran short, that she got her chance. For years, she was the only woman in a top post in a major pharmaceutical company.

Her research revolutionized both drug-making and medicine. Elion made organ transplants possible. Her drugs helped transform childhood leukemia from a disease that was invariably fatal to an illness that 80 percent of its young victims survive. She developed treatments for gout and herpes, which can be fatal for chemotherapy patients. She developed the first drug that attacks viruses. Her research laid the foundation for AZT, for years the only drug approved by the Federal Drug Administration for AIDS patients.

Even more important than developing these individual drugs,

Elion helped change the way drugs are discovered. Instead of the traditional trial-and-error method, she and her collaborator George Hitchings studied the subtle differences between how normal and abnormal cells reproduce. Then they developed drugs to interrupt the life cycle of abnormal cells while leaving the healthy cells unharmed.

Elion was only one of several strivers in her family. She was born to an immigrant family in New York City on January 23, 1918. Her father, Robert Elion, was descended from a line of rabbis that a relative traced back through European synagogue records to the year 700. Robert came to the United States from Lithuania at the age of twelve and worked nights in a drugstore so that he could graduate from the New York University School of Dentistry in 1914. He ran several dental offices, invested in stocks and real estate, and built blocks of houses in the Bronx. He loved music, and from the age of ten Trudy was his favorite date for Metropolitan Opera performances. He was a frustrated tourist, too, and made a hobby of using maps, train and bus schedules, and the like to plot imaginary trips. He was known as a wise and intelligent man, and fellow immigrants sought his advice about their problems.

Elion's mother, Bertha Cohen, had emigrated alone at the age of fourteen from a part of Russia that is now Poland. Like Robert Elion, she came from a scholarly family: her grandfather had been a high priest. It was the custom among Russian Jews to send first the oldest children to America to get established and then to arrive later with the youngest children. So when Bertha arrived in New York, she stayed with older sisters. She attended night school to learn English, worked in the needle trades, and married at age nineteen. Gentle and self-effacing, she had great common sense. She urged Trudy to have a career, any career, just so long as she could earn her own money and spend it as she liked. Like most wives at the time, she had to justify every expenditure to her husband. "Getting a little extra money was like filing a new grant application," Elion recalled. "You had to have an explanation, and you had to essentially go begging for it. You couldn't just go out and buy something."

When Trudy was three years old, her grandfather arrived from Russia. He had been a watchmaker, but his eyesight had faded.Now he had time to take his little red-haired granddaughter to the park and to tell her stories. A learned biblical scholar, he knew several languages, and he and Trudy spoke Yiddish together. Their close and loving relationship lasted thirteen years, until his death.

When Elion was six years old, her brother Herbert was born and the family moved to the Bronx, then an open suburb with large parks for playing. Herbert proved to be a tease. When Elion brought boyfriends home, he liked to hang around and turn off the lights. But when Herbert got stuck on his homework, he turned to Trudy. "She's inherently a teacher," Herbert states. "She had a beautiful way of getting to the heart of the matter." Their childhood relationship—love and gentle rivalry—continues to this day. Until Trudy won a Nobel Prize, the family considered Herbert—who owns a bio-engineering and communications engineering firm—the brighter of the two. Now their roles have changed, ever so subtly.

Trudy was a shy bookworm with an insatiable thirst for knowledge. "It didn't matter if it was history, languages, or science. I was just like a sponge." She idolized Louis Pasteur and Marie Curie—"people who discovered things"—and devoured popular science books like Paul de Kruif's *Microbe Hunters.* "Those books were so exciting," remembered Elion. "It was like reading a novel. It was a mystery story that they could solve, and they became people. They weren't just names." Her heroes had to be discoverers, but their sex did not matter.

In 1929, when Trudy was eleven, relatives frantically urged her father to sell his stock. But he held on. More a rabbi than a businessman, he thought selling out would hurt other investors. After the stock market crash in October 1929, he spent the rest of his life trying to repay his creditors.

Her father's bankruptcy changed Trudy's prospects radically. When she graduated from high school at age fifteen, she knew she wanted to go to college. She recalls, "Among immigrant Jews, their one way to success was education, and they wanted all their children to be educated. Furthermore, it's a Jewish tradition. The person you admired most was the person with the most education. And particularly because I was the firstborn, and I loved school, and I was good in school, it was obvious that I should go on with my education. No one ever dreamt of not going to college. That never came up. It was assumed you went to college."

The problem was money. Fortunately, the City College of New York was free. Competition to enter was fierce, but her grades were extremely high and she was accepted by Hunter College, then the women's section of the university. If Hunter had charged tuition, neither she nor Rosalyn Yalow (chapter 14) could have attended college.

Elion's father hoped she would study dentistry or medicine, while her English, French, and history teachers wanted her to major in their subjects. The only teacher who did not care what Elion studied was her chemistry instructor!

She chose her career when she visited her beloved grandfather in the hospital, where he was dying slowly and painfully from stomach cancer. "That was the turning point," she declared. "It was as though the signal was there: 'This is the disease you're going to have to work against.' I never really stopped to think about anything else. It was really that sudden." She never lost that shining goal. She chose chemistry as her major instead of biology, however, to avoid dissecting animals.

Elion graduated from Hunter with highest honors in 1937 during the Depression. Knowing that she had to have a doctorate to do chemical research, she applied for financial aid to fifteen graduate schools all over the country. Not one would give her a graduate fellowship or a scholarship or an assistantship. Later she realized that she should have spotted the reason: sex discrimination. "Although there weren't many fellowships, there were some, and I was willing to go almost anywhere in the country." She could not find a job either. Neither her honors nor her Phi Beta Kappa key nor her pretty, smiling face helped. "It didn't make a particle of difference," Elion learned. "There weren't many jobs, and what jobs there were, were not for women."

Looking back, Elion was amazed at her innocence. "I hadn't been aware that any doors were closed to me until I started knocking on them," she explained. "I went to an all-girls' school. There were seventy-five chemistry majors in that class. As it turned out, most of them were going to teach it. But women in chemistry and physics? There's nothing strange about that. So when I got out and found that they didn't want women in the laboratory, it was a shock. I really hadn't anticipated that. Of course, I have to say it was a very bad time to graduate. It was the Depression, and nobody was getting jobs. But I had taken that to mean that *nobody* was getting jobs."

A wonderful job interview opened her eyes. She was positive she would get the job—that is, until she heard the words, "You're qualified. But we've never had a woman in the laboratory before, and we think you'd be a distracting influence."

"I almost fell apart," Elion recalled. "That was the first time that I thought being a woman was a real disadvantage. It surprises me to this day that I didn't get angry. I got very discouraged. But how could I say, 'No, I won't be a distracting influence'? How could I know what the men were like?" Grinning, she points to a black-and-white photo. "I wasn't bad-looking. I was kind of cute." In desperation, Elion enrolled in secretarial school.

In the meantime, Elion had met the man of her dreams. Leonard was a brilliant statistics major at City College and won a fellowship to study abroad. When he returned, he and Trudy decided to marry. Then he became desperately ill. The diagnosis was acute bacterial endocarditis, a strep infection of the heart valves and lining. Today penicillin kills the infection immediately, but Leonard caught it a few years too soon and died. "It was a heartbreaker, and she never really fully recovered," according to her brother.

"I didn't think about marriage again for a long time," said Elion. "But that was never my intention, and nobody ever said anything about it. In fact, my family was kind of hoping I would get married. But they never really pestered me about it. That was unusual. All of my auxiliary family—my aunts, uncles, and cousins—everybody wanted to find somebody for me. But my mother and father left me alone."

As time went on, Herbert observed, "No one could match up to Leonard. And then, as it was more and more in the past, he became more and more bigger than life, and so the memory was just enshrined." Elion told future beaux right off that she did not have time to marry.

For seven years, Elion worked marginal and temporary jobs, trying to get experience, trying to inch her way up the chemistry ladder toward research. After six weeks in secretarial school, she taught biochemistry to nursing students for three months. Meeting a chemist at a party, she volunteered to work in his lab free—to learn. The company president told a new anti-Semitic joke each morning, not realizing that both Elion and her benefactor were Jewish. When she left the lab a year and a half later, however, she was earning the munificent sum of $20 a week. Furthermore, she had lived at home and saved $450—enough to pay for one year of graduate school.

Elion was the only woman in her graduate chemistry classes at New York University, but no one seemed to mind. Elion certainly did not. She worked as a doctor's receptionist half-days to pay for carfare and lunch. Then she took education classes on the side and became a substitute teacher in New York high schools. Nights and weekends, she worked on her master's degree at NYU. The university turned down the heat over the weekends so the lab was much colder than the water bath she needed for her experiment. She worked in a winter coat and warmed the room with Bunsen burners.

By 1942, the United States was at war and the number of male chemists available for industrial lab work was dwindling. The longer the war lasted, the better the job prospects were for women scientists. When Elion finally got her first lab job, she immediately quit substitute teaching. She was testing food products for A&P grocery stores, checking the vanilla beans for freshness, the fruit for mold, and the pickles for acidity—a long way from cancer research. At first, she learned a good deal about instrumentation. Then, when the work became repetitious, she announced to the boss in her quick Bronx accent, "I've learned whatever you have to teach me, and there's nothing more for me to do. I have to move on." Persistence kept her going. She adopted Admiral Farragut's motto: "Damn the torpedoes! Full speed ahead!"

By 1944, even research labs were hiring women. She got just the job she wanted at a Johnson & Johnson lab in New Jersey. Six months later, the unit closed, and she was offered a new position testing the tensile strength of sutures. She politely but firmly declined, "I don't think that is what I want to do, thank you very much."

It was her father who found her the job of her dreams. "What's this Burroughs Wellcome Company?" he asked one evening. The company had sent a sample of Empirin painkiller to his dental office. "I looked it up. It's not far from here. It's only about eight miles, right over the border in Westchester County."

"I'll call," Trudy agreed. "But I don't think they have a research laboratory."

But when she telephoned, the answer was, "Yes, we have a research laboratory."

"Do you have any jobs?"

"Yeah, we do," came a laconic answer.

"Oh, can I come for an interview?"

"Yeah, you can come for an interview."

"Can I come on a *Saturday*?"

"Yeah," the voice said. "We're open on Saturdays because of the war."

So Elion put on her best suit, topped her flaming red hair with a cute little hat, and went to Burroughs Wellcome. By sheer luck, George Hitchings was working that Saturday. He alternated weekends with an organic chemist who also had a job opening. If she had gone to work for the organic chemist, she would not have stayed long at Burroughs Wellcome. "He was only interested in *making* compounds, and what turned me on was what the compounds did," she declared.

Hitchings had already hired one young woman, Elvira Falco, and she advised Hitchings not to hire Elion. As Elion tells the story, "She looked at me, and said, 'She's no chemist. She won't get her hands dirty.' I was much too elegant."

Hitchings disagreed. Elion had "verve" and a collegiate record that was tops. "She wanted fifty dollars a week. I thought she was worth it," Hitchings declared. Elion was twenty-six years old and planned to stay at Burroughs Wellcome only as long as she was learning things. She never left the company because, after mastering organic chemistry, she moved on through biochemistry, pharmacology, immunology, and virology. "It was one new field after another, and the compounds were taking me there, and that was wonderful."

Burroughs Wellcome turned out to be a highly unusual company, a British firm owned by a charitable trust. Until 1986 when 25 percent of the company was sold for stock, Burroughs Wellcome was run solely to benefit the Wellcome Trust, which supports research laboratories and medically related activities like medical museums and libraries. Two American pharmacists, Silas Burroughs and Henry Wellcome, had founded the company in England in 1880. Wellcome wanted the firm to discover drugs to treat serious, incurable diseases. He promised his scientists, "If you have an idea, I'll give you the freedom to develop it."

The company's American lab was housed in a converted rubber factory in suburban Tuckahoe, New York. "Our conditions were not A-1," as Hitchings delicately phrased it. Elion, Falco, and a young English chemist named Peter Russell shared a large room without air-conditioning or adequate hood ventilation. A baby-food plant downstairs dehydrated infant formula year-round, and the lab floor's summer temperature registered more than 140 degrees Fahrenheit. To protect her feet, Elion wore thick rubber nurses' shoes.

Life in the lab was "very, very fun," Falco reminisced. Russell was an expert at raunchy jokes; Elion, a shy and well-brought-up young lady, invariably blushed deep red. Falco and Russell relished wash-bottle fights; blowing in one neck of a two-necked flask sent a plume of water streaming out the opposite neck. Elion was more serious. She and Falco became friends and competitors, and Hitchings sometimes had his hands full.

The lab quickly became the center of Elion's social life. For twenty-five years, she and Falco subscribed to the Metropolitan Opera and attended performances together. Whenever Elion scraped some money together, she spent it on travel—"not new cars, not new furniture. It was travel." And some of it was with Burroughs Wellcome colleagues. After the war, she and Falco drove west in Falco's car. When Elion took the wheel for the first time, she turned right into oncoming traffic and crashed. A few years later, she got her own car and took Falco for another drive, turned right, and ran into a lamppost. Elion promised never to drive Falco anywhere again. Elion also became close friends with Hitchings's wife and children, and they vacationed together, too. Decades later, she still visits Hitchings's children and their families.

Hitchings was an unusual man, Elion soon discovered. A graduate of the University of Washington, he had earned a Ph.D. from

Harvard University and taught before joining the company at age thirty-seven. Graduate work had turned him into a frustrated nucleic acid biochemist. Fortunately, the American subsidiary of Burroughs Wellcome allowed Hitchings to do what he wanted.

When Hitchings and Elion started work in the 1940s, little was known about nucleic acids. Biochemist Oswald Avery at the Rockefeller Institute in New York City had just discovered that DNA (deoxyribonucleic acid) is the carrier of genetic information. James Watson and Francis Crick would not discover DNA's helical structure for another decade.

As Hitchings explained to Elion, he disliked the traditional trial-and-error method of discovering new drugs. He wanted a rational, scientific approach based on a knowledge of cell growth. All cells require nucleic acids to reproduce, but the cells of bacteria, tumors, and protozoa require especially large amounts to sustain their rapid growth. They should be acutely vulnerable to any disruption in their life cycle, he hypothesized.

Hitchings divided the nucleic acids among his staff and assigned the purines to Elion. The purine bases—adenine and guanine—are the building blocks of DNA and RNA (ribonucleic acid). Arranged in varying sequences, they transmit hereditary data to cells.

Elion was fascinated from the beginning. Little was known about the biosynthesis of the enzymes involved, so she was exploring new frontiers. Each series of experiments was like a mystery as she and Hitchings tried to unravel what the microbiological evidence meant. An optimist, Elion happily worked long hours, never truly satisfied until she had explored a hundred or more variations. Falco was amazed at how long Elion could concentrate on a problem. "Like a male," thought Falco.

Hitchings let Elion follow her instincts. He could have told her to make the compounds and let someone else find out how they work. But the group was small, and everyone did a little of everything. And as soon as Elion knew how the chemicals acted, she was hooked. That was what she wanted to do.

Elion started publishing her findings within two years. From the beginning, Hitchings let her write her papers herself and list her name first. "That was a great thing for him to do," she said. "I'm very grateful he let me. He corrected them, helped me, and it's a tradition that I've carried on." Unlike many drug companies, Burroughs Wellcome encouraged its scientists to publish their findings once patents had been registered. Eventually, Elion published more than 225 papers.

When Elion showed Hitchings drafts of her papers, however, he never praised them. "Is it good?" she asked. "Yes," he replied. "Is it very good?" she pushed. "Yes, you know it is. You know that," he answered. Fifty years later, when he complimented her on a blue suit, she claimed it was the first compliment he had ever paid her.

Instead of praise, Hitchings gave her promotions and, after the publication of her twentieth article, arranged for her membership in the prestigious American Society of Biological Chemists. "I knew he thought I was good," she conceded, "but he wouldn't tell me."

Some of her ideas were hunches; others came from trying to figure out what puzzled her. She was constantly asking herself, "What does it mean?" and "Why did it happen?" She concluded that science is a constant process of deduction and intuition and trial-and-error and back-to-the-drawing-board and then, always, more questions. She even worked summer weekends at her parents' country cottage.

"Does your boss know how much you work weekends?" her mother demanded.

"Look, I'm not doing this for him. I'm doing it for me," Elion reiterated. Then one weekend, when she did not take work to the cottage, her mother fretted, "What's the matter? I know, you don't want to tell me. You're sick." Her mother finally understood what motivated her daughter.

The first time Elion presented a report at a scientific conference, a man in the audience questioned her conclusions. She argued back. "Do you know who he was?" whispered her shocked friends. He was a well-known authority from the Rockefeller Institute. Afterward, he asked Elion, "Can I take you to lunch?" There they discussed her work. In her ten-minute talk, she had not had time to give all her proofs. "I was right. It was my work. I had backup material.... I appeared shy when I gave a paper, but I wasn't shy. I never talked about something I didn't really know about," she emphasized.

From then on, Elion enjoyed scientific meetings. When she explained her results to university researchers, they shared their latest data with her. Soon she was part of a network of pioneers in purine research. Inside the company, some thought Elion was buttering up the brass. But her friends in academic research proved vital to her career; they intervened at crucial times to help. Without them, she would not have won the Nobel Prize.

For two years, Elion traveled by subway each evening after work, trying to earn a Ph.D. from Brooklyn Polytechnic Institute. Then, suddenly, she faced a critical decision. The dean of the school called her into his office and told her that she could not continue working part-time on her doctorate. "You're going to have to quit your job and go full time," he informed her.

"Oh, no, I'm not quitting that job. I know when I've got what I want," she retorted.

"Oh, well, then, you're not very serious," he shrugged. With great reluctance, she abandoned her dream of getting a doctorate. Only when she was awarded honorary degrees from George Washington University and Brown University did she finally decide that she had "perhaps" made the right decision after all.

Elion thought of 1950 as the WOW! year when she synthesized not one, but two, effective cancer treatments. The first was a purine compound that interfered with the formation of leukemia cells.

Tested on animals at the Sloan-Kettering Memorial Hospital in New York, diaminopurine looked so spectacular that the hospital tried it on two acutely ill leukemia patients. One was a twenty-three-year-old woman named J.B. For two years, J.B. was in such complete remission that her doctors decided their diagnosis must have been wrong. She married, gave birth to a child, and then had a relapse and died. She had indeed had leukemia. Today, J.B. would have been given larger doses over a longer period of time and probably would have been cured. During her two-year remission, however, she had received no treatment at all. Elion still weeps for J.B.

The drug put Elion and Hitchings on an emotional roller coaster. "We saw remissions that gave us joy, but almost all were followed by a relapse," Hitchings explained. It hooked them on cancer chemotherapy. Elion's compound, however, was too toxic and produced severe vomiting. So she began studying its biochemistry, figuring that if she knew how it worked, she could make other compounds like it. Eventually, she made and tested more than one hundred purine compounds.

Elion was just getting over the shock of J.B.'s death when she substituted a sulfur atom for the oxygen atom on a purine molecule. The new compound was 6-mercaptopurine, 6-MP for short. In

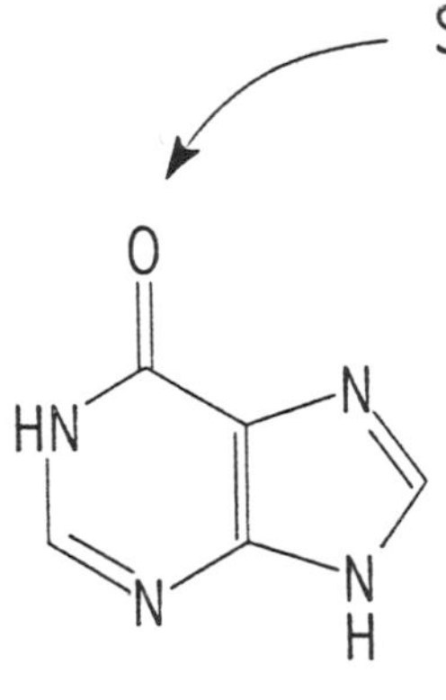

Figure 12.1. 6-MP. Gertrude Elion made 6-MP, her first antileukemia drug, by substituting a sulphur atom for the oxygen atom on a purine molecule.

animal tests, mouse tumors treated with 6-MP failed to grow. Even more important, the treated mice lived twice as long as mice with untreated tumors.

In 1950, half of all children with acute leukemia died within three or four months. Fewer than a third lived as long as a year. Yet children given test treatments of 6-MP experienced complete, though temporary, remissions. Columnist Walter Winchell reported the good news that George Hitchings had discovered a possible leukemia treatment. Within days, the Food and Drug Administration approved the drug for commercial release under the trade name Purinethol. The government's drug approval system was simpler before the thalidomide drug tragedy, so 6-MP remains the first and probably the only new compound ever released by the Food and Drug Administration before the bulk of the supporting data had become available. Later, Elion and Hitchings published the data and reported on it in scientific meetings.

By itself, 6-MP did not cure leukemia. The children treated with it eventually relapsed and died. Visiting them, Elion got a terrible, sinking feeling. Watching children live and die by her drugs, getting letters from their parents and reports from their doctors, Elion was as emotionally torn as she had been at her grandfather's death. As before, drug-making became a personal issue and engaged her heart as well as her mind. She told Hitchings she wanted to study the metabolism of 6-MP in the human body. She wanted to figure out how to make its effects last longer. She did not know anyone who had ever studied the metabolism of a drug before, but she spent six years trying to understand every detail of 6-MP.

"The disappointment with 6-mercaptopurine was that it wasn't good enough," Elion declared. "Until we learned how to use combinations for cancer therapy, we weren't curing anybody with single drugs. In a way, it was both an excitement and a real disappointment when it turned out that these kids who looked well relapsed. So there was this constant struggle: How to beat it? How do you get around it? Why are they relapsing? What can we do to make it better? And for eighteen years of my life, I tried to make 6-mercaptopurine better. I was insistent that this was going to work."

In 1950 Elion also synthesized a close relative of 6-MP called thioguanine. Once physicians learned to combine 6-MP or thioguanine with other drugs, they could finally treat childhood leukemia effectively. Today, children with leukemia are given 6-MP combined with one of a dozen or more drugs, and most of them will go into remission. Then, after several years of maintenance therapy with 6-MP and another drug, approximately eighty percent of the children are cured—a word that cancer therapists never dared use before. Thioguanine's primary use today is not for childhood leukemia; it is used instead for treating acute myelocytic leukemia in adults.

Elion was only thirty-two years old when she synthesized the revolutionary drug 6-MP. She had opened up an entirely new area of research in leukemia therapy. She had demonstrated that minute chemical changes in a compound will fool malignant cells. She also finally stopped worrying about not having a Ph.D. But most important, she started curing people. Watching children get well, Elion wondered, "What greater joy can you have than to know what an impact your work has had on peoples' lives? We get letters from people all the time, from children who are living with leukemia. And you can't beat the feeling that you get from those children."

"This is like you're the doctor in a way, and you're doing something directly for these people," she elaborated. "The intermediary is the doctor but the excitement is really yours because you know you gave them the tools. And so when the Nobel Prize came in, everybody said, 'How does it feel to get the Nobel Prize?' And I said, 'It's very nice but that's not what it's all about.' I'm not belittling the prize. The prize has done a lot for me, but if it hadn't happened, it wouldn't have made that much difference." Her reward was curing patients.

Still shy on purely social occasions, she felt completely at home in science. James Burchall, now molecular genetics chief at Burroughs Wellcome, concluded, "She lives in the world of science and drug-making and still finds this is the great challenge and fascination and the great focus of her life. It's her challenge and her joy."

Despite her success in developing a treatment for cancer, Elion could not protect her loved ones from the disease. During the 1950s, her mother became ill with cervical cancer. Today cervical cancer is almost 100 percent curable, but Bertha Elion's Central European upbringing had left her too embarrassed to consult a physician during the early stages of the disease. Elion was powerless to help. Bertha Elion's death in 1956 was one of the most painful experiences in her daughter's life. Like the deaths of her grandfather and her fiancé, her mother's illness brought home to Elion that what she did in her laboratory affected the lives of real people.

Elion soon developed a no-holds-barred strategy. Once she discovered a compound, she used it as a tool to find another. If a compound altered a step in the way nucleic acid was synthesized, she used it to explore that step and all its implications. "This kind of leverage technique, where every time you get a piece of information you use it as a tool to pull more information out—that was one of the key elements of her strategy," according to James Burchall. "You use everything you can get your hands on."

Elion was pushing the Hitchings approach to its logical conclusion, implementing it as far as the compounds would take her. Along the way, she discovered a remarkable series of drugs, each one attacking a different part of nucleic acid's life cycle.

By 1958, other researchers were following Elion's leads. In Boston, Robert Schwartz tested 6-MP on rabbits to see if it had any effect on their immune response. Fed 6-MP and challenged with foreign antigens, the rabbits were unable to produce any antibodies to the foreign substances. The discovery had great implications for organ transplants. Surgically, organ transplants had been feasible for years. The obstacle was the body's immune response, which rejected any graft within a few days and killed the animal. A young British surgeon, Roy Calne, had been experimenting with transplants in dogs. When he read about 6-MP, he decided to try drug treatment, too. When he gave daily doses of 6-MP to a dog with a transplanted kidney, the animal lived an amazing forty-four days.

Calne stopped by Burroughs Wellcome on his way to a fellowship in Boston and asked Hitchings and Elion if he could test some of 6-MP's relatives. They gave him a pocketful of vials, including 57-322, later known as azathioprine and marketed as Imuran. Imuran was a highly sophisticated version of 6-MP that Elion had synthesized. Tests showed that it might be effective as an immune-system suppressant.

A few months after his visit, Calne wrote Hitchings a brief note: 57-322 was "not uninteresting." Hitchings knew that, "translated from the British, he had a very exciting result." Hitchings and Elion hotfooted it up to Boston to work with researchers there. The heroine of the story was a collie named Lollipop. Calne transplanted a kidney into Lollipop and gave her Imuran; she lived 230 days and had a litter of pups before dying from an unrelated cause. By 1961, knowledge had progressed far enough for Dr. Joseph E. Murray to transplant a kidney between two unrelated people. Murray's third patient survived, and Murray received a Nobel Prize in 1990.

Elion's drug made organ transplantation possible. For the first time, patients could be given transplanted organs without their bodies rejecting them. The first heart-transplant patient took Imuran in 1967. It is still used to prevent kidney rejection. Of about one hundred thousand kidney transplants performed in the United States since 1962, most have used Elion's drug. It is also used to treat autoimmune lupus, anemias, hepatitis, and severe rheumatoid arthritis.

Sometimes, when Elion gives a talk somewhere, a stranger approaches her afterward to thank her for making his or her kidney transplant possible. As Elion observed with considerable understatement, "When you meet someone who has lived for twenty-five years with a kidney graft, there's your reward."

Asked to choose her favorite drug between 6-MP for leukemia and Imuran for kidney transplants, Elion cannot. She got a "high" from both: "It's hard to choose among your children. Each drug was wonderful, each one rewarding. It's been rewarding all along."

In the early 1960s, Elion was still trying to make 6-MP's effect last longer. She knew that an enzyme, xanthine oxidase, acted within minutes in the human body to break down 6-MP to another compound. What if she could inhibit the enzyme xanthine oxidase? Then 6-MP should last longer in the body. Burroughs Wellcome chemists had synthesized a compound that did just that. Tests showed, however, that although it protected 6-MP from destruction in the body, it did not prevent 6-MP from producing bad side effects. Elion was delighted to learn, however, that it did reduce the body's production of uric acid. The compound was called allopurinol.

Too much uric acid in the body is not only extraordinarily painful, it can also be fatal. Uric acid builds up in the body either because too much is produced or because the kidneys are not excreting enough. In either case, the acid precipitates into chalky crystals in the joints and causes the exquisite pain of gout. The deposits can also form painful urinary stones that block and damage the kidneys. Radiation and chemotherapy patients develop the same problem when the rapid destruction of their cancerous tissues makes uric acid accumulate. Before allopurinol, more than ten thousand gout sufferers died from kidney blockage each year in the United States alone.

A night watchman crippled by gout received the first allopurinal in clinical tests during 1963. His pain disappeared within three days, and he returned to work.

Ten years later, it was discovered that allopurinol is also an effective treatment for Leshmaniasis disease, a major problem in South America. Elion strongly encouraged the company to pursue the issue, irrespective of the profits involved. Allopurinol kills the protozoa responsible for the disease by interfering with their production of purine bases. Allopurinol may also be effective against another South American problem, Chagas' disease, in which childhood insect bites can trigger an autoimmune disease and death by the age of thirty. "She has a real social conscience," observed Thomas Krenitsky, now research vice president at Burroughs Wellcome. "In fifty years, Trudy Elion will have done more cumulatively for the human condition than Mother Theresa."

By the mid-1960s, Elion had developed her own identity, separate from her boss Hitchings. When Krenitsky first heard about her during the early 1960s, his Yale professor had said, "Oh, yeah, that fellow Elion has worked with George a long time." But by the time Elion was in her late forties and early fifties, she was well known in her own right. Krenitsky attended a national scientific meeting with her during the mid-1960s and watched as a prominent scientist greeted her with cordial deference. After he had turned away, she whispered to Krenitsky, "Five years ago that guy snubbed me." Suddenly, Krenitsky realized that Elion would never snub someone unimportant. "She's always herself. She's always Trudy. That guy wasn't always the same person with everybody, but she is, whether the other person is a student, a glassware washer, or the president of the company. She's egalitarian, and she lives it. She's not elitist, that's for sure."

Elion was collecting honors on her own, too. The American Chemical Society awarded her its Garvan Medal in 1968. The two-thousand-dollar-prize was prestigious, but until 1980, it was the only award that the American Chemical Society gave to women. Women could not compete with men for other ACS awards. Despite its drawbacks, the Garvan medal was Elion's first stamp of approval, and she was delighted. She told the society that she had had a "totally feminine" reaction—she cried.

Soon after, George Mandell, a professor at George Washington University, telephoned. He worked in purine chemistry and knew many of her one hundred papers. "Look, the kind of work you're doing, you've long since passed what a doctorate would have meant," Mandell told her. "But we've got to make an honest woman of you. We'll give you a doctorate, so we can call you 'doctor' legitimately." Standing on the platform at George Washington University, clutching her long-sought Ph.D., Elion thought only, "I wish my mother were here." Bertha Elion had wanted so desperately for her daughter to have a career.

When Hitchings retired from active research in 1967 to become vice president of research, Elion was on her own. She became head of the Department of Experimental Therapy and, for the first time, could show what she could do without Hitchings. During the twenty-three years that they had written papers together, not even Burroughs Wellcome insiders had been able to identify their respective contributions. Nevertheless, Elion says that she and Hitchings had had their differences for years. She considered them a team but doubts that Hitchings did. He was always the boss; he used "I" for their work while Elion used "we."

"He can be very patronizing," Elion remarked. "He perceives that he started it all.... But actually he was always willing to listen to suggestion. I said I wanted to study metabolism in 1953; nobody was studying drug metabolism.... He just let us do what we thought we should be doing."

The upshot was that "at fifty-five, I had had enough already of being junior," Elion remarked. "And then I had the opportunity to show what I could do on my own."

As always, that meant discovering an effective drug. "She's quite clear on what achievement in this area means," Burchall emphasized. "Achievement is finding new medicines to treat medically important indications that aren't currently being met, and there is no other bottom line. She doesn't do science in the abstract. She does it for a purpose, to advance a compound that may be useful to treat a disease. She's trying to find the crucial experiment or series of experiments that will decide a question."

In 1968, Elion revisited one of her failures—and turned it into one of her biggest successes. As she tells the story, "We decided to return to a path that had intrigued us as early as 1948, the path to antivirals." Scientists were convinced that no drug would ever be developed against viruses. They thought that any compound toxic enough to damage the DNA of a virus would harm the DNA of a healthy cell too. Consequently, no one had done much antiviral chemotherapy for years. Elion's first purine compound, the diaminopurine that had cured J. B. in 1948, for example, had showed some antiviral properties. But it was so toxic that Elion had put it aside and spent twenty years pursuing drugs to fight leukemia, organ rejection, and gout.

One of Elion's hallmarks, however, was learning from her mistakes. So when she heard that something similar to her 1948 compound showed some antiviral activity, she decided to look up her old friend again. She wanted a compound that would inhibit the multiplication of virus cells without affecting normal cell division. When she sent a sample of a related compound to England to be tested for antiviral activity, she got back an excited telegram: "This is the best thing we've seen. It's active against both the herpes simplex virus and the herpes zoster virus."

Herpes viruses include both herpes zoster, which causes shingles, and herpes simplex, which causes mouth and genital sores. Herpes infections can be fatal for patients with leukemia, cancer, and transplanted organs and bone marrow. When the immune system is suppressed by disease or chemotherapy, the virus can spread to internal organs or to large areas of the skin. A drug that attacked herpes would do far more than cure cold sores.

For four years, Elion's team studied related chemicals that had some effect on herpes. Then Howard Schaeffer, the head of Burroughs Wellcome's organic chemistry division, tried removing the sugar molecule perched on a purine sidechain. "Do you really need a whole sugar there or can you use a piece of a sugar to fool an enzyme?" he asked. Schaeffer had the idea of putting different sidechains on the purines, hoping that the viral enzymes would be confused. After Schaeffer and his associate Lilia Beauchamp further synthesized and modified the compound, it was one hundred times more powerful than anything seen so far.

Elion's team went to work, trying to understand its metabolism—how and why it worked, and why it was so nontoxic and so selective. Acyclovir turned out to be different from any other compound Elion had ever seen. It is so similar to a compound needed by the herpes virus for reproduction that the virus is fooled. The virus enters normal cells and starts to make an enzyme that helps it reproduce. This enzyme activates acyclovir and turns into something that is toxic to the virus. In short, acyclovir makes the virus commit suicide.

For four years, from 1974 to 1977, more than seventy-five researchers kept acyclovir secret. "We didn't tell anybody anything," Elion emphasized. "It was an amazing exhibition of what you can do when you're excited with what you're doing."

At cancer meetings, colleagues asked, "What are you working on?"

"Oh, you know, purines," she answered casually.

"Yeah, back in the rut," they replied.

But Elion knew, "We had to protect it from the other companies. We knew the minute it got out, everybody would jump on the bandwagon—and they did!"

Acyclovir went public at a 1978 scientific conference. Thirteen posters filled an entire alcove, explaining acyclovir from its synthesis to its activity, enzymology, metabolism, toxicology, and so on. Besides treating shingles, acyclovir is also effective against Epstein-Barr virus, pseudo-rabies in animals, and herpes encephalitis, a frequently fatal brain infection in children.

"Hey, wait a minute," people asked Elion. "Have you guys been working on this all these years? We never knew you were working on virology."

"Well, if I'd wanted you to know, I'd have told you," Elion retorted exuberantly.

Elion called acyclovir her "final jewel.... It was a real breakthrough in antiviral research. That such a thing was possible wasn't even imagined up until then." No one knew there were so many different enzymes specific to particular viruses. "After that, everybody went to work in the field, so in addition to being an important compound, it was an important landmark," Elion realized. If one virus had its own specific enzyme, maybe others did, too.

It took seven years before acyclovir was ready to market, but it was the pot at the end of rainbow. Marketed as Zovirax, it is Burroughs Wellcome's largest-selling product, with worldwide sales of $838 million in 1991.

Before acyclovir was discovered, Burroughs Wellcome had moved in 1970 from suburban New York to Research Triangle Park in the Piedmont region of North Carolina. The move to rural North Carolina was not easy for a born and bred New Yorker. But Elion managed. She stuffed a two-story condominium with travel souvenirs, family portraits, photos, statuettes, artwork, music, and plants. She maintained family ties with her brother's children by phone and plane. She kept her Metropolitan Opera subscription and flew to New York as often as possible for the opera. In fact, she attended every classical music concert in the Triangle, all the James Bond movies, and many of the college basketball games. She found a close friend in her neighbor, Cora Himadi, and together they traveled all over the world. Although Elion is no athlete, she will do anything for a photograph, even tramp up and down a mountainside. There is hardly a country that Elion has not visited in Europe, Asia, Africa, or South America. And before she knew it, she was a dedicated Tarheel.

Elion retired in 1983 and became a consultant to the company. Within a year, her former unit had used her approach to produce Azidothymidine, called AZT, which was the only drug licensed to treat the AIDS virus in the United States until late 1991. Elion is often credited with AZT's development but says she played no direct role. "The only thing I can claim is training people in the methodology.... The work is all theirs," she emphasized.

At 6:30 A.M. on October 17, 1988, Elion was washing her face when she got a phone call from a reporter: "Congratulations! You've won the Nobel Prize!" She assumed he was joking until she heard who the other winners were: George Hitchings and Sir James W. Black of the University of London. Black developed the first clinically useful drug for blocking beta receptors. The three shared $390,000.

It was the first Nobel for drug research in thirty-one years, and one of a few for cancer treatment. Elion was not only the employee of a pharmaceutical firm—a class of researchers rarely honored by the Nobel Committee—but she had no Ph.D. The prize, however, was not for a specific drug but for demonstrating the differences in nucleic acid metabolism between normal cells and disease-causing cancer cells, protozoa, bacteria, and viruses. As the Nobel Committee observed, "While drug development had earlier mainly been built on chemical modification of natural products, they introduced a more rational approach based on the understanding of basic biochemical and physiological processes." Elion was seventy years old, and eighty-three-year-old Hitchings was one of the oldest winners in many years.

Elion had expected Hitchings to get the prize, but not herself. "He expected it years ago, but when he didn't get it then, he thought he wouldn't get it at all," she explained. Later, Elion learned that several prominent academic scientists had nominated them as a pair. A Nobel Committee member wondered, though, why Elion should be included. "Has she really contributed?" he inquired.

One of Elion's longtime university friends replied, "Have you looked at the papers from the early days? She's first author." The first name listed on a scientific publication traditionally indicates who is primarily responsible for the information. Then the professor pointed to Elion's antiviral work, which had occurred after Hitchings' retirement. That tipped the balance. Without her university friends, she might not have won the prize.

The day the Nobel was announced, Hitchings was in New York, so he and Elion held separate press conferences. Despite the geographic distance between them, they emphasized the same point: The prize was just the icing on the cake. The real rewards had come from curing patients with their drugs.

At the Nobel ceremonies in Stockholm, Elion sat on the dais, listening to the orchestra play Mozart. Smiling just as naturally as if she were in her own living room, she gently tapped her foot and nodded her head from side to side with the music. She enjoyed it all. She brought her nieces and nephews along with their spouses and children, four of whom were under the age of five. Insisting that the children be allowed to attend the formal banquet, she told an astonished official, "I'm not going to bring them all the way to Sweden and then have them spend the evening in a hotel room. You put them at a separate table where they can see their parents and their parents can see them, and they'll be fine." And they were, enchanting press and hotel employees alike.

The prize did a lot for Elion. It turned her into a highly visible role model for women. It gave her a full-time secretary and a bigger office crammed with Nobel memorabilia and other honors. It got her the National Medal of Science, the nation's highest science honor, in 1991. She scandalized her brother by breaking a family rule never to turn down a previous invitation to accept a better one: she canceled a seminar at Tufts University to collect the prize from President George Bush. She was elected to the National Academy of Sciences, long after she had organized the campaign that got Hitchings admitted in 1975. (The equally prestigious Royal Society had also admitted Hitchings without her.) When Burroughs Wellcome gave Hitchings and Elion $250,000 each to donate to charity, Elion gave hers to her alma mater for women's fellowships in chemistry and biochemistry.

The prize has had its bittersweet elements. It strained her fifty-year-old friendship with Hitchings. Friends say that Elion's inclusion in the prize came as a shock to him and brought the undercurrent of their competition into the open. Elvira Falco, friend of both for decades, commented later, "They worked together as well as any two people worked together. But I have a feeling that when your assistant gets as much credit as you have, it may be a difficult thing."

Elion is working as hard as ever. Besides consulting at Burroughs Wellcome, she teaches research methods to medical students at Duke University. She has served on the national committee for reviewing procedures for the approval of new cancer and AIDS drugs; the National Cancer Advisory Board; World Health Organization committees on three tropical diseases, filariasis, river blindness, and malaria; and on boards of the National Cancer Institute, the American Cancer Society, and the Multiple Sclerosis Society.

Through it all, friends say, Trudy Elion is still Trudy—poised, unpretentious, and as intent on curing diseases as she was when she dedicated her research to her grandfather, her mother, her fiancé, J.B., and the children with leukemia.

One of her favorite awards is the simplest: the letter about Tiffany, the high school student with herpes encephalitis. In her letter, Tiffany's mother asked Elion for an autographed photo to tape to the front of the family refrigerator for all to see. Then she enclosed Tiffany's picture—"so you could see for yourself how great she looks, once again, thanks to you."

* * *

13

Rosalind Elsie Franklin

July 25, 1920–April 16, 1958

X-RAY CRYSTALLOGRAPHER

WITH A GLASS WAND, Rosalind Franklin gently coaxed apart the lintlike fibers of DNA. Pulling them as thin as spider silk, she bunched them in tiny packets of parallel strands. Then, delicately controlling the humidity, she beamed X rays at the little bundles and photographed the thread of life.

Quick, fierce, and fun-loving, Rosalind Franklin was a commanding leader, an idealist about science, and in her time the supreme experimentalist analyzing the molecules of heredity.

Fascinated by matter, what the world is made of, and how it formed, Franklin liked facts—indisputable, provable, hardcore facts—not high-flown theories or insubstantial speculation. As she declared, "Facts are facts." While still in her early twenties, she had uncovered data about coal that established her reputation as an expert experimentalist. The evidence she later revealed about viruses helped lay the foundation for structural biology.

In the early 1950s, Franklin almost discovered—by herself—enough information about the structure of DNA to explain the molecular basis of heredity. DNA, a molecule found in all living cells, is the coded blueprint for transmitting inherited characteristics from one generation to another. The facts she did uncover about the molecule helped James Watson and Francis Crick beat her to the Nobel Prize—data they used without her knowledge and without fully crediting her.

Once the structure of DNA was understood, the field of molecular biology exploded; it became the most significant scientific development of the late twentieth century. The most important technology of the twenty-first century is expected to be bioengineering or recombinant DNA, in which programmed DNA is injected into simple organisms like bacteria to produce a desired characteris-

Rosalind Franklin at Ciba Foundation conference in London on the nature of viruses. March 26–28, 1956.

Rosalind Franklin.

April 2, 1956. Rosalind Franklin in Madrid with (left to right) Ann Cullis, Francis Crick, Donald Caspar, Aaron Klug, Odile Crick, and John Kendrew.

tic in future generations. Insulin, growth hormone, and the clotting factor missing from the blood of hemophiliacs are already manufactured commercially with recombinant DNA technology.

Today, as the facts about Franklin's life and scientific prowess have emerged, they have cast a shadow on Watson's and Crick's achievement and reestablished Franklin's place in the sun.

Rosalind Franklin was born in London on July 25, 1920, the second of five children in a wealthy Jewish banking family. Her ancestors had lived in England since 1763, and her grandparents lived in upper-class English style. They had a large house in a comfortable section of London and a country home and, when they wintered in the Mediterranean, a retinue of English servants accompanied them.

Rosalind's father, Ellis Franklin, and her mother, Muriel Waley, were raised in a tradition of public service and philanthropy. Her banker father taught science as a volunteer at the Working Men's College and helped numerous Jews escape from Nazi Germany. Among her aunts were socialists and activists in women's causes and trade unionism. An uncle was such a strong supporter of women's rights that he served six weeks in prison in 1910 for taking a dog whip to Winston Churchill, then a prominent antisuffragist. Another relative was elected to Parliament three times before the government ruled that Jews could serve without taking an oath of office on a New Testament Bible.

As a child, Rosalind felt discriminated against because she was a girl. She thought her family did not understand her and remembered her childhood as a tense struggle for recognition. Because she did not like "let's pretend" games and detested dolls, her parents found her "practical and unsentimental.... literal-minded and not imaginative." She preferred making things—sewing, carpentry, and Meccano building sets. While her mother praised Rosalind's "exquisitely neat" embroidery and a "beautifully planed" coin cabinet, her analytical mind was harder to recognize.

But Rosalind needed reasons, proofs, and facts. She read through the Bible to find a reason for believing in God and concluded, "Well, anyhow, how do you know He isn't a She?" Quick, logical, and precise herself, she was impatient with slipshod, vague, and woolly arguments.

When Rosalind was eight years old, she caught a string of colds and flus, and the family doctor recommended a convalescent boarding school near the coast. With the best of intentions, her parents agreed. Somewhat hopefully, her mother regarded Rosalind's year away from home as "a neutral experience." As far as she could tell,

Rosalind was "a little homesick.... but never actively unhappy." Asked how she was getting on, the little girl replied laconically, "All right." In truth, Rosalind hated the school and always resented that year away from home.

Rosalind absorbed an unfortunate lesson from her boarding school convalescence. She decided that it was safer to ignore illness and pain than to seek help. When a needle became stuck deep in her knee joint, she walked for blocks to a hospital alone and in excruciating pain.

In London, she attended St. Paul's Girls' School, an academically rigorous day school for the daughters of well-to-do families. During part of one semester, she stayed in a Parisian pension to improve her French. She returned home an ardent Francophile with a zest for French dressmaking, cooking, and travel. From then on, she made her own clothes and raised and lowered her hemlines with each changing fashion.

Thanks to the excellent physics and chemistry classes offered by St. Paul's, Rosalind decided by age fifteen to become a scientist. An avid amateur astronomer, she followed star maps in the *London Times* and searched night skies for constellations. Hoping to study physical chemistry at Cambridge University, she took—and passed—the entrance examinations. She was in for a bitter disappointment.

Her father, who strongly disapproved of university education for women, refused to pay for her to attend Cambridge. He had once planned a science career for himself and would have been delighted if a son had pursued the same course. But women should do good works as volunteers; they should not be professionals. His refusal touched off the only crisis in her parents' happy marriage. Rosalind's favorite aunt, Alice Franklin, stormed over to inform her brother that she would personally send Rosalind to Cambridge. In the ensuing row, Rosalind's mother announced that she, Muriel, would pay for Rosalind's education out of her own family money.

Faced with three irate women in the family, Ellis Franklin backed down and agreed to pay for Rosalind's university education. His approval was grudgingly given and resentfully received. Rosalind loved her mother deeply, but she never entirely forgave her father, even though her virus work eventually made him quite proud. As she frequently told friends, daughters have special disadvantages.

In 1938, a year before the outbreak of World War II, Rosalind Franklin entered Newnham College, a women's college in Cambridge University. For a woman, Cambridge was much like a girls' boarding school. Before a Newnham woman could entertain a man in her room, she had to move her bed out into a public corridor. Women faculty members, most of them unmarried, seemed extraordinarily serious and formidable. Franklin decided that she never wanted to be like them. Years later, she almost turned down a job offer from Cambridge, rather than become a Cambridge woman don.

The outbreak of World War II in September 1939 precipitated another disagreement with her father. Ellis Franklin wanted Rosalind to quit the university and do volunteer defense work. Rosalind, on the other hand, was determined to continue her studies. Luckily, the government made it clear that all science students should complete their education.

One of the few blessings of the war was Rosalind's friendship with Adrienne Weill, a distinguished French woman physicist who had worked with Marie Curie and Irène Joliot-Curie at the Curie Institute. After Weill's escape to England, she worked in Cambridge where Franklin became her friend and for one year her boarder. It was Weill who found Franklin a job and a room in Paris after the war.

After graduating from Cambridge in 1941, Franklin spent a year doing research in physical chemistry with the future Nobel prize–winning chemist Ronald Norrish. Then she took an unpromising job that established her reputation as a research scientist. As her contribution to the war effort, she began to study the physical structure of coals and carbon for the British Coal Utilization Research Association. Rooming with a cousin, she bicycled fruiously through air raids across the exposed Putney Common each day to her job in South London. She never complained, but she was terrified. (During her last illness, she suffered delirious nightmares about cycling across the common and wondering if the war would ever end.)

In her laboratory, Franklin focused on a large and important wartime problem: how to use England's coals and charcoals more efficiently. In a series of elegantly executed experiments, she discovered the structural changes that occur when coal and carbons are heated and showed why some heated carbons turn into graphite as their molecules form parallel layers that slip and slide apart. She did the laboratory work herself, producing masses of experimental data. When the laboratory banned uncertified personnel from its machine shops, she simply turned its warning signs around and kept on working.

Between the ages of twenty-two and twenty-six, she published five papers on coals and carbons that are still quoted extensively today. Her research helped found the science of high-strength carbon fibers. It proved vitally important for both the old charcoal industry and for nuclear power, which uses graphite to slow the rate of fission. The work earned her a Ph.D. from Cambridge University in physical chemistry in 1945 and made her, at age twenty-six, a recognized authority in industrial chemistry. By today's standards, she was almost unbelievably young to have produced such important research.

Franklin soon realized that she needed to master the developing field of X-ray crystallography in order to understand matter—the material that the universe is made of. Crystallography, a branch of physics, is a powerful technique used to reveal the position of atoms within matter. Traditionally, crystallographers aim X rays at crystalline solids composed of atoms arranged in a regular and repeated pattern. The X rays enter the crystal; many of the rays pass completely through, but others are reflected inside the crystal and exit in different directions to strike photographic film or other types of detectors. (See figures 10.1 and 10.2, on pages 232 and 233.) By studying the intensity and angle of the spots on the film, researchers could figure out the positions of the atoms within a crystal. Crystallography was a British invention and specialty, and many women, like the Nobel Prize–winner Dorothy Hodgkin, achieved early prominence in it.

Franklin was never a traditional crystallographer, however. She never worked with regular, single crystals. Instead, she pioneered the use of X-ray diffraction to study disordered matter like carbons and complicated matter like large biological molecules.

When the war ended in 1945, Franklin wrote Adrienne Weill to ask if she knew any jobs in France for someone who knew a little about physical chemistry and a lot about the holes between carbon molecules. Through Weill, she found a job in Paris at the Laboratoire Central des Services Chimiques de l'État, beginning in 1947.

When Franklin arrived in Paris at the age of twenty-seven, she began the happiest three years of her life. A strikingly good-looking woman, she had clear olive skin, raven black hair, and brilliant eyes that could sparkle with amusement or flash with rage. Slim and quick-moving, she dressed fashionably in an understated European style. Her coworkers were young, many of them Communists from the wartime French Resistance. Together, they lunched in bistros, invited one another for dinners, spent weekends picnicking, and took group vacations climbing mountains, skiing, and camping. At first, she was shocked at such closeness; Cambridge women were not used to males and females sharing hotel rooms.

Speaking French, Franklin seemed to shed her British reserve. "She was a great deal of fun, not a heavy person at all," said Anne Sayre, her biographer and friend in Paris. "I thought she was very young for her age, slightly prankish, teasing. She was older than I was, but I felt like her aunt." Off work, Franklin could sparkle gaily with a slightly teasing, mischievous wit. Although "formal occasions" like banquets made her glum, she loved small dinners. She became an expert on the latest French slang and played elaborate French word games at top speed. She liked gossiping about friends' tangled love affairs and shopping in flea markets, street fairs, and department stores. Her conversation was light, quick, observant, and often amusing. "Gaiety was her note," crystallographer David Sayre remarked.

After the war years cooped up in England, Franklin gloried in being able to travel freely around the Continent. She plotted itineraries with minute precision, correlating maps, guide books, and international timetables to locate the most economical routes through the most mountainous scenery. She loved mountains and outdoor life, strenuous twenty-mile-a-day hikes, and bike tours—no matter what the weather. She could react in surprising ways to travel incidents, though. A friend, Vittorio Luzzati, complained sharply about their hiking in foul weather one day, and her eyes filled with tears. On another hike, Luzzati found an artillery shell from the war and showed it around; Franklin blanched, stiffened, and turned away. Something had happened during the war, he thought, that she could not discuss.

She could be happily married, but she did not want children, she confided to Luzzati's wife, Denise. Franklin loved children too much to hand them over to nannies, and her commitment to science prevented her from being a full-time mother. Nor did she like her parents' upper-class lifestyle. Her flat was simple, and although her family was well-to-do, she was a Socialist.

She worked in a nineteenth-century French army explosives laboratory that was flooded with light, coated with dust, and stuffed with old brassworks. While working, Franklin was intense, reserved, and private—even austere. She took science seriously and hated to waste time. Although she hated idle chitchat at work, she loved a good science argument, and she and David Sayre argued crystallography hammer and tongs. As far as she was concerned, passionate debates were part of the fun of being a scientist. She could be pitiless and make seminar speakers feel like fools. Coworkers who left the darkroom a shambles made her angry. She was unmarried in her late twenties, though, and the French in the 1950s dismissed her foibles as spinsterish.

By 1950, after three years in France, Franklin realized she had to get down to business. If she wanted a career in England, it was time to go home.

Her timing was excellent. Crystallographers already knew how to determine atomic positions in small, simple, and highly regular crystals. Now they were turning to extremely large and complex arrangements in biological matter. Borrowing techniques from physics, biologists and biochemists were solving one major problem after another.

Physicist John Randall, who invented the key to radar in World War II, formed an interdisciplinary team of physicists, chemists, and biologists to study living cells at King's College in the University of London. The team knew that DNA (deoxyribonucleic acid, to be precise) carries genetic information from one generation to another. It was also known that atoms of many proteins are shaped like a helix, that is, like a spiral staircase or an extended coil of springs. But no one understood DNA's structure or dreamed that it would explain heredity.

At King's, a graduate student named Raymond Gosling was taking X-ray photographs of DNA molecules. His photos were the best yet taken, but Randall decided an expert should analyze them. Randall went headhunting, heard about Rosalind Franklin, found her a fellowship, and hired her. Writing to Franklin to explain her new job, Randall made it clear that she would be working alone on a new topic, not the subject they had discussed earlier: "After very careful consideration and discussion with the senior people concerned, it now seems that it would be a good deal more important for you to investigate the structure of certain biological fibres in which we are interested.... This means that as far as the experimental X-ray effort is concerned, there will be at the moment only yourself and [the graduate student] Gosling, together with the temporary assistance of a graduate from Syracuse, Mrs. Heller."

Franklin arrived for her first day of work at King's College in 1951 and walked straight into a meeting fraught with consequences for her future. Randall's second in command, Maurice Wilkins, was away for a short holiday. Wilkins had been Randall's graduate student before World War II, had worked on the atomic bomb during the war, and was to play a crucial and controversial role in Franklin's life at the laboratory. In Wilkins's absence, Randall attended the meeting. And Randall turned DNA and Gosling over to Franklin "lock, stock, and barrel." No one in the lab had worked on DNA for several months, and Franklin assumed she was in charge. When Wilkins returned, on the other hand, he supposed she had been hired as a high-class technical assistant to supply the team with experimental data for it to analyze.

Gosling was caught in the middle. "I don't think Wilkins ever imagined that giving a problem to Rosalind meant that nobody else was going to work on it. The lab wasn't built like that, but Rosalind was built like that." Later, Gosling wondered if anyone had bothered to explain to Franklin the department's hierarchical command structure, Wilkins's position as its linchpin, or the fact that team members worked and published together. Certainly, King's was the only place where Franklin had much difficulty working with colleagues.

Although Gosling had been handed from one thesis advisor to another like a bale of hay, he quickly decided that Franklin was "terrific." She had a strong personality that people either loved or hated. She had strong opinions and high principles and did not compromise; if something was worth discussing, it was worth defending. She was dedicated, but she had a good sense of humor. As she told a friend, "What's the point of doing all this work, if you don't get some fun out of it?" Once, while trying to understand how radiation penetrates the skin of a sphere, she and Gosling peeled oranges. Giving up in frustration, they had a glorious orange fight, hurling fruit at each other across the lab.

Franklin needed a research partner so she could toss ideas around instead of oranges. As Francis Crick pointed out later, "It is one of the requirements for collaboration of this sort that you must be perfectly candid, one might almost say rude, to the person you are working with. It is useless working with somebody who is either much too junior than yourself, or much too senior, because then politeness creeps in and this is the end of all good collaboration in science." Gosling was too young and inexperienced to be the counterpoint Franklin needed. Unfortunately, there were few other prospects in sight.

Wilkins was the obvious candidate. He was interested in DNA, and he and Franklin got along at first. But Wilkins was "meditative, speculative, markedly indecisive," wrote the historian Horace Judson. He was "shy, passive, indirect...he could respond to vigorous disagreement only by turning aside." Wilkins thought carefully before speaking; Franklin was quick, decisive, and impulsive, and could snap at people. "She scared the wits out of me," Wilkins told colleague Aaron Klug. Between Wilkins's shyness and Franklin's lack of small talk, their meetings consisted mostly of staring at each other. Only later did their relationship deteriorate into antipathy and what Judson called "one of the great personal quarrels in the history of science."

At lunch, Franklin discovered that King's College was considerably more formal than Paris. A number of women scientists worked on the staff, but they were not allowed to eat with the men in the men's common room; women ate outside the lab or in the students' cafeteria. After work, the men visited a male-only bar for beer and shoptalk; the women were not invited. As a result, the men talked science casually among friends while the women operated in a more formal office atmosphere. Later, Franklin concluded that King's was also cool to foreigners and Jews.

Shut off from casual friendships at King's, Franklin developed a social life that was almost completely independent from the laboratory. She reserved her evenings and weekends for the theater and films, volunteer work for the Labour party, friends from St. Paul's School and Cambridge, weekly visits with her family, and trips to the countryside. At home, she gave small dinner parties where she introduced British friends to French delicacies like artichokes, new potatoes cooked in butter, and good wine. On holidays she snorkeled in Corsica, climbed mountains in the Alps and Yugoslavia, and toured Israel and Europe, snapping pictures wherever she went.

At the lab, Franklin and Gosling worked alone, collecting enough data about DNA to write five papers. Between her articles, reports, and the lab notes she kept in little red exercise books—all of which passed at her death to her friend and colleague Aaron Klug —it is possible to follow her work over the next few years.

First, she adjusted her X-ray camera to get a needle-fine beam. Then she worked on her DNA sample. Purified DNA looks like the fibrous lint from an old handkerchief, but no one had ever deciphered the molecular structure of such complex fibrous matter. Previous experimenters had tried, using thick DNA fibers. Franklin, who had already worked with amorphous coals and clay structures, knew how to deal with materials that were not fully crystalline. So she was able to invent a new and better method of aligning DNA's lintlike fibers.

With a glass rod, she pulled thinner fibers than had ever been made before and laid them parallel. Since a single fiber was too fine to scatter an X-ray beam, she bundled the gossamer threads together for bulk. Then she matched the optics of the X-ray beam to the diameter of the fibers to get a clearer picture. Finally, she studied how the fibers behaved in a humid atmosphere. Standing them over a closed container of saltwater, she measured the moisture concentration in the air and correlated it to the behavior of the fibers. As a physical chemist, she realized that humidity control was one of the keys to getting a clear picture. Water molecules, filling the spaces between the atoms of a crystal, hold the crystal erect and stable.

Soon, she could show that DNA molecules exist in two forms, A and B, depending on how much water they absorb. When the air surrounding the fibers reached a relative humidity of 75 percent, her X-ray photographs resembled the best pictures that Gosling had taken before her arrival. She called these photographs the dry, A-form of DNA. When the humidity rose to around 95 percent, the

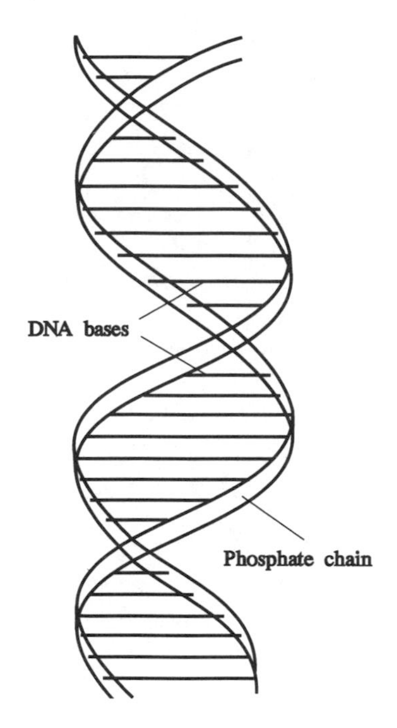

Figure 13.1. DNA.
The DNA molecule is shaped like a helix with base steps and outside phosphate chains going up and down.

molecules stretched 25 percent longer and actually popped off their stand. X rays scattered by these wet fibers produced fewer details on the photographic film. Instead, they made a simple cross shape, the characteristic sign of a helix, as Franklin knew. The cross indicated that a wet DNA molecule was shaped like a helix. She referred to the cross as the wet, B-form of DNA.

Franklin could actually make the DNA molecule shift from one form to another by changing the moisture of the air around the fibers. In only one year at King's College, Franklin had transformed the study of DNA. Her discovery that DNA existed in two forms gave her a big advantage. Other researchers worked with samples that were, unknown to them, mixtures of the two structures.

Because the molecule could absorb and give off water from the surrounding air so easily, she also deduced the location of the phosphate sugars known to be in DNA. In a stroke of intuition, she concluded that they are located on the outside of the molecule close to the surrounding water. The bases, tucked inside the helix away from the water, march up the helix like so many steps in a staircase. She was right in both respects.

She had discovered the first of four crucial points about the arrangement of the DNA molecule. Three more pieces of the puzzle had to be deciphered. She still had to learn that the molecule is shaped like a helix composed of two phosphate strands wound together; that the two strands are oriented in opposite directions so that one marches up and the other down; and that each base step is composed of a pair of two particular bases. Whoever grasped all four points about DNA could explain heredity. But neither Franklin nor anyone else knew that.

As Franklin began producing data, Wilkins became anxious to interpret it. "How dare you interpret my data for me?" she snapped. She thought Randall's and Wilkins's attitude toward her was, "Thank you for the pretty pictures. We'll analyze them." Wilkins told her that the simpler, crosslike B-picture indicated a helical structure, but she objected to his jumping to conclusions. Although she was proceeding on the assumption that both forms of DNA were helical, she wanted hard evidence to prove the point, not supposition. During the fall of 1951, she had a terrific argument with Wilkins that nearly resulted in her return to Paris. Afterward, they agreed to differ.

In November of 1951, Franklin gave a colloquium talk at King's College on her work to date. Staring at her from the rear of the room was a strange, skinny broomstick of a fellow with pop-eyes and wild hair. It was a young Midwestern geneticist, James Watson, who was working on DNA at Cambridge University with an English graduate student, Francis Crick. At this point, Franklin knew far more about the structure of DNA than either Watson or Crick. Watson could have learned a good deal from her talk, but he prided himself on not taking notes at lectures and he was so busy analyzing Franklin's physical appearance that he remembered her data incorrectly.

Later, in a bestseller that ridiculed Franklin's personality and scientific talent, Watson critiqued her lecture like a beauty contest. "There was not a trace of warmth or frivolity in her words," he complained. Franklin never wore glasses, but in his imagination, he put them on her and wondered "how she would look if she took off her glasses and did something novel with her hair." Watson thought the audience was afraid of "Rosy...Maurice's assistant," as he called her. "To be told by a woman to refrain from venturing an opinion about a subject for which you were not trained...was a sure way of bringing back unpleasant memories of lower school." Although Franklin's career was largely free of the legal and bureaucratic discrimination that hampered many other women scientists, more subtle problems did impede her progress. Watson's condescension, for example, immediately excluded him as a possible collaborator.

Thanks to his garbled version of Franklin's data, Watson and Crick produced a model of the DNA molecule and called in their friends to admire it. When Franklin saw it, she instantly spotted their mistakes and told them so. "Most annoyingly, her objections were not mere perversity," Watson admitted. "At this stage the embarrassing fact came out that my recollections of the water content of Rosy's DNA samples could not be right." Sir Lawrence Bragg called the model "the biggest fiasco" he had ever been associated with and forbade Watson and Crick from working on DNA. Horning in on a King's College project was not good sportsmanship—especially when you were wrong. So far, Franklin was still ahead of Crick and Watson. The next time they built a model based on her experiments, though, they might get her facts right.

By spring of 1952, Franklin was the only person working on DNA full time. During the previous eighteen months, she had also made the only significant progress toward solving the problem. That May, Franklin left her X-ray beam focused on extra-wet DNA fibers for an extra long time. After sixty-two hours of exposure, she had a magnificently vivid photograph of DNA, the simple cross reproduced in so many biology textbooks. Its cruciform pattern clearly originated from a helix-shaped molecule. The picture is regarded as one of the most beautiful X-ray photographs ever taken. But Franklin put it away in a drawer. She continued analyzing the A-picture of dry fibers. It had more details and promised to deliver more facts.

That spring, the United States State Department refused to issue a passport to the eminent American chemist Linus Pauling. He had been invited to speak at a protein conference in London in May 1952,

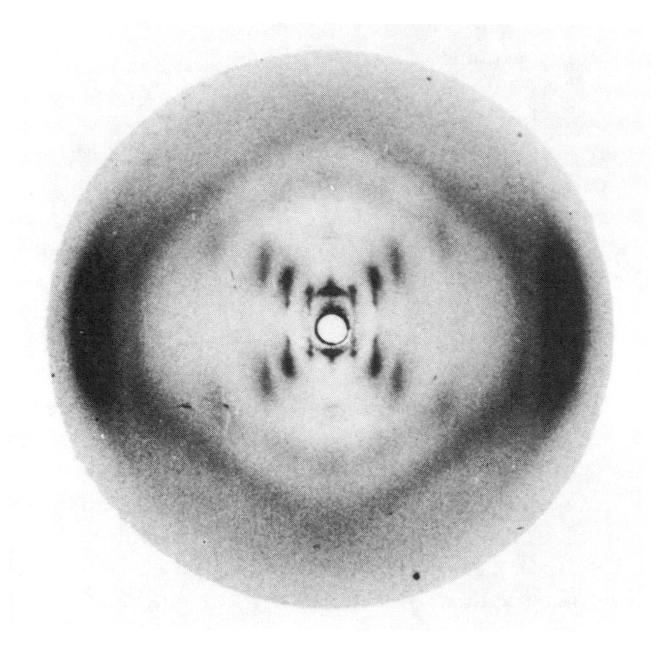

Rosalind Franklin's X-ray photograph of DNA.

but a congressional witness had accused him of being a Communist. Pauling denied the charge, but in the anti-Communist hysteria of the postwar era, he lost his chance to attend the conference. As Pauling realized later, the government's travel ban had prevented him from seeing Franklin's data and her X-ray photographs. Had he done so, Franklin and Pauling might have discovered the structure of DNA together, before Watson and Crick. If so, Pauling might have won three Nobel Prizes instead of two. This was the second time that Franklin had lost an opportunity to acquire a collaborator. She was still analyzing her data alone. As the historian Judson observed, "The situation [in Cambridge] was exactly the reverse: volatile collaborative enthusiasm and no data."

Meanwhile, at King's College, the split between Wilkins and Franklin was widening rapidly. Although both discussed their work with others in the lab, they rarely spoke to each other. Wilkins had begun duplicating Franklin's data as best he could, but her DNA sample and her technique may have been superior; for whatever reason, his photographs produced much less information than hers. He wrote Crick, "Franklin barks often but doesn't succeed in biting me. Since I reorganized my time so that I can concentrate on the job, she no longer gets under my skin."

Wilkins had been confiding his problems to Watson and Crick but rejecting their suggestions to build toylike models based on what he knew and could guess about the structure of DNA. Franklin, like the rest of the researchers of King's College, thought model building was pointless. "We all felt that you could build models 'till the cows came home,' but how could you tell which one was right? And why bother when you had the [X-ray] spots? It was a down-to-earth attitude," Gosling remembered.

On the advice of her friend Luzzati, Franklin began the complex mathematical calculations that crystallographers use to solve simple crystals. No one had ever used these so-called Patterson calculations to solve a fiber structure, however. The job was immensely complicated—much too complicated, as it turned out. She had begun to doubt whether the dry, A-form of DNA was helical. Ironically, Wilkins shared her doubts and actually published them in an agency report. Franklin did not publish her doubts; she had to be sure of her data before she committed herself.

Franklin joked about her hesitation that summer. She and Gosling hand-lettered a black-bordered funeral announcement for the dry, crystalline A-form of DNA: "It is with great regret that we have to announce the death on Friday 18th July 1952 of D.N.A. Helix (crystalline). Death followed a protracted illness." The "Death of the A-Helix" was a joke—but it had some steel in it.

Franklin was bogged down in mathematical calculations through the winter of 1952–1953. She still wanted to know whether the A-form, like the wet B-form, was helical. The question was quite legitimate at the time. "It's one of the quirks of perception in science that helical symmetry is now very obvious for any structure that is periodic and fibrous. But the idea was really very novel in the early 1950s," emphasized her friend, Donald L. D. Caspar, now a physics professor at Brandeis University. Then she misread another piece of data, and her doubts about the helix-shape of her dry A-form were reinforced. So she lavished time and energy on the A-form, when she could have been reaping the fruits of the wet B-form.

If she had asked Crick's advice, the two of them might have solved the structure together in a few months. At one point, Franklin and Crick actually met in line for tea, and Crick tried to give her some off-the-cuff advice. But Crick had a reputation as an eccentric, a flashy theoretician, and Franklin ignored him. Crick admitted, "I'm afraid we always used to adopt—let's say, a patronizing attitude towards her." Once again, Franklin had missed a chance to acquire a research partner.

Suddenly, early in 1953, the balance of power shifted from Franklin to Crick and Watson. In two fell swoops, facts that Franklin had painstakingly developed were given—without her knowledge or permission—to Watson and Crick in Cambridge. For the first time in two years, they would know more about DNA than she did. The race was quickening without Franklin's even knowing it.

Linus Pauling precipitated the race by writing a draft article about DNA and sending it to his son in Cambridge to critique. Peter Pauling promptly gave it to Watson, who took it down to King's College to show Franklin on January 30, 1953. She coolly pointed out that there was not a shred of evidence to prove that DNA was a helix. And she was right—she had the evidence and was not finished analyzing it. She must have been annoyed: Franklin, the woman with no time at work for idle chitchat, interrupted by Watson, the brash American who had no facts but who considered small talk "the essence of getting along with people." She had written Pauling's lab for information and received no reply, yet here was Watson waving a copy of Pauling's manuscript. She could tell at a glance that it was wrong; it was based on five-year-old photographs of DNA made long before she discovered its two different forms.

Watson transformed this scene into the dramatic climax of his bestseller, *The Double Helix.* In the book, Watson lectured her on helical theory and "implied that she was incompetent in interpreting X-ray pictures." When Franklin moved toward him, he became "fearful that in her hot anger she might strike me." Those who know

Franklin believe she must have been annoyed. But they also regard Watson's scene as a clever dramatic device bearing little relationship to reality. Watson was twenty-five years old and well over six feet tall; Franklin, at thirty-two, was five feet six inches tall. The idea of Franklin's physically attacking Watson seems more farce than fact.

Racing out of Franklin's lab, Watson met Wilkins in the hall. Commiserating, Wilkins went into the next room and grabbed a copy of Franklin's spectacular B-form cross. Without asking Franklin's permission or telling her what he was doing, he showed it to Watson. "Look, there's the helix, and that damned woman just won't see it," Wilkins complained.

The instant Watson saw the picture, his mouth fell open and his pulse began to race. He and Crick had been working with five-year-old photos of DNA and had had no inkling of its two forms, wet and dry. The picture told him the basic dimensions of the helix.

As Wilkins complained later, "They could not have gone on to their model, their correct model, without the data developed here. They had that—I blame myself. I was naive—and they moved ahead....We were scooped, I don't think quite fairly." Defending himself in 1992, Watson countered, "I didn't feel guilt. The picture was old. I'm sure Maurice wouldn't have shown it to me if it had been only two weeks old."

With the race heating up, Sir Lawrence Bragg lifted his injunction against Cambridge DNA work. If the American Pauling was working on DNA, Cambridge owed it to Britain to enter the fray. As a result, Watson and Crick were hard at work. Watson was coordinating information from friends and colleagues, weaving their data together as surely as if they were all part of an interdisciplinary team.

Once Watson had won the Nobel Prize, he stopped doing research and became an administrator and textbook writer. As he explained his role in DNA, he said, "Except for my writing, all my work has been getting other people to help me. If I have to use someone else to get the answer, I'll do it....The most important thing in science is getting the answer, not showing that you've done it yourself....It helps you doing science if you're very social." Thanks to Watson's attitude, Franklin was competing against Watson and Crick *and* all the experts Watson queried.

The second transfer of data from Franklin to Watson and Crick involved a government report. In it, Franklin had summarized the data that she had reviewed in her November 1951 colloquium talk—the lecture that Watson had remembered incorrectly. Her report was distributed to members of the agency's review committee in December. Max Perutz, a young crystallographer who headed a research unit at Cambridge, passed the report to Watson and Crick in early

February. Whether or not the report was intended to be confidential, normal etiquette would have been to ask her permission before handing her data around. Perutz apologized later, saying "I realized later that, as a matter of courtesy, I should have asked Randall for permission to show it to Watson and Crick, but in 1953 I was inexperienced and casual in administrative matters and, since the report was not confidential, I saw no reason for withholding it."

Thanks to the agency report, Crick and Watson finally had Franklin's colloquium facts accurately. They knew the water content of the fibers and the placement of the phosphate sugars on the outside of the helix. Even more important, however, Crick's reading of Franklin's report told him something that neither Franklin nor Watson knew. Elements in Franklin's data resembled horse hemoglobin crystals, which Crick had studied while writing his Ph.D. thesis. Thanks to his traditional crystallographic training and the equine connection, Crick realized that one of the outside chains of the DNA molecule must go up and the other down. That way, the molecule looks the same when it is turned upside down. Franklin was still struggling to grasp that point as she moved back and forth between the A and B forms. At this juncture, Watson and Crick finally had more data than Franklin.

Franklin brought her beautiful photograph of the wet B-form out of her drawer that same week. Starting on February 10, she began analyzing it and building models to help visualize her mathematical calculations. Sketching the A-model first, she almost figured out the key concept that Crick had already discovered: that the outside chains march up and down the outside of the molecule. In her lab book, she drew the dry A-form as a figure eight: one chain up and the other chain down. At this point she was not thinking in terms of a helix for the A-form, although the spiral S shape virtually assumes a helix, Klug noted.

Looking through the little red exercise books after her death, Klug broke away again. "Oh, it's awful. I can't bear to look at it. She's finally making the right connections between A and B. She's shuttling back and forth between the two things.... It's awful to see it." As Franklin confided later, "I could have kicked myself for not noticing it."

Franklin was still in the running, though. In fact, regarding the most important concept of all about DNA, both groups were on an equal footing. Neither Franklin nor Watson and Crick had yet discovered base pairing. The helix is visually elegant, but biologically the important point about DNA is the base pairing. It is the code that passes individual characteristics on to succeeding generations.

Crick's memory is that he suggested base-pairing on February 27.

But Watson's book claims that he, and no one else, figured that part out the next day. Using evidence uncovered by biochemist Erwin Chargaff, Watson knew that pairs of bases form the steps of the helical staircase. Building models of the molecule showed him that each step consists of a *particular pair* of bases: adenine with thymine or guanine with cytosine.

To reproduce itself, DNA simply divides in half longitudinally, leaving one outside chain attached to one of the bases; the complementary base is attached to the opposite chain. Finally, each chain makes its complement and recombines. This incredibly simple mechanism explains how genetic information can pass from generation to generation for thousands of years without change. Triumphantly, Watson and Crick showed their model to colleagues and wrote their friends. Strangely, neither told Franklin or Wilkins about it, despite the help they had received from Franklin's data.

Working on the B-form photograph in February, Franklin broke through the impasse that had blocked her for nine months. By February 23, she knew for sure that the wet B-form is helical and that its helix is made of two, not three, chains. Counting her deduction about the location of the phosphate chains on the outside of the helix, she now had two of the four vital points about DNA. She had not yet recognized the remaining two concepts: the one side down/other side up chains and the base pairing. Nevertheless, at the beginning of March, she and Gosling wrote a paper summarizing what they knew about the beautiful B-form photograph.

By the time Franklin got her manuscript typed, it was March 17, 1953. The next day, a *Nature* magazine editor called. Watson and Crick had solved the structure of DNA. They had submitted an article on March 6. The editor thought Wilkins and Franklin might like to contribute articles to accompany theirs. Hastily, Franklin revised her manuscript slightly to support Watson and Crick's hypothesis. They had won the race—even before Franklin knew they were competing.

Nature rushed the Watson and Crick article into print faster than it had published anything before. The article is scarcely one thousand words long, a mere one page. It offers a hypothesis without proofs. It cites no authorities or historical record. Nor does it credit the scientists on whose shoulders it was built. Crick and Watson could have published their theory jointly with Franklin. Instead, they merely thanked physical chemist Jerry Donohue for "constant advice and criticism." Then, in the next-to-last sentence they add ambiguously, "We have also been stimulated by a knowledge of the general nature of the unpublished experimental results and ideas of Dr. M.

H. F. Wilkins, Dr. R. E. Franklin, and their co-workers at King's College, London."

Could Franklin have solved the structure of DNA on her own? Her friends and supporters have debated that ever since. Franklin has become a patron saint of feminists, and for them the answer is clear. "Had Franklin not had her work secretly taken from her and had she thus been allowed enough time to use her data to solve her puzzle, there is hardly any doubt that she would have unraveled the helix—perhaps even before Crick and Watson. For, after all, Watson and Crick would then have had to have made their own unequivocal photographs of the DNA helix. This they had not succeeded in doing," charged G. Kass-Simon in *Women of Science: Righting the Record.*

Franklin's colleague Aaron Klug thinks she was only one and a half steps away from solving DNA on her own and that she would have done so eventually: one-half step for the opposite direction of the chains and a whole step for the base pairs. Klug's opinion is not to be dismissed lightly. He was Franklin's closest collaborator and friend for four years at Birkbeck. He won a Nobel Prize for chemistry in 1982 and directs one of the world's leading molecular biology centers, the Medical Research Council Laboratory of Molecular Biology in Cambridge. Moreover, he has studied her papers and notebooks extensively, more closely than anyone else alive.

"It is rather heartbreaking to look at these notebooks and to see how close she had come to the solution by herself," Klug observed. "Crick and I argue whether she was one and a half or two steps behind. She had two things to do: She didn't know that the chains ran in opposite directions; I maintain she was almost at the point of spotting that."

"The other thing was, how do you put the bases in? She knew they had to be on the inside; and she had talked about base interchangeability. The step from base interchangeability to base pairing is a quite long one but she was poised to make it," Klug asserted. "If you steep yourself in the notebooks as I have, you get the pace.... She didn't need intuition. She had facts. She wasn't highly imaginative like Crick or Pauling, but she was a superb experimentalist, a good analyst, and she'd have done it her own way."

Watson cites a curious reason for Franklin's failure to win the DNA race. In Watson's view, Franklin lost because she was interested in the steps more than the goal, she wanted to analyze the material herself without help, and she had "no patron, no one who cared for her." According to Watson, Lord Victor Rothschild, who then chaired the Agricultural Research Council, should have helped her. Why

Rothschild? Because he was a Rothschild and she was a Franklin related to the Samuels, an important Jewish family in England. "She came from one of the most prominent Jewish families in Great Britain. We had no idea who Rosalind was," Watson protested. "She just made life difficult for Wilkins."

By that time, Franklin was so disenchanted with King's College that she had decided to leave. She asked John Desmond Bernal if she could join his group at Birkbeck College, the graduate night school of the University of London. Bernal agreed, provided that Randall let her bring her fellowship. Randall and Bernal made a gentlemanly deal: Franklin could leave and become head of her own, larger research group—a considerable promotion—but she could not work on nucleic acids. Prohibiting a scientist to think about a problem she had been working on for years is unimaginable today. But at the time, British scientists were accustomed to dividing up the research world and allocating different projects to particular laboratories. Randall was clearing the decks for Wilkins to pick up his work on DNA, free of any competition from Franklin. She was not even supposed to help Gosling finish his Ph.D. thesis. Moving to Birkbeck in mid-March, Franklin ignored Randall and quietly helped Gosling get his degree.

At Birkbeck College, Franklin settled into part of two ramshackle townhouses at 21 and 22 Torrington Square. The buildings on either side had been bombed out during the war, and the 120-year-old remains needed major repairs. Her first office, under the roof, was artfully decorated with pots and pans to catch the leaks when it rained. Each evening, she opened an umbrella and placed it carefully over her desk to protect her papers overnight. Later, she moved to a downstairs office.

Despite the raindrops, Franklin finished her coal and DNA studies there. She produced two papers crammed with DNA information that, in some respects, scientists are just catching up with. By forcing Franklin's move to Birkbeck, Randall had actually done her a favor. Over the next five years, she published seventeen articles on viruses. She established a reputation as the world's finest experimentalist for dealing with helical structures. Bernal considered her one of the "major founders of biomolecular science." Sir Lawrence Bragg said he had not believed it was possible to discover as much about viruses as she did.

Franklin led a four-person research team at Birkbeck. Besides Klug, she had two graduate students: Kenneth C. Holmes, now professor at the Max Planck Institute for Medical Research in Heidelberg, and John T. Finch, now at the MRC Laboratory of Molecular Biology at Cambridge. Klug joined the group after he met

Franklin on the stairs, heard about her work, and switched research topics. He became her first and only collaborator. Known as an extraordinary theoretician, he enjoyed debating with Franklin. Together they developed "marvelously delicate techniques for securing new and beautiful X-ray data," recalled crystallographer Dorothy Hodgkin.

Franklin proved to be a commanding leader with presence and even an aura of authority about her, recalled Finch. "She knew what she wanted to do scientifically, and she knew experimentally how to get there."

She was also deeply devoted to scientific research at a very high level of performance and could be single-minded and fierce. "She could be very pleasant, and she had a sense of fun. But in the lab, she was actually quite tough. She could snap at people," Aaron Klug recalled. "It would have gone quite unremarked if she had been a man. But she stood up for things. She was rather persistent. She wasn't the saintly, nunnish figure portrayed in the BBC film *Road to the Double Helix.*"

Brandeis University professor Donald Caspar, who also worked with Franklin on viruses, remembers, "The most negative things I can think about her are still admirable qualities....She wouldn't put up with nonsense. She was a very vital human being who didn't indulge in speculation."

Reminiscing years later, Holmes said, "She had charisma. She was a fascinating, very attractive woman, and she affected all of us in a very deep way. Her friends and students have great difficulty thinking about her because it's so painful."

Had Franklin been a man in charge of the research team, she might have been called the "strong silent type." She did not suffer fools graciously. She was not soft and gentle like Hodgkin, and she did not approve of her research assistants getting distracted by romance or hobbies. At this stage in her career, she strove for results.

And as always, she resented boring social functions. At dinner in Birkbeck's common room one evening, she sat silent all through the meal as others chatted. Franklin was a bit choosy about where she put her effort and, if she did not think the occasion was important or interesting, she did not try. But late in the dinner, she decided to make a contribution. Thus, when there was a lull in the small talk, she pronounced loudly: "It's a good year for mushrooms." And that was all she said. Unaccustomed to seeing her socially, her students decided that her only flaw was an incapacity for small talk.

Otherwise, she was a good mentor. "Go on, you do the first draft," she told Holmes when they wrote their first paper. Then she turned his totally inadequate draft into a good paper. "That was one of the nicest aspects about her," Holmes recalled. "She didn't control you on a day-to-day basis."

Having studied DNA at King's College, Franklin began working on RNA (ribonucleic acid) at Birkbeck. She decided to work on viruses that are composed of both RNA and protein, RNA being the infective part of the viruses. By understanding RNA's structure, she hoped to explain how a virus particle, which is not in the full sense alive, can grow and reproduce in other cells. In Franklin's five years at Birkbeck, her group outlined the general molecular structure of several RNA-containing viruses and helped lay the foundation of structural virology. At the time, her group was the world's leader in using X-ray diffraction to uncover the molecular structure of viruses.

Like other virus researchers, she concentrated on tobacco mosaic virus (TMV). TMV was to viruses what corn and fruit flies were to genetics—the model used to establish basic scientific principles. TMV is stable, easy to handle, and abundant. She particularly liked the way TMV's long, rod-shaped particles produced detailed X-ray diffraction patterns with a wealth of information about molecular structure. She was intrigued with TMV for two other reasons too. She was convinced—correctly—that structural studies of TMV would help scientists understand the organization of other regular virus particles, including the polio virus and the common cold virus. Second, TMV's fibrous structure was even more technically challenging than DNA. Franklin's DNA research had made her the world's expert in fiber diffraction, so she was intrigued by all the difficulties involved.

Watson had hypothesized that TMV is constructed in a helix, but a different type of helix than DNA. Franklin swiftly confirmed his conclusion. But when she measured the helix, she discovered that he had underestimated the number of small protein subunits that form each turn of the helix. She also located the long single strand of RNA—the carrier of the virus's genetic information and hence the source of its infectivity. She showed that it exists—not in the helix's central cavity—but buried deep between the subunits of the virus's protein coat. For the first time, it was possible to understand the structural relationship between protein and a nucleic acid and how they fit together.

Watson and Crick actually met occasionally and exchanged information with Franklin and Klug on the virus structure project. Seeing them, Franklin was cheerful and ebullient and there was no sign of animosity among them. Franklin had great respect for Crick's ability. She became close friends with Crick and his French wife, Odile, and together they traveled through southern Spain one summer. She was more reserved about Watson, referring to him as "the horrible American."

Franklin had a rousing argument with the director of her funding agency in 1956. Storming back from the meeting, her eyes filling with tears of rage, she complained angrily that "the ARC refuses to support any project that has a woman directing it." Fortunately, friends helped arrange for a three-year grant from the U.S. Public Health Service to continue her work at Birkbeck.

During 1956, Franklin reported on her results at conferences in London, Madrid, and New England, and visited labs in Berkeley, Los Angeles, Pasadena, St. Louis, and New Haven. At Berkeley, where Franklin worked for a month with the Nobel Prize–winner Wendell Stanley, she had trouble getting a ride to a lab picnic. Watson's stories about "Rosy," the temperamental bluestocking, had preceded her. Afraid to tangle with her, the young people in the lab wiggled out of giving her a ride. So Stanley himself drove her. At the picnic, the students discovered that Franklin was actually lively and fun, and they were forced to revise their opinions of her. Later that summer, when she climbed in the Rocky Mountains with other Americans, she befriended them as well. Nevertheless, Franklin was still an outsider in the scientific establishment.

Several episodes of terrifying pain that summer sent her to an American physician, who told her to see a specialist as soon as she got home. The diagnosis was ovarian cancer. Over the next two years, Franklin had three operations and experimental chemotherapy. She was irritated by doctors, nurses, and surgeons who refused to answer her questions. On the other hand, she refused to talk about the illness with her friends or relatives. Only her close family and research group knew much about it.

After her first bout of illness, the cancer went into remission for almost ten months and she resumed tennis and mountain climbing, theater going, and work. At one point, she convalesced with the Cricks; they did not know what her operation had been for or how serious it was, but she felt easier with friends who knew nothing. When Crick suggested that Franklin and her group move to what became the MRC Laboratory of Molecular Biology in Cambridge, her main fear was that she might become a spinster professor like those she had hated as a student. Nevertheless, she decided to move with her group.

By accident, Frederick L. Schaffer's laboratory at the University of California at Berkeley had crystallized some polio virus, the first crystals ever formed from an animal virus. Schaffer's wife agreed to try to take a thermos of the crystals to Franklin for analysis. British customs officers questioned Mrs. Schaffer about the contents of the thermos. Polio caused as much fear and hysteria in the mid-twentieth century as AIDS does today, and the tiny crystals in the thermos were fully infectious. But she replied jauntily, "It's polio virus. But it's all right. It's crystalline." She sounded so trustworthy that customs let her through.

When Franklin got the thermos, she waved it at her mother. "You'll never guess what's in here," she teased. "Live polio!" Then she opened the family refrigerator and slipped the virus in.

Working with an infectious virus in a dilapidated, dirty laboratory without proper safety equipment was dangerous. The Salk vaccine had been available for only three years, and many people had not yet been vaccinated against the disease. Franklin, on the other hand, knew by now that she was dying. Watching the work, Bernal's secretary thought Franklin was a modern Marie Curie. Soon after Franklin's death, the polio work was halted because of the risks.

Franklin's work on small plant viruses attracted such widespread interest that both the august Royal Society of London and the Royal Institution of Great Britain requested material to exhibit. In 1957, the Brussels World's Fair committee asked her to build two models of virus molecules. The request was a great honor. During the 1950s before jet travel and television, world's fairs were glamorous and exciting events. She was the first scientist to know enough about the structure of a virus to build a realistic model. A revolution in biology was just beginning. For fair visitors, her model would be their first glimpse of biology in terms of the molecules that make up all living organisms.

A small plastics company produced dozens of white plastic, shoe-shaped pieces for a six-foot-tall model of TMV. Each shoe represented one molecule of the protein coat surrounding the virus. When they arrived, however, they were all slightly too big. So Franklin, Klug, Finch, and Holmes spent a day filing them down by hand. When they assembled the model, they omitted a few shoes in order to show the single strand of RNA winding around like a bracelet near the hollow section of the molecule.

Watson had thought that TMV would "self-assemble" by replicating protein subunits over and over again at the end of the growing helix. But when Franklin and Klug started putting the model together, they discovered that getting the assembly process started was actually quite difficult. Some kind of special mechanism must be involved, they concluded. Unraveling that process occupied the next decade. Scientists were so interested in the models that the Royal Institution of Great Britain displayed them before they were shipped to Brussels.

Rosalind Franklin's model of the TMV virus molecule which she built for the 1958 World's Fair, on exhibit at the Royal Institution in London before transport to Brussels.

During the last year of her life, Franklin seemed softer and easier to approach. Aware that she was dying, she worked on TMV until a few weeks before her death, putting her data in order. The day after organizing a supper party for her parents' fortieth wedding anniversary, she checked into a hospital for the last time. By her bedside, she kept an invitation from a Venezuelan laboratory to spend a year in Caracas.

On April 16, 1958, within a few minutes of the time that her last scientific paper was due to be read at the Faraday Society, Rosalind Franklin died. She was thirty-seven years old. She had made crucial contributions to one of the most important discoveries of the twentieth century. Her work on two other major biological problems and the techniques for solving them helped lay the foundations of structural molecular biology. And that summer, her virus models went on view before 42 million visitors to the Brussels World's Fair.

* * *

In 1962, four years after Franklin's death, the Nobel Prize for medicine was awarded to Francis Crick, James Watson, and Maurice Wilkins. On the basis of what the three winners said in their Nobel Prize lectures, no one would have known that Franklin had contributed to their triumph. Their three Nobel lectures cite ninety-eight references, none of them Franklin's. Only Wilkins included her in his acknowledgments.

If Franklin had lived, would she have won the Nobel Prize? Most scientists today believe that she deserved it. Nobels are given only to living persons, however, and each prize can have no more than three winners. Would the committee have known about her contributions? And if it had, would the committee have been willing to give a third of the prize to her and not to Wilkins? Or would the committee have awarded two prizes, one in medicine and the other in chemistry, and split them among four winners?

"The Nobel committees have sometimes made quirky awards, omissions and downright mistakes, but we cannot doubt that the value of her work was known," the historian Judson concluded. Everyone in Randall's unit at King's College knew her work; so did Crick. Bragg, a crystallographer, would have understood the importance of her published articles. It was he who insisted that King's College in the person of Wilkins share the Nobel Prize. Furthermore, when the Nobel Committee studied the publications of the four scientists, they would have realized that Franklin's papers contained by far the most hard data. Had she lived, she might well have shared the Nobel Prize for one of the twentieth century's greatest scientific achievements.

Six years after Watson got a Nobel Prize, he wrote *The Double Helix,* a breezy account of his DNA experiences. He catalogued everyone's foibles and idiosyncrasies, from his first sentence ("I have never seen Francis Crick in a modest mood") to Sir Lawrence Bragg ("I quietly concluded that the white-mustached figure of Bragg now spent most of its days sitting in London clubs like the Athenaeum"). The manuscript raised a storm. As drafts were passed around and subjects complained about their treatment, Watson softened and modified some of his portraits—except Rosalind Franklin's. She was dead and could not argue.

In Watson's book, Franklin plays the role of "Rosy," the wicked stepmother. She is both Watson's central rival and the stereotypical old maid who keeps the plot line moving. Besides denigrating her personality, Watson attacked her scientific abilities, accusing her of being categorically "anti-helical" and opposed to model-building. Fortunately, Watson could not let a good story go to waste, so he also related how he and Crick used her data from the funding agency report and her X-ray diffraction photograph of DNA.

In response to complaints from Franklin's friends, Watson added an epilogue to the book. It stated that his "initial impressions of her, both scientific and personal...were often wrong." But he did not change his portrayal of her in the book. His fictionalized stereotype of a woman who has abandoned her femininity for science made the book more readable and exciting.

Some scientists praised its lighthearted rendition of scientific research. Others were outraged. "He has carelessly robbed Rosalind of her personality," Anne Sayre protested. Dismissing Watson as a case of retarded emotional development, the Nobel Prize–winner André Lwoff charged that "his portrait of Rosalind Franklin is cruel....At the very least, the fact that all the work of Watson and Crick starts with Rosalind Franklin's X-ray pictures and that Jim has exploited Rosalind's results should have inclined him to indulgence." Robert L. Sinsheimer complained that the book is "unbelievably mean in spirit, filled with the distorted and cruel perceptions of childish insecurity." "It was a mean, mean book," observed Nobel Prize–winner Barbara McClintock. Watson is an excellent writer but arrogant and a well-known antifeminist, commented Nobel Prize–winner Rita Levi-Montalcini. David Sayre believed that Watson's book lowered the moral tone of scientific research by glorifying "the big grab for credit." To restore Franklin's scientific reputation, Klug wrote two papers outlining her DNA contributions for *Nature* magazine in 1968 and 1974.

The controversy continues to this day. As late as 1989, Anthony Serafini's biography of Linus Pauling stated, "There are so many actual and possible degrees of unethical behavior that it is difficult to draw the line. Sometimes, of course, the case is clear, as when James Watson made use of Rosalind Franklin's data without crediting her in the famed DNA race....Certainly Watson and Crick would not have gotten the Nobel Prize had they not stolen her data."

Despite such criticism, Watson asserted early in 1992 that if he were writing the book again today, he would write it the same way. "Because that was the way it happened. I told it like it was. But you get into trouble when you tell it like it happened."

In the short run, the book enhanced Watson's reputation as a brash and brilliant young scientist on the move. In the long run, it contained a time bomb. His admission that he had used Franklin's data without her knowledge has tarnished not only his brilliant achievement but Crick's as well. And his fictionalized portrayal of her personality and scientific achievements has made her the martyred saint of feminists and women scientists. The oddest element of the entire story is that it was Watson himself who brought the facts of her contributions to light. He cast a shadow on his own achievement and shone the sun on hers.

EPILOGUE

In January 1992 the English Heritage society placed a historical marker outside Franklin's apartment at 22 Donovan Court, Drayton Gardens, in the Kensington neighborhood of London. The plaque is inscribed: "Rosalind Franklin, 1920–1958, pioneer of the study of molecular structures including DNA, lived here 1951–1958."

* * *

14

Rosalyn Sussman Yalow

July 19, 1921–

MEDICAL PHYSICIST

Nobel Prize in Medicine or Physiology 1977

When Rosalyn Sussman Yalow's brother was a first-grader, his teacher smacked his hand with a ruler. He promptly burst into tears and threw up. Five years later, when Rosalyn entered first grade, the same teacher hit Rosalyn with a ruler. Rosalyn struck back. Marched to the principal's office for questioning, Rosalyn explained that she had been waiting for years to avenge her older brother.

Amused and proud, her parents encouraged Yalow's combative spirit. They staged a triumphant photograph in the park: Rosalyn, a tiny five-year-old wearing enormous man-sized boxing gloves, looms over her brother. He lies sprawled on his back, looking as if she had knocked him out in a ferocious fight. The snapshot has become crinkled and faded with time, but Yalow keeps it handy in her desk. Smoothing the photo, she says, "That's the attitude that made it possible for me to go into physics."

This is the Yalow—part protective Earth Mother, part Aggressive Warrior—who helped invent a technique so sensitive that it can detect a teaspoonful of sugar in a body of water sixty-two miles long, sixty-two miles wide, and thirty feet deep. Thanks to their daring, she and her scientific partner, Solomon A. Berson, developed the radioimmunoassay (RIA) procedure. Thanks to their determination, they convinced the scientific community of its value. In 1977, after Berson's death, Yalow won the Nobel Prize for their discovery. She was the first American-born woman to win a Nobel Prize in science.

RIA revolutionized endocrinology—the study of ductless glands and hormones—and the treatment of hormonal disorders like diabetes. For the first time, doctors could diagnose conditions caused by minute changes in hormones. Thanks to Yalow and Berson, dwarfed children can be treated with human growth hormones; newborns are tested to prevent retardation caused by underactive thyroids; blood banks are screened for deadly diseases; infertile couples are tested for insufficient sex hormones; fetuses are checked for serious deformities like spina bifida; athletes are tested for drug abuse and crime victims for poisons, and on and on. Yalow and Berson's work—a spectacular combination of immunology, isotope research, mathematics, and physics—also started a new science, neuroendocrinology.

Yalow's scientific virtuosity was self-made. Neither of her parents had set foot in a high school, much less a university. Her mother, brought from Germany to the United States as a small child, quit school after sixth grade. Her father was born on the Lower East Side of New York City, the melting pot for Eastern European immigrants. He quit school after the eighth grade and became a streetcar conductor before opening a one-man paper and twine business. Rosalyn was born on July 19, 1921, in the South Bronx—"the part that's now a disaster area." Except for three and a half years spent at the University of Illinois, she has lived her entire life in New York City.

Few adults in her Jewish neighborhood had attended more than elementary school. Nevertheless, as "people of the book," they valued learning. "It was also characteristic of the times that the people were self-educated," Yalow explained. Her father, Simon Sussman, read the *New York Times* and kept the financial records for his business in beautiful Spencerian handwriting with each rounded letter parallel and slanting to the right. His wife, Clara Zipper Sussman, read every school book her children brought into the house. Rosalyn herself learned to read before kindergarten. By the time she was five years old, she and her brother were making weekly trips to the local public library. "I remember the rules for joining the library. You had to be able to read a statement and sign your name. Every one of my friends aspired to do this as early as possible. By the time we were five or six, most of us joined the library."

Yalow believes she was born to be independent. "Perhaps the earliest memories I have are of being a stubborn, determined child," Yalow explained. "Through the years my mother has told me that I was fortunate that I chose to do acceptable things, for if I had chosen otherwise, no one could have deflected me from my path." When she was three years old, her mother took her to see a movie and afterwards stopped off at an egg store. When she came out, Rosalyn wanted to go home one way, and her mother wanted to go another. Rosalyn promptly sat on the pavement. Unable to carry Rosalyn without breaking the eggs, her mother gave in. Yalow grins as she

Rosalyn S. Yalow, 1977.

tells the story. It was probably the last time that she dealt with a problem by sitting it out.

"She was so independent, but she was nice about it," her mother recalled. "I let her do what she wanted—as long as I thought it was right." At age ten, Rosalyn enrolled in an all-girls junior high school. To make sure that Rosalyn got off at the right trolley stop, her mother followed her onto the trolley and sat in the next car. As soon as Rosalyn spotted her mother, she got off the car and waited for the next one. That night, she told her mother, "Don't follow me. My friends will think I'm a baby." When her brother was sick, Rosalyn dressed in a homemade nurse's uniform and insisted that only she could give him pills. When she wanted braces, she helped her mother at home, turning collars for a neckwear factory. "Who had the money? If you wanted something, you worked for it. It didn't keep me from doing my homework," was Yalow's attitude.

Her father in particular encouraged her to do whatever boys could do. "I never really got the message that girls were not as important as boys," Yalow admitted. "I was very close to my father. He took me to baseball games. I can tell you all about the 1934 Yankee team."

By the time Rosalyn was eight years old, she had decided to get married, have children, and be a "big-deal scientist. . . . I liked knowing things. The thing I like is logic, and this characterizes all of science," she explained. Her friends wanted to be scientists too, but only Rosalyn wanted a family in addition. As she told them, "If you think now about how you're going to get married, you'll do both."

In junior high, Yalow narrowed her science goals to medical research. In an accelerated class, she completed three grades in two years. At a parent conference, a teacher confided to Mrs. Sussman, "You know, your daughter is a genius." Mrs. Sussman thought, "Genius? I don't want a genius. I want a normal child." She was thinking of Albert Einstein. "I never met the man, but I had heard he was a little peculiar," she said.

When Yalow graduated from high school at age fifteen in 1937, her parents advised her to become an elementary school teacher. "That's what bright Jewish girls did in the thirties," Yalow explained. "Boys were urged to become doctors and lawyers." She could not afford tuition, but her grades won her admittance to Hunter College, the highly competitive women's college. Tuition at city colleges was free. Hunter, where Gertrude Elion also studied, now charges tuition.

Yalow stood out, even at Hunter. One of her physics classes was right after lunch and, to wake up students who were groggy, the professor announced that he would make two mistakes in his lecture. Yalow found three.

Choosing a major was easy. "In the late thirties when I was in college, physics, and in particular nuclear physics, was the most exciting field in the world," Yalow explained. "A few people would sit around and talk to each other. One of them would have a great idea, go to the laboratory, and work for days or weeks, and make the discovery worthy of a Nobel Prize. It was an absolutely fantastic time."

In 1939, fission was discovered by Lise Meitner, Otto Hahn, and Fritz Strassmann. When Enrico Fermi, the legendary Italian-American physicist, gave a colloquium at Columbia University on the new discovery, every physicist in New York wanted to attend, and Yalow competed desperately for a seat. She wound up hanging from the rafters in the top rows of the lecture hall, listening to Fermi explain the unbelievable news that an atomic nucleus could split and release energy. As Yalow is quick to explain, nuclear fission resulted, not only in atomic bombs, but also in the radioisotopes that she eventually used in medical research.

To top off Yalow's enthusiasm about nuclear physics, Ève Curie published a biography of her mother, Marie Curie, in 1938. "Every woman scientist read that book twenty thousand times. We were all going to be like Madame Curie." The story had special meaning for Yalow. "For me, the most important part of the book was that, in spite of early rejection, she succeeded. It was in common with my background, with my being aggressive." Five years later, when Yalow was in graduate school, she was enthralled all over again by the Curie story when the Hollywood film *Madame Curie* appeared. "'Til my dying day, I will remember Greer Garson and Walter Pidgeon… coming back to the laboratory at night and seeing the glowing that meant they had discovered radioactivity. That was exciting!"

Yalow blames Hunter's "old maid" women professors for discriminating against her because she wanted to combine science and marriage. "In my career, I got help from men, not women. Neither of the two women professors I had in the physics department at Hunter did anything to get me along in physics," Yalow declared.

The woman who chaired the department decided that Yalow was not serious about science because she wore lipstick and dated boys. The woman had a difficult personality and regularly lost her temper and resigned during faculty meetings; later, she would retrieve her letter of resignation from a secretary's desk before Hunter's president could act on it. "Fortunately, my senior year, one of the faculty members got a hold of the letter before she could remove it and it got to the president in time," Yalow declared. The department chairman would never have recommended Yalow to a graduate school.

Despite Hunter's "old maids," Yalow is grateful that she attended girls' schools. "Very often it's better for women to be protected until they're further along. If I'd been in a coed school, maybe I'd have gotten less attention. If the men professors had had twenty men students, they wouldn't have paid attention to me."

Yalow wanted desperately to go to medical school but knew she would not be admitted. American medical schools would not admit Jewish men, let alone women. In 1937, fewer than 3 percent of the pre-med students from the City College of New York were admitted to a medical school in the United States; the college was predominantly Jewish at the time.

Even if Yalow had been accepted, she could not have paid the tuition. She tried her second choice, physics, because teaching assistantships were available. Anti-Semitism was rife in graduate schools, too, though. Purdue University responded to her application, "She is from New York. She is Jewish. She is a woman. If you can guarantee her a job afterward, we'll give her an assistantship." With no guarantee possible during the Depression, Yalow had no chance.

Asked about discrimination, Yalow replied, "Personally, I have not been terribly bothered by it. I have understood that it exists, and it's just one other thing that you have to take into account in what you're doing.... If I wasn't going to do it one way, I'd manage to do it another way."

During her senior year, she took a part-time secretarial job in Columbia's medical school with the understanding that, in return for learning stenography, she could take some science courses at Columbia. Thus, in January 1941, at age nineteen, after graduating from Hunter College with high honors in physics and chemistry, Rosalyn Yalow entered secretarial school.

Fortunately for her, World War II was about to start and the draft was draining American graduate schools of their male students. For a few short years, before the men returned as veterans, graduate schools admitted women students rather than close their doors. Partway through shorthand class, Yalow got an offer for a teaching assistantship in physics at the University of Illinois, the most prestigious school she had applied to. "It was an achievement beyond belief," she recalled. She tore up her steno books, stayed on as a secretary until June 9, and took free physics courses under government auspices at New York University that summer.

In the fall of 1941, she took a train to Champaign-Urbana. Her father wanted to pay her fare but she retorted, "I'll pay my own way and I'll pack my own valise. I don't want my parents to stand on the platform and say, 'Our baby's going away.'"

But what if World War II had not occurred? Yalow replied

confidently, "I'd have taken graduate courses while I was a secretary, and someone would have recognized me. I'm a very good secretary."

The dean of engineering at the University of Illinois congratulated Yalow when she arrived. She was the first woman permitted in the engineering school since 1917, during World War I. Evidently, the engineering program, which included physics, only admitted women immediately before and during world wars. As a woman, she was not considered qualified to teach engineers; she could only teach less advanced pre-med students. Despite such subtle discrimination, her seventy-dollar monthly salary and free tuition made her feel like a rich woman. She was no longer "a stubborn, determined child, but rather a stubborn-determined graduate student."

Yalow had taken fewer physics courses than any other first-year graduate student. To compensate, she audited two undergraduate physics courses and enrolled in three graduate classes, in addition to teaching half-time. "Like nearly all first-year teaching assistants, I had never taught before—but unlike the others, I also undertook to observe in the classroom of a young instructor with an excellent reputation so that I could learn how it should be done."

Only one of Illinois' physics professors would accept a woman teaching assistant: Robert Payton recognized Yalow's excellent mind and her courage and follow-through. She succeeded so well that the department gave assistantships to two more women. When they quit as soon as their husbands received their Ph.D.'s, the department complained, "Look, we take them in, they go so far, and they don't finish."

After the attack on Pearl Harbor, on December 7, 1941, physicists left the university to do defense work, and army and navy students arrived for government-sponsored training. With fewer instructors and more students, the department finally let Yalow teach engineers.

By then, Rosalyn had met her future husband, Aaron Yalow, a rabbi's son from Syracuse and another aspiring nuclear physicist. Both did their Ph.D. research under the guidance of Maurice Goldhaber, later director of Brookhaven National Laboratory, where Chien-Shiung Wu's husband worked. Goldhaber's wife, Gertrude, also advised Rosalyn. Gertrude Goldhaber was a distinguished physicist who had no university position because of Illinois' nepotism rules against two relatives working for the same university.

The same rule prevented Rosalyn and Aaron from marrying as long as they were both teaching assistants and technically faculty members. When Aaron got a fellowship, he was no longer considered a faculty member and they were able to marry on June 6, 1943. Aaron was "enormously supportive of me and my work. You must choose a husband in keeping with the life-style that you want," Rosalyn advised.

Yalow lives for her work and her family. Aaron accepted that she would not be a homemaker partner, but he did not help around the house either. When Rosalyn traveled, she cooked his meals ahead of time for him to warm. Neither she nor her parents had kept a kosher kitchen, but she did it for Aaron so carefully that distinguished rabbis have celebrated Passover at her house. When she dines out without Aaron, though, she sneaks shrimp and bacon. Had Aaron interfered with her professional life, however, "That would have been a matter of principle." As for Aaron, he said, "I've served as an extra pair of hands and eyes and another brain." Yalow's mother added, "Believe me, she married the right man. He did a lot of things for her. If he wasn't such a good person, she couldn't have gotten as far as she did."

Asked who the better physicist is, Rosalyn Yalow replied that her aggressiveness sometimes made her better than Aaron. Illinois' department chairman, for example, so disliked Rosalyn that he asked Aaron to prove his comprehensive examination problem twelve different ways. So Aaron did. When the chairman tried the same trick on Rosalyn, she snapped back, "Goldhaber and Nye taught it to me this way, and if there's anything wrong, you better talk to them about it." The chairman walked right out of her exam and did not return. "I knew I was right, and I wasn't going to be troubled by that guy."

As soon as Rosalyn received her Ph.D. in nuclear physics in January 1945, she hurried back to New York City. Both she and Aaron needed physics jobs, so they had to live in a city with several universities. Unable to find a position in nuclear physics, she became the first woman engineer at the Federal Telecommunications Laboratory, the research lab for the International Telephone and Telegraph Corporation. When her research group folded a year later, she returned to Hunter to teach. Aaron, who joined her in New York in September 1945, eventually became a physics professor at Cooper Union.

Hunter had no research facilities, however, and Rosalyn was getting nowhere in nuclear physics. The alternative—high-energy physics—was big-machine, big-team science and she preferred working with small groups. So she looked for another field. Radioisotopes were just coming into use in medicine, and Aaron suggested medical physics. Through him, she met "The Chief," Dr. Gioacchino Failla, the dean of American medical physicists. After talking briefly with Yalow, Failla picked up the phone, dialed, and told a colleague, "If you want to set up a radioisotope service, I have someone here you must hire." As Yalow noted in an unusual spirit of obedience, "Dr. Failla had spoken."

When Yalow reported for work at the Veterans Affairs Medical Center in the Bronx in 1947, she quickly turned a janitor's closet into one of the first radioisotope labs in the United States. Nuclear reactors, built during the war to make atomic bombs, were producing radioisotopes—radioactive forms of chemical elements—and making them available for scientific research. The Veterans Administration, the federal government's agency for veterans' programs, thought that radioisotopes would be a cheap alternative to radium for cancer treatment. Yalow realized otherwise.

She had been reading a book by the 1943 Nobel Prize–winner George Hevesy that showed how radioactive isotopes could be used as tracers in chemical and physiological processes. Short-lived radioactive isotopes give off detectable particles as they move through the human body. Hevesy's book had an enormous influence on the twenty-six-year-old Yalow, and she considers him one of the scientific progenitors of her career, along with Henri Becquerel, the discoverer of radioactivity, Marie and Pierre Curie, and Irène and Frédéric Joliot-Curie.

Still teaching full-time at Hunter, Yalow developed a radioisotope service for the hospital and started several research projects with its physicians. Her engineering experience proved helpful because commercial instrumentation was not available yet and she could make or design much of her equipment. Within two years, she completed eight publications.

Working at a veterans' hospital kept Yalow far from the scientific mainstream. Most scientific research in the United States is conducted in universities. She was a female Ph.D. in a hospital dominated by male physicians and military officers. "The only way she could make her point was to be very precise, definitive, and assertive; otherwise nobody would have listened to her," her student Mildred S. Dresselhaus of the Massachusetts Institute of Technology noted. "Sometimes people see her brusque side. But to be noticed in the world of science, she had to be that way. She was an outsider in every way. She was working in a new field of physics, and she didn't have the right credentials in medicine. So she had to let them know she was for real."

Outsider status did not scare Yalow. "I had survived the University of Illinois physics department and the ITT research laboratories, and I was confident that I was needed and could do a good job," she remarked. "I've always been well organized. I've always considered what I wanted, been prepared to work for it." There were "hints of discrimination" during her early years with the Veterans Administration, and she was "mad as the devil" when she was told not to attend an out-of-town conference because there were no hotel facilities for women. But on all substantive matters, local VA authorities supported her.

Yalow believes that "women, even now, must exert more effort than men do for the same degree of success." On the other hand, she is also convinced that "the trouble with discrimination is not discrimination per se but rather that the people who are discriminated against think of themselves as second-class." As far as she was concerned, "There was something wrong with the discriminators, not something wrong with me."

In January 1950, she cast her lot with medical physics and resigned from Hunter College. She taught one course for one more semester only because she had discovered an outstanding young student and wanted more time with her. Mildred Spiewak Dresselhaus was the first student taken under Yalow's wing. Thanks to Yalow, Dresselhaus switched her major from elementary education to physics. She went on to become an MIT professor, a member of the National Academy of Engineering and the National Academy of Sciences, and a president of the American Physical Society. Yalow was a "decisive, no-nonsense, inspirational teacher," Dresselhaus says. Yalow mothered her students, turning them into protégés. "It comes completely natural to her. It's that instinctive. She doesn't know she's doing it." Dresselhaus went into a different field but whenever she gave a talk nearby, Yalow showed up with Aaron and her canvas shopping bag. "I didn't even realize she was well known in her field until she got the Lasker prize the year before her Nobel Prize. But she knew what was happening in my career step by step. She was like a parent focusing on her child," Dresselhaus said.

Recognizing that medical physics required an interdisciplinary approach, Yalow began looking for a collaborator to complement her strengths. In the spring of 1950, she met Solomon A. Berson. Berson, a young resident physician in internal medicine at the VA hospital, was a brilliant Renaissance man. An accomplished violinist, he played multiple games of blindfolded chess and loved art and philosophy. He was an elegant writer and speaker and a talented physician and biologist. His personality was intense, charming, and likable. Like Yalow, he was an outsider and had faced academic anti-Semitism. The son of a Russian Jew, he graduated from City College of New York in 1938 and was rejected by twenty-one medical schools. It was three years before he was accepted by the New York University Medical School. Talking over their possible collaboration, Berson gave Yalow a series of mathematical puzzles. She liked his sense of humor and began working with him.

Within two years, they had formed such a magnificent partnership that it lasted twenty-two years. "Stick with me, and I'll have your name up in lights," Berson told her. He supplied the biological insights, physiology, anatomy, and clinical medicine; Yalow provided the physics, mathematics, chemistry, engineering, and rational muscle. Each learned the other's field. Yalow ended up knowing more physiology than most leading physiologists and became one of the few nonphysician members of the Association of American Physicians. Yalow taught Berson mathematics and physics so well that a Yale physics professor lectured to his class from Berson's classical-mechanics notes and Berson thought about teaching mathematics. Their scientific styles complemented each other too. Berson was broad and sweeping; Yalow was logical, quick, and precise.

Like a long-married couple, Yalow and Berson developed shortcuts to communicating with each other. "It was a kind of eerie extrasensory perception. They didn't have to waste time talking. Each knew what the other was thinking. One could start a sentence, and the other could finish it. Each raved about the other. Each had complete trust and confidence in the other," recalled Stanley J. Goldsmith, a physician in their lab who became director of physics and nuclear medicine at Mount Sinai Medical Center in New York City.

"They didn't waste time with small talk or polite sentence structure," Goldsmith remembered. Instead, they asked, "Where's the data? What does this mean? When did this happen? How can you say that?" When Yalow continued their waste-no-time language after Berson's death, she terrified the uninitiated. "At scientific meetings, they were known as tough, and either one of them would get up and say it doesn't make sense," Goldsmith continued. "But they were the same with each other. It wasn't hostility. It was honesty. They felt they were on a mission. The mission wasn't a Nobel Prize. The mission was understanding the phenonema."

Newcomers often thought that Yalow and Berson were married to each other, even though each was happily married to someone else. Yalow did the "woman's" work. As team administrator, she made plane reservations and arranged for manuscript typing. They planned experiments together, but she generally set them up. Dashing upstairs one morning, she told a friend, "Oh, yeah, I forgot to make Sol's lunch." Outside the lab, they were inseparable, whether running a conference or chatting at cocktail parties. But occasionally at parties Berson told her to stay with the wives. "It was a characteristic view at the time," Yalow explained. "Also, I was living in the medical world, and I didn't have a medical degree."

Although Berson dominated everybody, including his bosses, he and Yalow shared credit equally. They alternated first authorships of papers. When he accepted a prize, it was with the understanding that she would get one soon too. Although they considered each other equals, Berson captured the attention. The breadth and depth of his knowledge dazzled their colleagues. Outside the lab, he was the front man, writing their articles and delivering most of their speeches. When he and Yalow had a beer with a colleague, Berson talked ninety percent of the time.

The Yalow and Berson professional family extended to many of their young postdoctoral fellows. Yalow told postdocs that she was their professional mother, Sol their professional father, and other fellows their brothers and sisters, and that they should share information with each other but be cautious with outside competitors. "We were a protected species," according to Seymour Glick, now at the Ben-Gurion University of the Negev in Israel. "Anyone who dared attack us unfairly had Sol and Ros in the full array of their most aggressive stances to contend with." The team did not have much money, but Berson personally paid the travel expenses of postdoctoral fellows going to meetings. In addition, he was extraordinarily supportive of women scientists, according to Johanna Pallotta, a Harvard Medical School endocrinologist, and others. "He was unbelievably wonderful to me and to all the women," Pallotta emphasized.

Berson and Yalow operated at ninety miles an hour, eighty hours a week. "They would come into a room and argue and bounce things off each other and throw ideas around. Then they'd go running out of the lab to test whatever it was, wildly flying around," recalled Dr. Eugene Straus, who worked with both.

"I'm a morning person," Yalow noted. "When I'm stewing about something I'm working on in the laboratory, I'll wake up at two or three in the morning and by daybreak everything will fall into place and I know the experiment that has to be done the next day. Usually there are a lot of experiments that have to be done." Yalow worked through fevers, flus, and several years of anemia. She typed her children's papers before work, dashed home for ten minutes to put a turkey in the oven, stayed up until two A.M. with chemical assays, and returned to work by eight the next morning. Once they worked until 4 A.M., Berson caught a nap on her porch, and they were both back at work before 8 A.M.

Yalow and Berson invented the radioimmunoassay (RIA, for short) by accident—as an offshoot of their insulin research. To test how long insulin remained in a diabetic's system, Berson and Yalow had injected patients with radioactively tagged insulin. Then they took frequent blood samples and measured how fast the hormone disappeared from each patient's system. To their surprise, the tagged insulin took *longer* to disappear from diabetics than from nondiabetics who had never had insulin. Why did adult diabetics retain insulin longer than other people? Was the effect caused by the hormone or the diabetes? One group of people had taken insulin even though they did not have diabetes. These were schizophrenics who had undergone insulin shock therapy. Berson and Yalow figured that if schizophrenics also retained insulin for an abnormally long time, the effect would have to be a result of their insulin treatment; it could not have been caused by diabetes, because they did not have the disease. And sure enough, the schizophrenics also retained tagged insulin longer than nondiabetics who had never had insulin.

Berson and Yalow concluded that people who took insulin developed antibodies to the insulin molecules. During the 1950s, insulin was obtained from the pancreases of pigs and cattle. Although animal insulin was almost the same as human insulin, the human immune system produced antibodies to fight the foreign insulins. The antibodies inactivated the insulin, making it more difficult for the body to process the hormone. A few diabetics actually became resistant to the insulin they needed to survive; Berson and Yalow discovered that they had a much higher concentration of antibodies in their blood than other diabetics. Today, manufactured insulin is genetically engineered to be precisely the same as human insulin.

Berson and Yalow had overturned one of the long-held traditions of diabetes: that an insulin molecule was too small to stimulate an antibody. The prestigious journal *Science* rejected their article outright, and *The Journal of Clinical Investigation* refused to publish it until the words "insulin antibody" were deleted from the title. Twenty-two years later, Yalow was still fuming at the *Journal of Clinical Investigation*.

Ironically, Berson and Yalow were not the only investigators to discover that people who had been treated with insulin process the hormone more slowly than others. "It was luck for both sets of investigators to discover that insulin disappears more slowly from one group of patients than from another.... But it isn't by accident that you interpret the observation correctly," Yalow emphasizes. "That's creativity.... I still think discovery is the most exciting thing in the world."

Berson and Yalow could have spent the rest of their careers studying insulin. Instead, they came to a startling realization: Although their technique measured the antibodies to a hormone, its inverse would measure the hormone itself. As Yalow put it, "Once you saw it one way, you saw it the other way." Later, other researchers confessed that they had had the same idea but done nothing about it. Berson and Yalow did. They were indefatigable.

They called their method a "radioimmunoassay" (RIA) because it used radioactively tagged substances to measure antibodies produced by the immune system. In principle, the test was simple: First, Yalow and Berson put a natural hormone from a patient into a test tube. Second, they added its antibody. Third, they mixed in a small amount of the radioactive form of the hormone. Then they waited a few days or hours. While the solution incubated, the radioactive hormone and the natural hormone competed for the privilege of combining with an antibody molecule. Neither hormone had an advantage; the antibody did not care which it bound to. By measuring how much of the radioactive form succeeded in bonding with the antibody, Yalow and Berson could tell how much of the natural hormone had been present in the patient's body.

If a patient's blood contained a large amount of natural insulin, for example, the natural insulin captured almost all the antibody, leaving little left over for the radioactive hormone to bind to. If the patient had only a small amount of natural insulin in his blood, most of the radioactive insulin was able to bind to the antibody.

Although RIAs are automated and computerized today, Berson and Yalow worked for days preparing solutions in two thousand or three thousand test tubes. Clearing their schedules for at least 24 hours, they ran tests straight through, day and night. With no technicians to help, they did all the work themselves. And as Berson smoked, they communicated in their quick, terse, hybrid language.

Their technique had several remarkable advantages. It was almost unbelievably sensitive. An RIA can detect a billionth of a gram—and there are about thirty-two grams in an ounce. In addition to its fabulous sensitivity, it was a test-tube operation. No radioactivity entered a patient's body. Further, tiny amounts of a substance could be tested. Until the RIA, a diabetic had to give one hundred cubic centimeters of plasma—roughly a cup—for each blood test. With RIA, a tenth of a cc was sufficient. Moreover, RIA worked for virtually every hormone and a variety of other biologically important substances. Finally, different substances could be tested simultaneously in any laboratory equipped to measure radioactivity.

Yalow and Berson published their idea for the radioimmunoassay in 1956. Then they spent three years developing the concept into a practical test, applying her physics and chemistry knowledge to the biological problem. They made few immediate converts, however. The technique was too new, complicated, and unbelievable to catch on. When Berson gave a speech at the University of Illinois, one person in the audience thought he would win a Nobel Prize and the other twenty-nine thought he was crazy.

Yalow and Berson were used to fighting the medical establishment, though, so they did not give up. Instead, they began training scientists to use RIAs. Other laboratories published only a few papers using RIA before 1965, but by the end of the decade, the field had exploded and RIA was a phenomenal success. The battle left its legacy; the years of striving reinforced their feeling of being outsiders. It also proved to them the value of questioning everything, accepting no givens, and forcing every fact to defend itself.

The first RIA revolution occurred in diabetes. By using RIAs, physicians could separate diabetics into two groups: those who typically lose the ability to make insulin during childhood and those who produce plenty of insulin, but during adulthood lose the ability to use it. Today the second type of diabetics is treated with diet and, if necessary, pills—but not injections.

By the 1960s, Berson and Yalow were analyzing masses of data with total self-confidence. "They took it upon themselves to rethink all of modern endocrinology, all of the hormones," observed John T. Potts, Jr., a world leader in RIA. "They had the boldness and quickness to move into one biological field after another. They learned a great deal about the critical experiments and the details of the new fields. Then they reassessed the endocrine physiologies, how the hormones really worked. For the first time, anyone could sample a particular body compartment by an accurate method and really have reliable data—and get masses of it. They could administer a stimulus or a suppressant to the hormone and study different kinds of diseases associated with too much or too little of the hormone. It was an awesome decade in intellectual productivity, and it took really fierce people. They'd work all night every other night and then go to another field, become the experts, and turn that field on its head based on their technique. They were both very opinionated; they weren't afraid of saying what they found."

Armed with self-confidence and careful experimentation, Berson and Yalow made daring and sweeping generalizations about how the hormones were made and handled in the body. "They deduced the general picture from an enormous amount of data taken from a very small arena; for example, from thousands of blood samples measuring the hormone in one body compartment. Then they could predict what must be happening somewhere else in the body to explain what they saw there," Potts explained.

Before RIA, little was known about human growth hormone. Essential for bone development and growth, it is secreted by the pituitary gland to stimulate the liver to produce growth factors. Dwarfism in children can be caused by insufficient growth hormone or by a variety of other factors. A dwarfed child with little or no growth hormone in his plasma shows little response to small amounts of insulin. When insulin is given to children who are short from other causes, they secrete more growth hormone. In this way, it became possible to determine which children can be helped by growth hormone treatments. Conversely, the RIA shows whether acromegaly—abnormal bone growth—is caused by the oversecretion of growth hormone.

From insulin and human growth hormone, Yalow and Berson swept on through a host of other substances, including parathyroid hormone, hepatitis B-antigen, and ACTH. In 1970, hepatitis B-antigen was a dreaded complication of blood transfusions; contaminated blood caused liver infections several months after surgery. Thanks to Yalow, Berson, and their co-worker John Walsh, transfusion hepatitis virus has been virtually eliminated from North American blood banks.

Perhaps the most important public health use of RIAs is to prevent mental retardation in babies with underactive thyroid glands. The symptoms of hypothyroidism are undetectable until a baby is more than three months old. By then, brain damage is irreversible. A heel prick and a few drops of blood on filter paper diagnoses at-risk newborns in time to prevent damage. For a few dollars a year, they can be treated so that their brain development will equal that of their siblings. As Yalow puts it, "What a bonus to these children, their families, and society!"

Today radioimmunoassays are used to measure the concentration of hundreds of hormones, enzymes, vitamins, viruses, and drugs within the human body. They are used to diagnose conditions caused by hormonal excesses and deficiencies. They identify people with peptic ulcer disease to determine whether ulcers should be treated medically or surgically. They show whether high calcium levels in the blood, which often lead to kidney stones, are caused by too much parathyroid hormone. Berson and Yalow produced a theory about the biosynthesis of ACTH that was later proven to be correct. RIA is useful for studying hypertension and for detecting hormone-secreting cancers and other endocrine-related disorders.

RIA made endocrinology one of the hottest areas of medical research. RIA had an enormous impact on clinical medicine, making major advances in the diagnosis and treatment of the body's major hormone systems, including thyroid function, growth, and fertility. It virtually started a new science—neuroendocrinology, the study of chemical messengers used by the brain to control the body's major hormone systems.

RIA also pinpoints the concentration of drugs in the blood—from heroin in drug abusers to steroids in athletes and antibiotics in the sick. Many drugs must attain a certain concentration in a patient's blood in order to be effective; if not enough drug is given, it is ineffective, and too much can be toxic.

Like the Curies, Yalow and Berson decided not to patent their discovery. Numerous commercial laboratories made enormous sums of money performing RIA. But Yalow explained curtly, "In my day, scientists didn't patent things. You did it for the people. Unfortunately, now is not the way life was." Furthermore, she wondered, "What would we have done with the money except pour it into research?.... I never had a research grant in my life. It was all VA money.... If I had five million dollars a year for research, it would be necessary for me to supervise a hundred scientists. It would be impossible for me to talk to each of them every day.... I'm psychologically adjusted to 'mom and pop' science."

During these high-flying RIA years, Yalow bore and raised two children. Benjamin was born in 1952, when Yalow was thirty-one years old. Elanna was born two years later. Yalow planned her family and worked five years at the hospital until she was established enough to defy VA regulations. According to the rules, all women employees resigned during their fifth month of pregnancy. The rule was enforced; while Yalow was pregnant with Benjamin, a pregnant veterinarian was forced to resign. Yalow, on the other hand, had made herself too valuable. When Benjamin was born, she joked that he was history's only eight-pound two-ounce, "five-month" baby. She took seven days off for his birth, fired the pediatrician who disapproved of her working and nursing, and nursed Benjamin for ten weeks. Then she trained him to sleep days and play nights when she was at home. Nine days after Elanna's birth in 1954, Yalow gave a lecture in Washington, D.C. By then the Yalows had moved from an apartment around the corner from the hospital to a house about a mile away in Riverdale, too far for her to nurse Elanna. The Yalows could have moved to a more elegant suburb, but she wanted to be able to lunch with the children.

"Life was a lot easier then—there was more household help," Yalow emphasizes. She had a sleep-in housekeeper until Benjamin was nine years old, and Yalow's mother "just happened" to drop in almost every afternoon. Mrs. Sussman disapproved of her daughter's working full time with small children, but Yalow never doubted that she was doing what was right for her. When five-year-old Benjamin faced a problem, she assured him, "You're a big boy now. See if you can deal with it. If you can't, I'm here for you. If you can, well, it's good you've learned how to take care of it."

When Elanna entered elementary school, Yalow switched to part-time help and still found time to volunteer for class trips. Weekends, she took the children to her office to play with the equipment and the rabbits, mice, and guinea pigs. She cared for the animals over weekends and vacations.

Young researchers soon learned that the only way to get a breather at work was to inquire about Yalow's children. Then Yalow would stop work to report that Elanna had had difficulty choosing between pink and coral lipstick that morning and that Benjamin was still involved in science fiction and computers. Benjamin and Elanna attended Bronx public schools, the prestigious Bronx High School of Science, and earned doctorates. Benjamin is computer-center director at the City University of New York, and Elanna directs a California company that establishes day-care centers.

In 1968, Yalow received some shattering news. After eighteen years as her full-time research partner, Sol Berson wanted to take another job. He would become the professor in charge of internal medicine at the Mount Sinai School of Medicine of the City University of New York. Yalow tried to talk him out of it. Both were convinced that eventually they would win the Nobel Prize—"the Big One," she called it. When that happened, the VA would give them their own research institute, she argued. But as Berson told his wife, Miriam, the discovery of RIA was essentially complete and he wanted to teach medicine, philosophy, or mathematics. He was ready for something new.

Berson assumed that he could supervise a large hospital department and continue research with Yalow on the side. Each Tuesday and Thursday, he put in a full day's work at Mount Sinai before coming to Yalow's lab at the VA to work all night—without remuneration. Sitting across from him at their partners' desk, Yalow caught him up on lab projects. At ten P.M. they met with their research fellows and started an experiment or an assay that would last through the night until he went to Mount Sinai the next morning to begin another day's work. The pace was unrelenting, and even Berson had to slow down. After a while, he came to the lab only one night a week.

In March, 1972, Berson had a slight stroke. Confidentially, he asked Eugene Straus, a young physician at Mount Sinai, to work with Yalow. A month later, in April 1972, Berson was elected to the prestigious National Academy of Sciences and went to Atlantic City, New Jersey, for a scientific conference. Alone in his hotel room, he suffered a massive heart attack and died. He was fifty-four. His death sent shock waves through the small endocrinological community.

Yalow telephoned Boston to tell Johanna Pallotta. Come to New York immediately, she said. "I'll stay at the lab until you get here." Yalow's world had fallen apart, but she took Pallotta home for the night at her house. "She held the group together with her strength when we were collapsing," Pallotta remembered gratefully.

Berson's death was a tremendous blow to Yalow. Friends say she was devastated for more than a year. For twenty-two years, they had worked together, shared ideas and work, spoken their own private language, invented a remarkable technique, and fought for its acceptance. But with Berson dead, the medical grapevine predicted that her career was over. She was told she could not win a Nobel Prize without him. RIA was unquestionably worthy of the prize; for many years, half the papers at the Endocrine Society were based on RIAs. But Nobels are not given posthumously, and they had never been given to a surviving partner. Furthermore, there was widespread feeling—which persists to a lesser extent today—that Berson had been the creative brains of the team. In her despair, Yalow thought she needed the status of an M.D. degree. She considered going to medical school at the University of Miami and then decided against it.

Finally, Yalow decided that if she had to prove herself all over again, she would. So, instead of an eighty-hour week, she worked one hundred hours. Renaming her laboratory the "Solomon A. Berson Research Laboratory" so that her papers would continue to bear his name, she took over Berson's editing and speaking engagements. She kept young Eugene Straus on as her research partner, and the lab published sixty articles between 1972 and 1976. She won a dozen medical awards in her own right. She showed that human antibodies to insulin can differentiate between pig, dog, and whale insulin even though the insulins are made of almost identical amino acid sequences. Then she and Straus showed that cholecystokinen, a well-known hormone that helps digest fats in the small intestine, is also a synaptic transmitter in the brain, communicating information from one neuron to another. It was the first time that a gastrointestinal hormone was found to have a dual role as a neural transmitter. It is still the classic example of the body's frugal use of one chemical to perform two completely different jobs.

Nobel Prize winners are announced each October, and every fall Yalow hoped to be among them. She put champagne on ice and dressed up a bit each Nobel announcement day—just in case. Each time she was passed over, Aaron Yalow said, "Her reaction was just, 'What do I have to do to win?'"

Pressure was reportedly building on the Nobel Committee to give her the prize, despite Berson's death. With every passing year, the power of the RIA technique grew more impressive. In 1975, she was elected to the prestigious National Academy of Sciences, three years after Berson. In 1976, she became the first woman to win the ten-thousand-dollar Albert Lasker Basic Medical Research Award, often considered the harbinger of a Nobel Prize. Yalow brought roast turkeys from home, made potato salad during a lab meeting, and threw a party for one hundred people.

At three o'clock the morning of October 13, 1977, Aaron and Rosalyn woke up. Unable to sleep, Rosalyn was at work by 6:45 A.M, when the telephone rang. After years of hoping, she had won the Nobel Prize in Medicine or Physiology. Aaron advised her to race home and change her clothes before reporters descended. She was back in the lab by eight. At her press conference, Aaron answered so many questions that she interrupted, "Aaron, Aaron, let me talk!"

Yalow was the second woman to win a Nobel Prize in medicine. Gerty Radnitz Cori had won in 1947. Yalow was also the first American-educated woman to win a science Nobel Prize. An outsider from a VA Hospital in the Bronx, she is one of eleven Nobel graduates of the City University of New York. She was also the first survivor of a research partnership; the Nobel Committee broadcast through its private network that no more exceptions would be made.

Combining family and science as usual, she and Aaron flew to California for Elanna's wedding and then two days later to Stockholm. Elanna and her new husband spent their honeymoon in Sweden. Yalow also took three Bronx students to Stockholm—one from her junior high school. one from Walton High, and one from Hunter College—to write for their school newspapers.

Yalow shared the $145,000 prize with two men who heartily disliked each other. Roger C. Guillemin of the Salk Institute and Dr. Andrew V. Schally of the Veterans Administration Hospital in New Orleans won the prize for discoveries concerning the hypothalamic area of the brain, which controls endocrine functions. Standing on either side of Yalow, Guillemin and Schally studiously avoided looking at each other. Seldom had such frank and open ambition shared the Nobel stage.

Nobel festivities are filled with rituals, one of which involved a Swedish student coming down the long banquet table to escort Yalow back up to a podium to speak. The student became confused. Proceeding down the wrong side of the table, he carefully came to attention behind Aaron Yalow. Across the table, Rosalyn realized what had happened. Presented with the choice of two Dr. Yalows, the student had bet on the man. Standing straight and tall in her long blue gown, Rosalyn Yalow marched herself up to the podium as the hapless aide accompanied her along the opposite side of the table. Such trifles never bothered Yalow. She got where she was by keeping her eye on important issues, not trivialities.

The Nobel Prize opened new and enjoyable vistas for Yalow. "Before Nobel, nobody had heard of me. Now I'm much more in public view, and I can do things I've never done before. It should happen to everybody," she joked. She hosted a five-part television series about her heroine Marie Curie. Before putting her gold medal in the bank vault, she had a small copy made to wear around her neck. She has received forty-seven honorary degrees from universities and colleges and decorated her office with several of their elegant ceremonial hoods.

Asked what she planned to do with the Nobel prize money, she replied, "I can't think of anything in the world that I would want that I haven't had.... I have my marriage, two wonderful children. I have a laboratory that is an absolute joy. I have energy. I have health. As long as there is anything to be done, I am never tired."

What about hobbies? she was asked. "What hobbies?" she answered. "What am I going to do? Ride a horse? Play tennis? This is where the excitement is.... After *x* years of playing tennis, do you really feel the next day there will be a dramatic change and you will revolutionize tennis?" In 1988, she received the National Medal of Science, the nation's highest science award.

A few months after Stockholm, Yalow used her Nobel to settle an old score. When her Nobel Lecture was reprinted in *Science* in 1978, she reproduced a portion of the *Journal of Clinical Investigation*'s letter that had rejected her report on insulin antibodies twenty-two years earlier. The letter appeared without the editor's name, but she had attacked him by name in a *New York Times* interview a few months earlier. Such personal attacks are rare in science and did not make her popular among her colleagues. Besides the editor, several anonymous experts in the field had reviewed the article and unanimously advised its rejection. Only the editor's name was public knowledge, however. In reproducing the letter in *Science,* only the portion about the size of the insulin molecule was mentioned. Omitted was the editor's reference to their "incoherent" writing.

Yalow worked in her VA lab for fifteen more years. For a number of years, she also held professorships at Albert Einstein College of Medicine in Yeshiva University, Montefiore Hospital and Medical Center, and Mount Sinai School of Medicine.

The aggressiveness that got Yalow through the first grade and into physics is still alive. But it has seemed less benign than in her early years. She is unpopular among many colleagues in her own field and among scientists in other subjects who have worked with her on professional projects. She is as logical, outspoken, and direct as ever and holds strongly to her opinions, but the attacks on competitors' work sometimes seem more personal than in the Berson days. A colleague complained that, when Berson argued, he expressed his point of view; when Yalow argues, she believes she has the answers.

Friends call her manner "detached and scientific"; critics call it "excoriating" and "dour." When Yalow publicly questioned the data presented in a talk by a young scientist from another field, the speaker felt as if a tank had run over her. Yalow still nurtures her protégés, but her scathing reviews of competitors' grant applications and journal articles could be devastating. One of her lab's eminent graduates and one of Berson's closest women friends—two colleagues she might ordinarily have been close to—came under particular attack.

As Yalow grew in confidence and stature after Berson's death, the memory of his contributions to their partnership seemed to fade too. "She's become awesomely full of herself with a kind of a viciousness of style for dealing with people," a colleague commented sadly. "She's become the object of wariness bordering on fear, depending on whether you are equal to or below her." When a young scientist argued with her in a public meeting, an older professor took him aside and warned him, "Don't argue with her. She can be vindictive."

She still does her homework though. Sitting on an eminent committee about radiation levels, she argued with a fellow panel member, quoting from his own articles. He had forgotten the details but she had not.

Yalow retired from the VA hospital in 1991, a victim of Nobel-style burnout. Delighted to close her lab, she complained, "Labs are terribly overregulated now, and it's not worth the hassle. I'd do it all over again in my generation. If I were starting out now? The whole modern system isn't as much fun as when I did it. One, small science isn't as easy. Two, it's completely overregulated. For thirty-five years my rabbits have lived in the same room with my guinea pigs. Now I must have a separate room for each species." An inspector who examined her laboratory complained that her RIA manual was inadequate. Yalow retorted, "I wrote the book," but she ended the dispute by assigning a technician to write a manual that only the inspectors have read.

When she retired, she took on a public service career as a science activist, lecturing about causes dear to her heart. Using her prestige as a Nobel Prize winner, she calls for more high-quality child care and more science and physics in education. Above all, she speaks out against what she believes is the public's unwarranted fear of small amounts of radiation. "People tend to confuse nuclear medicine with nuclear reactors and nuclear bombs.... Radiation from a nuclear plant is less than the amount of radiation coming from a coal plant," she declares. She adds that, "Nuclear war is more of a threat because of our dependence on foreign oil. If we were self-sufficient...and this is possible with nuclear power, then there would be fewer reasons for fighting."

As for the need to improve American science education, she notes that the United States lags far behind other countries in terms of Nobel Prize winners per capita. Excluding American prize winners who were born and educated abroad makes the United States fall even further behind. "There is at present in the United States a powerful activist movement that is anti-intellectual, anti-science, and anti-technology. If we are to have faith that mankind will survive and thrive on the face of the Earth, we must depend on the continued revolutions brought about by science." She points out that science and technology have raised the life span of people in the developed world from forty-five years in 1900 to almost seventy-five in 1992.

In speeches all over the country, she emphasizes the difference between men and women: only women bear children. Proud of her record as a superscientist, supermother, and superwife, she says, "You can have it all!" The United States, however, must pay more attention to child care, Yalow insists. "It's a tragedy for society when talented women do not have children." And women scientists, in particular, need good day care, she argues. "It is difficult in a field that changes as rapidly as science to drop out for a number of years and then hope to return without major retraining.... When I go out to the universities I ask, 'What have you done in the way of day-care centers?'" When Johanna Pallotta had her first child, Yalow flew to Boston and insisted that she get live-in help and continue her career full time.

Yalow believes strongly in equal access and equal opportunity for women. But she opposes women-only awards and affirmative action programs for women. She turned down the Federal Women's Award in 1961 and the *Ladies Home Journal* Woman of the Year prize in 1978. When she became president of the Endocrine Society, a professional organization of six thousand endocrinologists, she opened her 1978 presidential address with a blast at its women's caucus. Calling herself a "non-Establishmentarian," she said she felt secure enough to be "one of the few chairpersons who dares to be a chairman."

As Rosalyn Yalow explained later, "You go where the power is. You don't isolate yourself. You don't need to be protected. It's part of my aggressive approach to things."

* * *

THE NEW GENERATION

15

Jocelyn Bell Burnell

July 15, 1943 –

ASTRONOMER AND PHYSICIST

"ARE YOU TALLER or shorter than Princess Margaret?" the reporter asked.

Grinning, Jocelyn Bell thought, "We have quaint units of measurement in Britain."

But the next question—about how many boyfriends she had at a time—was no better.

The photographers were much the same. They posed her sitting, standing, reading, and running and yelled at her, "Look happy, dear! You've just made a discovery!"

"Archimedes doesn't know what he missed," she thought.

Jocelyn Bell, a twenty-four-year-old graduate student, had discovered pulsars, an entirely new class of unimaginably dense, burned-out stars. When reporters learned that Little Green Men from outer space and an attractive young woman were involved in the most dramatic scientific event of the decade, they headed straight for Bell. She would provide a soupçon of sex and excitement to spice up an already delectable story.

Bell's discovery provided vital clues to the evolution and death of stars, opened up new areas of astronomy, and provided physicists with giant new laboratories for the study of superdense matter, superstrong magnetic fields, general relativity, and gravitation. Yet no sooner had she discovered pulsars, been interviewed and photographed by the press, and finished her doctoral thesis than she abandoned competitive, world-class research. She married, changed her name to Burnell, turned her back on radioastronomy, and followed her new husband from job to job as he moved up his own career ladder. While their child grew up, she took part-time, temporary jobs in astronomy, taught adult education on the side, and furthered her lifelong commitment to social justice and to her religious faith. For the discovery of pulsars, her thesis adviser won a Nobel Prize.

Susan Jocelyn Bell had been born in Belfast, Northern Ireland, on July 15, 1943. She grew up in "Solitude," a country house on a large and rather remote piece of land with her younger brother, two younger sisters, and parents G. Philip and M. Allison Bell. Her father was an architect, so Jocelyn had plenty of Meccano building sets and could spend hours constructing elaborate houses and automobiles for her dolls. In summertime, the family went sailing. On Sundays they attended Quaker meeting in the country, and on weekdays the children went to school in the next village.

An idyllic childhood, except for two problems: nannies and schools. With few friends nearby, the Bell children spent their time playing together, supervised by a succession of nursemaids. Inevitably, the nurses preferred whoever was baby at the time. As the eldest child, Jocelyn was irritated by the obvious injustice. And as a girl, she was rankled when she overheard one of the nursemaids telling her colleagues how wonderful it was that Mrs. Bell had finally had a son. Her parents did not subscribe to that view, but Jocelyn absorbed the message that in Ireland women and girls did not count.

When Jocelyn was eleven years old, she flunked her eleven-plus. This now-defunct British school examination divided children irrevocably into those entitled to a college-preparatory secondary education and those destined for various degrees of vocational training. "I failed the exam for three reasons," Burnell explained in her typically organized, no-nonsense fashion. First, she was a late developer. Second, she took the examination when she was relatively young and it was scored so that young candidates had to do better than older candidates. The theory was that young students had time to take and pass the examination again. And third, her small country school was "not great."

Jocelyn's failure set off an intense family debate. She needed a second chance, but it would be unfair to give one child a boarding-school education and leave the others in inferior, country schools. Four tuitions would be expensive.

Another factor entered into the debate too. As Quakers, the Bells were particularly sensitive to the religious tensions dividing Protestants and Catholics in Northern Ireland. The Bell family had emigrated from Presbyterian Scotland to Northern Ireland two hundred years before and settled among the Catholic population. At some point, the family had joined the Society of Friends, a small Protestant denomination better known as Quakers. Quakers have been pioneers in coeducation, universal education, the abolition of slavery, and women's rights; they have opposed militarism, war, and brutality in prisons and insane asylums. The Quakers' social action

Jocelyn Bell Burnell at the time of the pulsar discovery.

Jocelyn in a dance costume.

Jocelyn Bell Burnell and her son.

Jocelyn Bell Burnell at the 4.5-acre radiotelescope.

arm, the American Friends Service Committee and the Friends Service Council in Britain, won the Nobel Peace Prize in 1947.

Friends' schools often emphasized science as well as social justice, and Quakers have made disproportionate contributions to science. Famous Quaker scientists include the chemist John Dalton, the geneticist Francis Galton, the anthropologist E. B. Tylor, the discoverer of antisepsis Joseph Lister, and the astronomer Arthur Eddington. It has been estimated that an English Quaker has almost fifty times as great a chance as the average English citizen to be elected to the prestigious Royal Society of scientists. As Burnell admitted later, "Those who know me well say my Quaker heritage explains much."

During the 1940s and 1950s when Jocelyn was a child, Ireland's religious strife—"The Troubles"—was not as obvious to outsiders as it became later. Nevertheless, her parents thought that Northern Ireland had become, as Burnell puts it diplomatically, "a bit more introverted than they wanted it to be." Sending their children abroad to boarding school might broaden their perspective.

Above all, Jocelyn needed a second chance. So in 1956, at the age of thirteen, she was sent to the Mount School, a Quaker girls' boarding school in York, England, just across the Irish Sea from Belfast. Her brother and sisters eventually followed her to York for their schooling, too. As an adult, Burnell dates the beginning of her scientific career from her failure at the eleven-plus. "I started by failing," she says.

"Going to boarding school gave me a new start. It was a good idea, and I did very well," Jocelyn recalled. Nevertheless, her six years there were not problem-free. "The Troubles started up again while I was away at boarding school. I found that very difficult to handle because it wasn't reported properly in the British newspapers. Because of official English press policy, there were just two-line snippets, and I could read between the lines."

In addition, the school staff was not well trained in physics, and its physics laboratory was short of equipment. She had become interested in astronomy, a physics-based science. One of her father's architectural contracts was designing a modern addition to Northern Ireland's eighteenth-century Armagh Observatory. Jocelyn visited the observatory with him and got to know its staff, who encouraged her to become a professional astronomer. Soon she had read all the popular astronomy books in her father's library. The school's weakness in the physical sciences, however, left her unsure of her ability to build instruments and do experimental physics. Characteristically even-handed, she does not blame the school. "Although criticism of the science teaching in girls' schools may be justified," she admits, "it can also be argued that women should have rectified recognized deficiencies by age forty and that failure to do so may point to a weakness in those women."

After graduating from high school in 1961, Burnell enrolled at the University of Glasgow in Scotland. Convenient to Ireland, it was one of the few places in Britain that offered a degree in astronomy. No sooner had she enrolled than she changed her mind about majoring in it. "I decided that would be burning my boats a little too soon," she explained. British astronomers tend to be university faculty members, and with a limited number of professorships, the field was highly competitive. "One has to be academically tops," she concluded. "So, in the end I did a physics degree, which still kept options open to do an advanced astronomy degree and left me other options too."

Physics at Glasgow challenged Burnell's intellect *and* her good humor. By the end of the first year, she was the only woman in her class left in physics. She soon realized that women in the physical sciences were "out of role." Students treated her as if she were "Jocelyn from Jupiter"—slightly unnatural. Luckily, Burnell has a well-developed ability to laugh at herself. When students learned that she was in the honors section and the teasing escalated, common sense told her it was time to learn not to blush. She simply ignored well-meaning women in her dormitory who advised her to quit physics and to leave the university with a three-year general degree instead of a four-year honors degree. Married women do not need that much schooling, they told her. Only when she was misguided enough to score tops in a mathematics exam was the teasing less than good-natured. In any event, she earned a bachelor of physics degree—with honors—in 1965.

Happenstance sent Burnell to Cambridge University for graduate school. According to legend, her graduate school application to Jodrell Bank at the University of Manchester, the best-known radio-astronomy center in Britain, fell behind a professor's desk. After Burnell discovered pulsars, the professor claimed with great embarrassment to have lost her application. Since she did not expect to be admitted to her other choice, Cambridge University, she applied to a department in Australia and started planning for the long trip.

"Then I got into Cambridge after all," Burnell said. "One of the ironic twists is that, as a result of the discovery of pulsars, so many people wanted to get in [to the Cambridge astronomy department] that they raised the entrance boundaries, and I could not have gotten in at all." So, having always supported Oxford University in the Oxford-Cambridge boat race, Jocelyn Burnell suddenly found herself studying astronomy at Cambridge.

Women have been active in astronomy since the middle of the nineteenth century. An American, Maria Mitchell, won the King of Denmark's medal for discovering a comet with a telescope in 1847. After Harvard University discovered in the 1880s that women could perform detailed computations for less money than men, observatories began hiring women to make tedious calculations and photographic analyses.

Female astronomers, however, had to overcome an indisputable fact: They often worked late at night with men in remote areas. To avoid such scandalous improprieties, some universities directed women into solar research because it was a daytime occupation. When Burnell was at Cambridge in the mid-1960s, American women astronomers were not permitted to use the important telescopes at Mount Wilson and Palomar Mountain in California. At Jodrell Bank, women were required to observe in pairs at night and were not allowed to drive home after work. Only in the past twenty-five years have women been permitted to apply for time on all of the world's major telescopes.

"Women astronomers of my generation feel (have been made to feel) a little unusual. 'Freak,' however, is too strong a word," Burnell decided. Nevertheless, women have contributed a large share of the most exciting recent developments in astronomy, especially in extragalactic space. They include E. Margaret Burbidge, Vera Rubin, Sandra Faber, Neta Bahcall, Margaret Geller, Jacqueline van Gorkom, and especially the late Beatrice Tinsley.

Although only about 7 percent of American astronomers are women, they were assigned in 1986, according to scientific merit, to one-third of the observation time at facilities like Kitt Peak Observatory outside Tucson, Arizona, and Cerro Tololo Observatory in Chile. The fact that astronomy has attracted more women than physics puzzles Burnell "because astronomers usually go through all the math and physics that are supposed to be the stumbling block to women going into physics."

When Burnell arrived in Cambridge in 1965, radioastronomer Anthony Hewish was preparing to build a large radio telescope to search for scintillating, or twinkling, radio sources.

Some celestial bodies can be detected by the visible light they emit or reflect back to Earth. Other bodies give off waves in different regions of the electromagnetic spectrum—from X rays and gamma rays at the short end of the wavelength spectrum to visible rays, infrared rays, and, at the long end, radio rays. Each region of the spectrum has its own story to tell. And each region must be studied with a different type of equipment built to receive that particular wavelength. Burnell did not know it when she arrived in Cambridge, but eventually she would study almost every area of the electromagnetic spectrum. At the time, she intended to investigate radio waves.

Radioastronomers do not look through ordinary telescopes, because radio waves are invisible to the human eye. In principle, they could listen to a simple home or car radio. But radio waves from space are so weak that radioastronomers must build giant receivers with enormous antennae and sophisticated amplifiers to collect and magnify the waves. The antenna can be a wire; any wire inundated with radio waves will develop an oscillating electrical current that can be amplified, recorded, and quantified. In Hewish's case, he planned to use 120 miles of wire. The longer the wire, the more electromagnetic waves it can intercept and the bigger the electrical current it develops. The currents are then fed into a feeder wire, which sends them to the amplifier.

Signals from cosmic radio sources are usually strongest at low frequencies. Accordingly, Hewish planned to build a radio telescope that would operate at 81.5 megahertz, close to the FM radio band, to catch radio waves 3.7 meters (approximately twelve feet) long. Anyone with a four-acre FM radio antenna could have heard roughly what Hewish hoped to hear—the waves emitted from quasars out in space. His telescope would scan a large part of the sky each week to plot the positions of scintillating radio sources on a sky chart.

In the grand old tradition of graduate student as slave laborer, Jocelyn spent her first two years in Cambridge helping Hewish build his radio telescope. To capture as much radio energy as possible, the telescope was basically a giant radio arrangement of wires stretched over 4.5 acres, roughly the size of fifty-seven tennis courts. An array of one thousand nine-foot-tall poles supported the 120 miles of wire, two thousand copper feeder wires, cables, and cord. Burnell and five other students built the telescope with the help of several enthusiastic visitors who sledgehammered posts into the ground during their summer vacation and actually seemed to enjoy it. Burnell is small and slim, but she learned to wield a twenty-pound sledgehammer with professional aplomb. She was in charge of installing the network of cables, attaching 350 connectors to them, and constructing 200 transformers. To keep costs down, Burnell made the cables herself, cutting and bending them into shape outdoors all winter in a canvas shelter.

Unbeknown to Burnell, she was not just building Hewish's radio telescope to scan for scintillating sources. She was also constructing a device that was ideal for discovering pulsars. It had a wide angular collecting area and repeatedly scanned the same area of the sky. It collected data on Hewish's quasars billions of light-years away and information on waves originating from much closer sources in our own galaxy. It was also the first radiotelescope designed to look for rapidly changing astronomical sources.

When the telescope was switched on in July 1967, Burnell shifted from brawn to brain work. At twenty-four years of age, she was in charge of operating the telescope and analyzing its data, under Hewish's supervision. The telescope scanned the entire sky every four days and spewed out data with four three-track pens that covered nearly one hundred feet of chart paper daily. Burnell analyzed the data by hand. The telescope was new and no one was familiar with its behavior, so the analysis was not computerized. If a computer had been programmed to look for specific kinds of data, Burnell might never have discovered pulsars.

After studying the first few hundred feet of charts, Burnell could distinguish between scintillating sources and background interference, such as French television, aircraft transmissions, and automobile ignition systems. Mapping the radio signals, she discarded those from manmade sources. "Radio telescopes are very sensitive—they have to be to detect the weak cosmic radio signals—but this means they are easily swamped by local radio interference. Fortunately, scintillation and interference usually look different on the charts, and one soon learns to distinguish between them," she explained later.

Although the telescope performed well, analyzing the charts was an overwhelming job. For the next eight or nine months she struggled to keep up with the paper spewing out of the pen recorders each day. By October, she was one thousand feet behind; by November, a third of a mile behind. In six months, she would be buried under 3.5 miles of the stuff. "The temptation to cut corners on the analysis must have been considerable," as science writer Nicholas Wade commented later in *Science* magazine.

In October, less than two months after the telescope started up, Burnell spotted the first "bit of scruff." It covered a mere half-inch of four hundred feet of chart. "The first thing I noticed was that sometimes within the record there were signals that I could not quite classify. They weren't either twinkling or manmade interference. I began to remember that I had seen this particular bit of scruff before and from the same part of the sky. It seemed to be keeping pace with 23 hours, 56 minutes, i.e., with the rotation of the stars." The scruff was keeping sidereal time, or star-time. Although Earth rotates every 24 hours with respect to the sun, it rotates every 23 hours and 56 minutes with respect to the stars.

"The nub of the discovery lay in that single instant of recognition," Wade commented. She had seen the scruff at the same time of day in previous charts, and it was "keeping sidereal, not terrestrial time." In short, it was coming from a source that rises and sets each day with the stars, not the sun.

"Unclassifiable things are too disturbing to be easily forgotten," she thought, so she remembered it. It occurred in a part of the sky that was transiting overhead in the middle of the night, when normally there should be little or no scintillation. Burnell could not find any cataloged radio source in that part of the sky either.

She told Hewish about the mystery, and they agreed to watch for it. Unfortunately, they had to wait until a high-resolution extra-fast chart recorder was installed. When it started up in November, Burnell hoped she would be able to see the scruff in more detail. For a month, she went outside every night to check the fast recorder. But the signals had disappeared completely. At first, Hewish thought it might have been a flare star. "It has died and gone, and you've missed it," he chided. She joked, "It's always the research student's fault!"

The night Burnell played hooky to attend a lecture in Cambridge, the scruff came alive again. Two nights later, on November 28, 1967, it reappeared again. As Burnell watched, the pen traced a series of regular pulses. They appeared to be equally spaced. As soon as the signal stopped, she ripped the paper off the chart recorder and measured carefully. The pulses were indeed equally spaced: one and one-third seconds apart.

Telephoning Hewish immediately, she explained that she had just recorded a regular series of pulses, one every one and one-third seconds. "Well, that settles it," he said. "It must be manmade." Burnell was puzzled by Hewish's reaction. "What I didn't know (but should have) was that it would be difficult to get such a rapid variation from a star, galaxy, or any other type of cosmic object then known."

Intrigued, Hewish came out to the telescope the next night to watch the recording. As if on cue, the signals came one and one-third seconds apart. They were lucky. It is now known that this particular pulsar can be fickle and does not often perform on demand.

The rapid, one and one-third second interval between signals was the biggest puzzle. "We had terrible trouble trying to sort out that conundrum," Burnell confessed. The fastest variable star then known signaled only three times daily; a star that pulsed every one and one-third seconds was nearly inconceivable. Yet the signal could not be manmade because it kept star-time, not Earth time. Hewish also ruled out radar signals bouncing off the moon, peculiarly orbiting satellites, and, after a poll of astronomers, a late-night scientist on Earth.

Hewish's other collaborators measured the pulses and discovered that they signaled every 1.3373011 seconds. They were accurate to within an astonishing one-millionth of a second. No other astronomical source signaled so regularly. Estimates put the source some two-hundred light-years away, far beyond our solar system, but still within our galaxy.

In honor of its regularity, Burnell nicknamed the scruff "Belisha Beacon" after the flashing orange globe that warns English motorists of pedestrian crossings. Others dubbed it LGM, for Little Green Men. They were only half-joking.

"One evening just before Christmas I went into Tony's office to discuss something and found I had walked into a high-level conference about how to present these results," Burnell said. "I was slightly peeved when I...realized they were discussing it." But the meeting apparently "just happened," and once she had arrived, no one told her to go away. So she stayed.

The discussion was about LGM. "We did not really believe it was Little Green Men, but on the other hand, we had no proof that it was not, nor had we any convincing alternatives to suggest. What was the responsible way to proceed in such circumstances?" Burnell wondered. After all, radioastronomers realize that they will probably be the first people on Earth to detect a signal from other civilizations in the cosmos.

As Hewish explained, with no satisfactory terrestrial explanation for the pulses, "We now began to believe that they could only be generated by some source far beyond the solar system, and the short duration of each pulse suggested that the radiator could not be larger than a small planet. We had to face the possibility that the signals were, indeed, generated on a planet circling some distant star and that they were artificial." In other words, Little Green Men. Was that why the signals appeared manmade but moved like a star?

Hewish wanted to keep the signals a secret until he was sure of the facts. "I felt compelled to maintain a curtain of silence until this result was known with some certainty. Without doubt, those weeks in December 1967 were the most exciting in my life," Hewish recalled.

Luckily, there was a simple way to test the LGM theory. If the signals originated from a small planet circling a sun far out in space, the signals would show evidence of a Doppler shift. The Doppler shift is the familiar sound effect that occurs as an ambulance siren moves toward and then away from you. As the ambulance approaches, the sound waves are bunched close together and sound high-pitched. As the ambulance moves away, there is a rapid change of tone as the sound waves space themselves out behind the ambulance. In the same way, when a planet orbiting around its sun comes toward Earth, its radio waves are compressed. As a planet moves away from Earth, the waves spread out again.

Hewish's team found no evidence of a Doppler shift, however, ruling out the possibility of LGM. At almost the same time, Burnell settled the question of the signal's source a different way. She found another.

After the LGM meeting in Hewish's office, Burnell went home "really rather annoyed, thinking, 'Here am I trying to get a Ph.D. out of a new technique and some silly lot of Little Green Men have to choose *my* frequency and *my* aerial to try signaling us.'" A quick look at her desk, piled with two-thousand feet of unanalyzed charts, sobered her up. After supper, Burnell returned to the laboratory to study them. The last chart she looked at before the lab shut for the night at 10 P.M. covered a section of the sky that was always badly contaminated by Cassiopeia A, one of the strongest radio sources in the sky. On the chart, during a break in the usual confusing mess, she spied something that looked remarkably like scruff. "Only this time," she realized, "it was from a different part of the sky.... With only minutes to go before I would be locked in the lab for the night, I checked previous records of that bit of sky, and sure enough—there it was on several occasions."

The lab was closing and Burnell was going home to Ireland to celebrate Christmas and announce her engagement to be married. The scruff was in a patch of sky that transited overhead at one A.M. Late that night Burnell went out to the telescope. The equipment did not operate properly in cold weather, and she had to flick switches, swear, kick, and blow hot air on it before she got it working. But for five crucial minutes, it behaved. A string of equally spaced pulses appeared, each 1.2 seconds apart. The second signal was even faster than the first!

Burnell dumped the chart on Hewish's desk the next morning and left for the holidays, much happier. "It was improbable that there would be two lots of Little Green Men simultaneously choosing the same unlikely frequency to signal the same planet Earth," she told herself.

Over Christmas Hewish kept the telescope operating, the recorders full of paper, and the pens well inked. He even checked on Christmas Day. He piled the unanalyzed charts on Burnell's desk.

The first day she returned, she found two more bits of scruff in different parts of the sky. "I wondered if I had not had too good a holiday!" she said. The fourth was much faster than the first three. With a period of only a quarter second, it was sometimes so strong that it drove the recorder pen against its end stops. "It was an awesome experience, watching a pen rushing across the paper and back again four times a second, and hard to believe that a star could behave like that."

In mid-January, Hewish decided it was time to write a paper announcing the discovery of "pulsars." He sent the paper—with S. J. Bell listed as the second of five authors from his group—to *Nature*, a prestigious British scientific journal. *Nature* published the article in only two weeks, an indication of its importance. A few days before it appeared on February 9, 1968, Hewish gave a seminar in Cambridge to announce the discovery. Every astronomer in Cambridge attended, giving Burnell her first inkling of the importance of her discovery.

An uproar occurred when the press learned that astronomers had actually considered the possibility of LGM from space. And when reporters discovered that S. J. Bell was female, the excitement turned to frenzy. For weeks, Burnell had her photograph taken, sitting and standing on embankments and sitting and standing over phony charts. A science reporter on an English newspaper invented the jazzy name "pulsar"; Burnell thought it was "a ghastly name," but it stuck. All through the excitement, she was trying to write her Ph.D. thesis and get her degree. The thesis detailed the angular diameters of about two hundred scintillating radio sources. She put pulsars in an appendix.

"It's impossible to overemphasize the excitement of that first year or so after pulsars were discovered," according to Burnell's friend, Joseph Taylor, astronomy professor at Princeton University. Astronomers who had signed up months before for telescope time were besieged by colleagues eager to do pulsar experiments immediately. Some creative dickering ensued: one night on a telescope soon equaled one week months later.

Although Hewish's team did not know whether the pulsars were white dwarfs or neutron stars, the latter seemed more likely. Neutron stars are unimaginably dense, burned-out stars only ten to fifteen kilometers (or five to ten miles) across. Astronomers had theorized in 1933 that neutron stars were mathematically and theoretically possible. But until the discovery of pulsars, there had been no evidence of their existence. Yet only a super-dense star could spin as fast as a pulsar. If an object as big and as heavy as Earth revolved every second, everything on its surface would fly off into space. Centrifugal force would pull the ground apart and hurl it into the galaxy; Earth's core would disintegrate.

A living star exists in equilibrium between gravity, which tries to pull all the matter in the star toward the center, and nuclear energy, which generates enough heat and pressure inside the star to keep it from falling in on itself. When a small star runs out of nuclear fuel, it is a burned-up ember, a white dwarf. When a larger star runs out of nuclear fuel, its matter falls in on itself with extraordinary violence. The collapsing matter may form a superdense neutron star at the center or, in the case of a heavy star, a black hole that swallows up the matter. During these violent transformations, some matter escapes and releases the energy of a supernova. Viewed from afar, the supernova looks like an explosion.

As a star dies and becomes a neutron star, it can shrink from a diameter of a million kilometers to ten to fifteen kilometers across. Its surface gravitational field becomes 100 billion times greater than that on Earth; the enormous pressure crushes matter and turns it into a thick soup of neutrons. The star becomes as dense as the nucleus at the center of an atom, a hundred million million (10^{14}) times denser than ordinary matter. As its mass contracts and compacts, it spins faster and faster, like the skater who brings her arms close to her body, reducing her effective size and increasing her density. As the star spins faster, it produces superstrong magnetic forces a million times more powerful than those on Earth. The twirling magnetic fields produce electric fields, and together they generate powerful radio waves that stream out of the north and south magnetic poles of the star. The star becomes a whirling lighthouse. When the beam of radio waves swings toward Earth, we detect them; when it swings away from us, the beam appears to shut off. Thus, the neutron star appears to observers on Earth to pulsate.

Radioastronomers and physicists rushed to study the process. Since a neutron star is as dense as an atomic nucleus, it was like a giant laboratory of nuclear matter ten kilometers across. Each neutron star has far more nuclear matter than exists in all the atoms on Earth. Physicists interested in extremely dense matter were fascinated because, on a density scale, neutron stars are superdense, stopping just short of black-hole status. Physicists interested in general relativity, gravitation, and superstrong magnetic fields could use neutron stars to test their theories experimentally. When even faster pulsars were discovered during the 1980s, the excitement flared up all over again. It is estimated that one hundred thousand active pulsars may exist in our galaxy alone.

And where was Jocelyn Burnell during all the excitement that she had created? At another university, in another city, quietly learning another specialty. As others leapt into the field she had discovered, she walked out. No one tried to convince her to stay. In 1968 she married, changed her now-famous name to Burnell, and with some regrets left radioastronomy. Had she not married, she might have gone to Jodrell Bank to continue work in radioastronomy, which was her first love. But her new husband, Martin Burnell, had a local government job in southern England, and Jocelyn wanted to be near him. "That move was the first of several that proved to be quite difficult. A lot of the moves I've made and the equally drastic changes of field within astronomy have been because I've moved as my husband moved around the country as a local government officer."

Burnell did not realize the magnitude of what she was giving up. She was just beginning to realize that many scientists thought of her as a discoverer of pulsars. "It's only over the years when one sees the impact it's had on the whole subject of astronomy and physics that you realize what you've started," she admitted.

Burnell taught at Southampton University for five years, doing gamma ray astronomy before moving in 1974 to the Mullard Space Science Laboratory in rural Surrey. By then, she had a one-year-old son and was working half-time in order to spend afternoons with him. In her spare time, she enjoyed walking, sewing, and knitting, and volunteering in local and national Quaker activities.

"I had originally intended to give up work completely to look after the child, but soon found that I missed the intellectual stimulus too much," Burnell said. "I was unwilling to have a nanny...[and] found part-time work a reasonable compromise." She liked the flexibility of choosing to do a job or not, depending on her home commitments. Practically speaking, it seemed better as a responsible parent to work part-time.

Her spectacular early success in astronomy gave her an advantage as she moved around the country trying to find a job somewhere near her husband. "If I hadn't had that behind me, I doubt if I'd have managed to stay in the field," she admitted. Professors and research directors were willing to bend the rules or create jobs for her when she moved into their area, so that she always had congenial and enjoyable work. She was frequently given more responsibility than her position warranted. But, as she noted, "My curriculum vitae does not look too good."

The Franklin Institute of Philadelphia provided the first clue that the excitement in Burnell's scientific life was not over. The institute, which prides itself on thoroughly investigating its prize candidates, awarded its prestigious Albert A. Michelson medal jointly to Hewish *and* Burnell in 1973. Although the institute refused to comment on its awards process, an officer indicated that "we are probably recognizing the recipients for equal efforts."

The Franklin award immediately fed speculation that Burnell might share a Nobel Prize with Hewish. Four physicists have won Nobels for their doctoral theses; Marie Curie is sometimes included as a fifth, but her radioactivity work was done for a University of Paris degree that is considerably more advanced and demanding than the Ph.D.

The Nobel Prize, reflecting Alfred B. Nobel's instructions, is given in only three scientific fields: physics, chemistry, and medicine or physiology. The Swedish experimental physicists who controlled the physics prize in its early years decided almost immediately to limit the prizes even more by excluding from the physics prize astrophysics, astronomy, and geophysics. But in 1974, the Nobel committee broke its own rules excluding astronomy and awarded the physics prize to astronomers Sir Martin Ryle and Anthony Hewish of Cambridge University. Hewish was cited "for his decisive role in the discovery of pulsars." If Burnell was disappointed, she never let on. Instead, she said diplomatically that the discovery of pulsars had brought her "enormous enjoyment, some undeserved fame." Little more was said until the following spring, March 1975, when Sir Fred Hoyle charged that Hewish had "pinched" the Nobel Prize by failing to give Burnell proper credit. (Hoyle is a noted British astronomer, the author of many of the popular astronomy books that Burnell had read as a child, and something of a gadfly in British astronomy.) His views started the joke that Nobel means No-Bell.

Asked to comment, Burnell replied, "It's a bit preposterous, and he has overstated the case so as to be incorrect.... My background in astronomy wasn't as good as Hewish's, and I didn't appreciate all the risks.... I continued to think it was a star until someone pointed out to me how fast [the pulse rate] was."

A few days later Hoyle wrote that his views had been "only crudely represented." The crucial step, he said, was Burnell's discovery of the signals and her finding that the source changed position with the stars. "There has been a tendency to misunderstand the magnitude of Miss Bell's achievement, because it sounds so simple—just to search and search through a great mass of records. The achievement came from a willingness to contemplate as a serious possibility a phenomenon that all past experience suggested was impossible.... I would add that my criticism of the Nobel award was directed against the awards committee itself, not against Professor Hewish. It seems clear that the committee did not bother itself to understand what happened in this case."

Cornell astronomer Thomas Gold, the first to offer an accepted explanation of pulsars, reiterated Hoyle's point. Burnell deserved a major share of the honor if she was the first to recognize that the pulsar was keeping sidereal time. "That realization would have been the first firm indication that the signals were coming from beyond the solar system and represented the true moment of discovery." In fact, at least one other astronomer did see the same bit of scruff that Burnell spotted—and ignored it as a machine malfunction. "You have to get him well lubricated to tell the story," Burnell laughed. "I can understand how it happened."

Hewish neither discovered nor explained pulsars, noted Princeton University astronomer Jeremiah P. Ostriker. If the Nobel Prize is given for a discovery, as Alfred Nobel intended, the prize should have included Burnell. "It's a pity that they got the Nobel and she didn't. They were just the people who'd raised the money for the laboratory in which she did her work."

At issue is the fine line between "full-fledged scientific collaboration and supervised research assistance, between irreplaceable and replaceable scientific contributions to prize-winning research," as sociologist Harriet Zuckerman defined it in her study of the Nobel science prizes. Hewish pointed out that Burnell had used his telescope, under his instructions, in a sky search that he had initiated. Later he told *Science* magazine more informally, "Jocelyn was a jolly good girl but she was just doing her job. She noticed this source was doing this thing. If she hadn't noticed it, it would have been negligent." Burnell herself contended that "Nobel Prizes are based on long-standing research, not on a flash-in-the-pan observation of a research student. The award to me would have debased the prize."

After investigating the controversy for *Science* magazine, Nicholas Wade concluded that a case can be argued for or against Burnell's inclusion. And George B. Field, astronomy professor at Harvard University, complained, "It's precisely questions of this kind that make one skeptical about the whole business of the Nobel prize."

As a Quaker from Belfast, Burnell knew the dangers of dissension in a small community like British astronomy, and she has worked hard to avoid it. She and Hewish are both "very careful with the relationship. The attention that has been focused on our relationship has strained it.... I think it's careful more than relaxed because we dare not be relaxed about it."

In any event, several other prize-giving groups recognized Burnell. She received the J. Robert Oppenheimer Memorial Prize from the Center for Theoretical Studies in Miami, Florida, in 1978. She was the first recipient of the Beatrice M. Tinsley Prize presented by the American Astronomical Society in 1987. The Royal Astronomical Society awarded her the Herschel Medal in 1989.

After the Nobel Prize fracas, Burnell stayed on at Mullard Laboratory. Still part-time, she was in charge of a small team analyzing data from a highly successful British X-ray satellite. Instead of receiving visible light waves or radio waves, the satellite zeroed in on celestial bodies that emit energy in the X-ray portion of the spectrum.

"It was an enormously successful satellite. We kept tripping over discoveries. Before we finished with one discovery, there'd be another one. It was tremendous fun," Burnell reminisced.

During her six years at Mullard, Burnell also worked part-time for the Open University. The Open University is a second-chance institution, offering correspondence instruction to adults who missed out on higher education in their youth. "It happens on a large scale, " Burnell said, "and I'm impressed and sobered at the very able students who work evenings for a degree for six or eight years. It's quite spoiled me for regular students." For Burnell, who also would have missed out on a university education without a second chance, the Open University was a natural.

Founded in 1969, it is now the United Kingdom's largest university, graduating eight percent of all UK university students. Open to all adults in Europe, irrespective of their high school records, the Open University enrolls the highest proportion of women and students from blue-collar families of any British university. Most of its students are between the ages of twenty-five and forty-five. Although four out of five do not have the high school qualifications to enter other universities, more than half graduate. They study at home from specially produced printed materials, early-morning and late-night television and radio programs, audio and video cassettes, home experiment kits, home computing, and short-residence schools. Tutors offer special help, in person or by telephone.

"It's very rewarding," Burnell said. "They're so keen, very committed. They're putting a lot of time and money into it. We do tutorials on weekends or, in Scotland where the population is so thinly spread, there are conference-call tutorials. Explaining vectors over the phone is quite demanding! I've had students on oil rigs, lighthouse keepers, in the Outer Hebrides. They're mostly doing it to get better jobs; others are mums out of work." Signed on as a tutor, she eventually became a consultant and guest lecturer.

In 1982, Burnell made her last job change in order to follow her husband. She took a part-time job at the Royal Observatory in Edinburgh, Scotland, taught part-time for the Open University, and enjoyed attending choral concerts during the Edinburgh Music Festival.

At the Edinburgh observatory, she gradually worked her way into an administrative job, managing from Scotland the James Clerk Maxwell Telescope, a large international telescope in Hawaii. Again, the project was in a new field for her: the telescope was designed to detect submillimeter radio wavelengths, in size between infrared waves and radiowaves. Submillimeter waves are absorbed by the water vapor in Earth's atmosphere, so the telescope was built 4,200 meters above sea level where the air is clear and dry. Burnell discovered she enjoyed administration but she was soon "grossly overloaded." In 1990, she switched to managing a new infrared satellite project.

Her moves had taken her from radio astronomy in Cambridge to geophysics and gamma rays at Southampton; to X-ray astronomy in Mullard; and to infrared and submillimeter astronomy at Edinburgh. In addition, she had worked in ground-based, satellite-based, rocket-based, and balloon-based astronomy. Few other British astronomers have such a broad background, according to Burnell. On the other hand, broad backgrounds are not the way to get ahead in research science, which puts a premium on depth of knowledge. Some astronomers her age who had leapt on the pulsar bandwagon were now full-time professors in prestigious universities.

Over the years, Burnell had become increasingly interested in women's issues. When she arrived in Edinburgh, she was surprised to discover that astronomy classes had the same percentage of women as in 1892, the year that the university's astronomy department first admitted women. "What is the block?" she demanded, concluding reluctantly that, "Womenfolk play the larger part in the decision. It has been my experience that it is other women who ask (too frequently) whether one really enjoys doing physics. It seems that attitudes are dictated largely by our sisters, and our cousins, and our aunts."

When churches moved to eliminate sexist language, she was irritated at first. "But gradually I have been alerted.... For example, my position is that of 'fellow' and the effort to complete a project is estimated in 'man-years.' Does this signal perhaps subconsciously to young women that this is not for them?"

"I believe the overlap between the sexes is much greater than the differences between them. I am unhappy with the assertion that women are better in particular areas of work and men in others. It seems more accurate to say that some *people* are more skilled in certain areas than others."

In Edinburgh, Burnell also began to have serious doubts about part-time, temporary jobs. "Looking at my female colleagues at the Royal Observatory, it seems that either we have made some unusual domestic arrangements or we have tried working part-time, hoping for good luck, obliging colleagues, and accommodating directors; *both modes seem far from satisfactory*.... The problem with part-time work is that it assumes that domestic and child care remain with women, and part-time and lower status jobs have often gone together traditionally." After years of part-time work, she declared, "I have some reservations about the emphasis on part-time jobs."

When her son was ten years old, he became seriously ill. Diagnosed as a juvenile diabetic, he was told that his pancreas had stopped functioning and that he would be kept alive with daily insulin injections for the rest of his life. A well-meaning friend sent the child a cheery get-well card, saying, "Hope you will soon be fit and well and good as new!" He was no fool and the card was kindly meant, but he knew he would never be "good as new" again—and he said so. He set his mother thinking about the role of the person who is not whole, who lives with hurts that do not mend, problems that do not disappear, crises that never end—people who are "broken for life."

Uncomfortable with modern society's emphasis on success, health, wealth, and achievement, Burnell struggled against "the shallow optimism where everything is always terrific and great and that couldn't admit of anything else." She was reacting in part to a diabetic child; in part to Northern Ireland, which refuses to mend quickly; and perhaps in part to the tension of combining career and marriage, an increasingly unsatisfactory compromise from the scientific point of view.

Yet problems like failure, old age, illness, poverty, injustice, and violence do not "go away." Her first fleeting thought, when her father died after an operation, was, "Well, bring him back to life again!" Then the truth sank in. Some pain never disappears.

After three and a half years struggling with her feelings, she put them together in a lecture for the Society of Friends' Yearly Meeting in Aberdeen in 1989. The lecture was subsequently published by the Quaker Home Service in London as a booklet entitled *Broken for Life*.

"Can you find a wholeness that includes pain and a readiness to suffer?" she asked. If God is a loving, caring God in charge of the world, why is there suffering? And why does so much of it fall on innocent people?

In her book, she offers a possible resolution to these ageless questions. Although she was loath to abandon the idea of a kindly God, perhaps God is not running the world. "If the world is not run by God, then the calamities that occur cannot be blamed on God. Perhaps God decided that we are responsible adults who should be given a free hand and allowed to get on with life without interference.... God would still exert influence on the world, but only through people, through their attitudes and what they do, through their healing and reconciliation."

As a physicist, Burnell found such randomness comforting. "It actually ties in very well with the randomness or uncertainty that modern physicists know is at the heart of everything and seems to be one of the 'givens' of this world." In fact, she found the idea liberating, "releasing one from the constraints of rewards and punishments, just and unjust, cause and effect."

"Sometimes religion appears to be presented as offering easy cures for pain: have faith and God will mend your hurts.... [But] healing so as to eradicate all trace of the encounter is not part of the package," she concluded. Brokenness is an essential ingredient in life. "Suffering can mature us and make us more sensitive to others; then through small deeds and kind actions we can interact with empathy, reassuring and helping others.... But pain is not part of a Grand Design and will not come to a purposeful ending unless we work at it to ensure that it does."

The byline on *Broken for Life*—"S. Jocelyn Burnell"—betrayed no trace of the Jocelyn Bell who discovered pulsars. That part of her life seemed irrelevant. Soon after she wrote the book, however, Burnell's marriage collapsed. As she joked, "I handled my maiden name badly. I discovered pulsars as Bell and got married. I wrote a book as Burnell and got divorced."

Rebuilding her life and looking for her first full-time job in two decades, she wondered briefly if she had been right to work part-time all those years. Would she do it over again? "I don't think I can handle that question right now," she responded.

In 1991, Burnell got a new job and left Edinburgh. She moved to Milton Keynes, a new English city, to become professor of physics at the Open University. It was her first permanent, full-time job in more than twenty years. The position, more physics than astronomy, represented a slight change of direction, but no more than that. Once more, she would be helping others get a second chance—or the first chance they never had. Fittingly, the job gives her another second chance as well. Her part-time work for the Open University, her many moves, and her broad knowledge of astronomy had turned out to be advantages. They may not have prepared her for a purely research career—she had always shied away from its commitment in order to leave time for teaching, social action, and people—but they had prepared her for a professorship with the Open University. She is the third woman to become a physics professor in the United Kingdom.

* * *

16

Christiane Nüsslein-Volhard

October 20, 1942–

DEVELOPMENTAL BIOLOGIST

Nobel Prize in Physiology or Medicine 1995

"THERE WAS A TIME when I was almost totally discouraged, and I was really, really distressed. I cried almost every night," Christiane Nüsslein-Volhard said, recalling her struggles as a student with a recalcitrant thesis topic. Even after switching projects, she experienced a month of failures and mishaps.

Then suddenly, she pulled herself together, saying, "I'll do it. I feel I have the strength."

"I had this sense of power," she remembered. "I wasn't being recognized as much as I wished, but I knew I wanted to do it. It was important for me." Returning to work, she glowed with happiness; her shyness had disappeared.

Nüsslein-Volhard's colleagues see her as strong, tough, and authoritative, but she maintains, "I'm not a very self-confident person usually."

One of the most important developmental biologists of all time, Nüsslein-Volhard has helped explain one of life's great mysteries: how a single cell becomes a complex creature like a fruit fly, fish, or human being. When she began studying the genetics of fruit-fly embryos during the 1970s, no one dreamed that their genes might also guide the growth of human embryos. Her studies of fruit flies and zebrafish have helped explain the genetic origin of human health problems from birth defects to cancerous tumors. Nearly half of all pregnancies end in spontaneous abortion, and one in twenty-eight infants has a birth defect. Many of these accidents stem from flaws in the genes that regulate the early development of embryos, genes that Nüsslein-Volhard and her research partner, Eric

Christiane Nüsslein-Volhard.

Left: Christiane Nüsslein-Volhard as a schoolgirl. *Below:* Christiane Nüsslein-Volhard, third from the left, vacationing with her parents, brother, and three sisters, about 1955.

Above: Christiane Nüsslein-Volhard and Eric Weischaus, right, and Trudi Schüpbach and Jamos Szabad at the time of the Nobel Prize–winning project in Heidelberg. *Below:* Christiane Nüsslein-Volhard and Eric Weischaus at a conference in Madrid in 1987.

Wieschaus, discovered in the fruit fly.

Nüsslein-Volhard is a study in contrasts. As a high-level woman scientist in Germany, she is virtually unique in her generation, yet the German women's movement looked askance at her embryology and genetics. She is a loner by inclination and worked for years outside the scientific mainstream, but she builds smooth-running research teams. She is a world-class scientist who prefers small-town life. A woman without mentors, she adopted the German poet Goethe as her hero. One of her scientific strengths is her artist's eye. Colleagues call her a blend of ambition and self-criticism, of brusqueness and generosity. She believes her temperamental moods are a mild and harmless version of the manic-depression that has more seriously affected some members of her family.

Christiane Nüsslein-Volhard grew up among artists in the pleasant Frankfurt suburb of Sachsenhausen. Her parents, Rolf and Brigitte Haas Volhard, married in 1939, early in World War II. Christiane, the second of their five children, was born on October 20, 1942, and was almost immediately nicknamed Janni, or "Yanni," as it is pronounced in German. She remembers her father, an architect and painter, as a charming, temperamental, and even charismatic man. Captured by U.S. forces after the war, he supported his young family by painting pinups and portraits of American officers. Canvas was scarce, and if a pinup didn't sell, he flipped it over and painted his children on the back. To help make ends meet, Janni's mother illustrated several children's books.

Evenings, while their mother played the piano, the children accompanied her on instruments. Janni played the flute and sang, belting out operatic arias with such gusto that a neighbor complained. Gluing reproductions of famous paintings to wood, the Volhards jigsawed them into puzzles, a game that sharp-eyed Janni continues to enjoy. Two Volhard children became architects, and two others studied art and music. As adults, they play chamber music together. "We take our art seriously," Nüsslein-Volhard said. "I'm always surprised that there are so many people around who don't have [art and music in their lives], and I wonder how they can survive. Whenever I go to the States, the first thing I go to is the Metropolitan Museum of Art."

Vacationing as a small child in the farm village where her grandparents had taken refuge during the war, Janni was thrilled by nature. She helped harvest crops and feed cows and horses. Although her earliest memory is of oatmeal so full of husks that it scratched her throat, in the country she savored ham and potatoes boiled in their skins. The little girl was also intrigued by her maternal grandmother, Lies Haas-Moellmann. She was a woman of strong discipline and character, and she painted in an impressionistic style that Janni greatly admired. Haas-Moellmann had abandoned her dream of studying art in Paris when she married a lawyer and always regretted her lack of education. But she gave her granddaughter an abiding respect for nature, good food, and career women.

Growing up in her family's Frankfurt home, Janni Volhard spent hours with her eyes to the ground in the yard and nearby woods. Obsessed by nature, she watched plants, staring inside their flowers, trying to understand what was happening. She collected snails, memorized plant names, elaborated theories, and planned complicated projects. The Volhard name is a distinguished one in German academia; a great-grandfather was an eminent organic chemist and scientific biographer, and a grandfather was a prominent professor of medicine. But within her immediate family, "No one took much notice of my interest in science. Intelligence didn't matter in my family. What counted was artistic accomplishment, wittiness, beauty." By the age of twelve she knew she wanted to be a biologist. "There was a certain feeling of loneliness in this."

Janni attended Frankfurt's Schiller School, a rigorous gymnasium for girls where her mother and aunts had studied. Her teachers were unmarried women, university graduates who taught high school because German universities hired almost no women. Senior year, her biology teacher discussed modern developments in genetics, evolution, and animal behavior. Janni Volhard developed her own theory of evolution, reported on research conducted by the Austrian animal behaviorist Konrad Lorenz, and memorized reams of nature poetry by the Frankfurt-born Romantic Johann Wolfgang Goethe. A month studying hospital nursing reinforced her conviction that she belonged in science, not medicine.

In a country aching to forget, Janni's teachers insisted on classroom discussions of anti-Semitism, genocide, guilt, resistance, and Hitler's personality. A graphic film made by the American government in connection with the Nürnberg war-crimes trials horrified Janni when it was shown in class.

Several years later, the Volhard children talked with their mother about the family's activities during the war. Two grandfathers lost their jobs because they were not Nazi party members, an aunt was tried and almost executed for impolitic remarks, and three uncles died. Her father, an air force pilot, crisscrossed Central Europe while transporting supplies to German forces at Stalingrad; Janni believes he must have seen evidence of the Holocaust and been deeply shocked. Although her family was not Nazi, neither were they in the Resistance, she noted. While her grandfather, the physician, tried to help individual Jews in small ways, nothing was done "in the grand manner.... I wouldn't say my family was particularly heroic. People were just struggling hard to have some normality."

After the war, her mother helped Jewish concentration camp survivors from abroad, finding them lodging in Frankfurt and accompanying them to courtrooms where they testified about German war crimes. Later, at their invitation, she visited several in Poland and Israel.

At school, Janni drifted through boring classes but plunged headlong into any subject that interested her. Entranced by a Goethe poem or a mathematical formula, she discussed them with her father; displaying her skills, she tried to win his approval. Although by now her family and teachers had grown accustomed to her scientific ambitions, she realized even as a young girl that whatever she accomplished she would have to do on her own. She rarely did homework, however, and passed her final examinations with mediocre grades. She almost failed English. Teachers praised her talent for scientific investigation, but regretted her "strong display of self-will." They thought her passionate, emotional, and motivated by curiosity, not ambition.

She would need all the "self-will" she could muster because the day she took her final examinations in 1962 her beloved father died suddenly of a heart attack.

When Janni Volhard entered Frankfurt's university a few months later, the odds were stacked overwhelmingly against her becoming a research biologist. As late as 1996 in Germany, men still outnumbered women two-to-one in university science courses and in science degrees in Germany. At the Ph.D. level, the gap widened to three-to-one. In the United States, men outnumber women Ph.D. recipients in science two to one.

In addition, Germany is a winner-take-all scientific community. Top researchers direct departments or institutes; others work at their direction. Unlike the United States, there are few intermediate possibilities. Moreover, German universities require a second lengthy thesis, called the *Habilitation*; researchers who pass this hurdle in their early forties must then find permanent jobs at other universities and relocate there. At the top of the career ladder in Germany today, women occupy only 5 percent of all professorships, including those in the humanities and social sciences, where women are more strongly represented than in science. In the United States in 1995 women accounted for 23 percent of all tenured, that is, senior, faculty members in four-year institutions of higher education and an estimated 15 percent of tenured science and engineering faculties.

Despite Janni Volhard's high hopes, university life in Frankfurt disappointed her. The anonymity of Johann-Wolfgang-Goethe-University, one of Germany's largest, forced her to confront her shyness. She had none of the small talk that makes meeting strangers easy; she said what she had to say and then fell silent. Planning her curriculum was complicated, too. Frankfurt's biology was too elementary, and when a zoology professor gratuitously decapitated a frog during a lecture, she switched her major to physics and chemistry. Eventually, those subjects became too abstract for her, and she ran out of options in Frankfurt.

Fortunately, the University of Tübingen was starting Germany's first biochemistry major. There she could study biology and get a solid grounding in the basic, quantitative sciences. But she and a young Frankfurt physics student, Volker Nüsslein, had fallen in love. Tübingen was approximately one hundred miles to the south, near the Swiss border. "I don't see myself making decisions," Nüsslein-Volhard explained later. Her opinions form gradually, until suddenly, "It's very clear. I have to do it." Most German women would have stayed in Frankfurt, she knew. But she could not. Abandoning Frankfurt, she moved to Tübingen in November, 1964. Volker Nüsslein supported her, saying, "Sure, you do that. There's no question."

"In retrospect, it was hardly a decision at all," Nüsslein-Volhard said. "We wrote letters every day. It was not that we were not in love. We took that very seriously. But this was a very difficult situation, and we tried to compromise."

Janni Volhard—later Janni Nüsslein-Volhard—has spent most of her career in Tübingen or in Heidelberg or Freiburg, two other small, medieval university towns within a seventy-five-mile radius. Tübingen—her favorite—is the smallest with fewer than 85,000 residents. Tübingen's university was founded in 1477; the astronomer Johannes Kepler and the philosopher Georg Wilhelm Hegel studied there. In the nineteenth century, a Tübingen study of snails provided the first demonstration of Darwin's theory of evolving species.

Nüsslein-Volhard may like novel and far-reaching perspectives in science, but she prefers to live in cozy and familiar locales filled with picturesque charm. As late as 1998, the longest she had ever been in a non-German–speaking place was three weeks.

Nüsslein-Volhard stayed ten years in Tübingen, starting as an uncertain twenty-two-year-old student and emerging as a full-fledged research scientist of thirty-two. At first she ignored boring classes and earned mediocre grades, but in her final year as an undergraduate, she discovered genetics and microbiology, protein biosynthesis, and DNA replication at the Max Planck Institute for Virus Research in Tübingen. Although she could barely understand the lectures, they excited her. And it was there that she met Heinz Schaller, her thesis advisor for her master's and doctor's degrees and for one year of postdoctoral studies.

Even as a young woman, Nüsslein-Volhard struck Schaller as original. After she married Volker Nüsslein and he switched to molecular biology in Tübingen, the couple moved into one of the town's oldest buildings. It was a drafty, sixteenth-century house that leaned against the wall of a medieval Cistercian abbey just outside town. The cluster of buildings at Bebenhausen, converted into a royal hunting lodge during the seventeenth century, is today preserved as a historic site. But in the 1960s, it was far from fashionable.

Nüsslein-Volhard loved it and immediately planted a flower garden. "A garden is artificial," she conceded. "You can't avoid that. But it's the closest you can come to nature." Puttering around the flowers, she said, "I look, and I try to understand.... It's a form of biology." The Nüsslein's little house became the headquarters for their friends' social activities. Volker Nüsslein was a postdoctoral fellow with Friedrich Bonhoeffer at the Max Planck Institute. Janni revered Bonhoeffer's family; three of his uncles died resisting the Nazis. Schaller's wife, Chica, now a professor at the University of Hamburg, was a graduate student in biology. So the three couples—the Schallers, Bonhoeffers, and Nüssleins—formed a close scientific and social circle.

Nüsslein-Volhard's determination to pursue a scientific career as a married woman set her apart from most of the women around her. On a practical level, German life discourages wives from working outside the home. Until recently, all West German stores closed during non-working hours: at lunch time, after 6:30 P.M., and on weekends from Saturday at 1 P.M. to Monday morning. Clothes washers required manual adjustment with each new wash or rinse cycle. Elementary schoolchildren are sent home for the day at lunchtime. Child-care centers are scarce and frowned upon. Women who allowed others to care for their childern were called Raven Mothers because, according to German folklore, the birds neglected their young. Early in the twentieth century, when Nüsslein-Volhard's grandfather tried to marry as an assistant professor, he was told that young scientists had to remain single. Today many German women still think that families and careers cannot be combined. Futhermore, postwar Germany provided few examples of prominent women in science; Lise Meitner and Hertha Sponer fled the Nazis, and Maria Goeppert Mayer married an American and worked in the United States.

Ignoring these difficulties, Nüsslein-Volhard convinced Schaller to change her research topic. Then she plunged joyfully into her new project. She got her first real laboratory training and learned to think in quantitative terms. Between 1969 and 1977, her name appeared on six articles, including a letter in the prestigious journal *Nature*. But according to Nüsslein-Volhard, someone else was made first author on one paper "because he was a young man and would have a family to support." At the time, she didn't view the decision as gender discrimination. "As grad students, we didn't talk about being discriminated against, though in retrospect I probably had more difficulties [than men]." But she didn't think about those things then. She believes her lack of awareness was actually an advantage. She wasn't distracted by constantly asking, "Am I treated badly?" or, "Do they take me seriously?"

Today Schaller agrees, "Janni should have been first author on the *Nature* paper, since she did most of the experiments and also participated in the writing. However, probably due to my lack of experience in leading a group (these two were my first Ph.D. students), B. H. was chosen to be the first author, since the major topic of the paper was supposed to be the subject of *his* thesis. There was no disagreement within the group about this decision."

In any event, by the time Nüsslein-Volhard finished her thesis on gene transcription in bacteria, she was an experienced molecular biologist. She could have expanded her thesis with more details, but she was bored again. She craved the stimulation of new and broader problems. Few German biology students thought about their long-term research interests. But Tübingen's Max Planck Institute was changing the direction of its research from viruses to developmental biology, and she attended its exploratory seminars, plied its researchers with questions, and combed the scientific literature.

Dropping molecular biology, she decided to focus on one of biology's biggest issues: How does a single cell develop into a complex living being? In other words, how do generations alternate between one-celled simplicity and multi-celled complexity? As the egg cell divides, it creates an organism with millions of cells, each armed with specialized information as to what type of cell it must become and where it should be located, whether in nerve, muscle, blood, or brain. So what information does that first, single cell contain? In particular, Nüsslein-Volhard wondered how many shape-making components are already present in the egg, where are they localized, what is their molecular nature, and how do they function? Over and over she asked herself, "How does something so complex develop out of something so simple?"

None of these questions could be answered in 1970. To solve the puzzle, she needed to artificially alter the informational content of the egg cell. Ideally, she could remove or displace individual genetic components and visibly change some part of the embryo. Biologists already knew that an organism's development is steered by genes; altering any of those small packets of inherited information causes mistakes in the organism's growth.

Searching for an experimental animal, Nüsslein-Volhard read about several bizarre fruit-fly mutants. They were poorly understood, primarily because no one had the tools to look at embryos. But she already knew some techniques and thought she could invent more. She concluded that the ideal experimental animal would be *Drosophila melanogaster*, the little red-eyed fruit fly that multiplies in overripe fruit and the dregs of wine glasses.

Although few German scientists studied fruit flies at the time, almost every breakthrough in mammalian genetics in the twentieth century had started with fruit-fly research. Geneticists maintained breeding lines of flies with odd eye colors, abnormally shaped body bristles, and the like. Knocking out or altering a gene with chemicals or radiation, they watch what defects result. Such experiments require enormous numbers of flies, which obliging female fruit flies produce quickly and cheaply in little jars. Their embryos develop conveniently outside the female's body. And within twenty hours of fertilization, a fruit-fly embryo has formed a larva encased in a cuticle that is divided into broad segments and richly decorated with bristles, hairs, and other visible features. Each detail is a landmark waiting to be mutated. Merely by observing a larva's cuticle, Nüsslein-Volhard thought she could spot subtle changes created by genetic mutations.

Eager to conquer new worlds, she parted from her thesis adviser "not on the best of terms," though they reconciled soon after. She had received a long-term fellowship from an international research consortium, the European Molecular Biology Organization, called

EMBO. Armed with it, she would make her way alone, without a mentor. And like a virtuoso, she would make the little fruit fly reveal more about development in the natural world than anyone thought possible.

To do so, she switched from trendy molecular biology to déclassée embryology, genetics, and fruit flies. Virtually every young man being groomed for scientific success in Germany went to the United States for postdoctoral training. But Walter Gehring, an established geneticist and embryologist, had started a top-flight biology laboratory in nearby Basel, Switzerland. Meeting Gehring at a scientific conference in 1973, Nüsslein-Volhard summoned up her courage and asked to work in his laboratory.

"She is certainly a very original kind of woman," Gehring immediately thought. She topped her curly hair with a big, brownish hat, like a man's but more oval and elegant, Gehring remembered years later. Nüsslein-Volhard made it clear to Gehring that she planned to combine genetics with developmental biology in Basel whether or not her husband stayed in Tübingen. "She was very determined...She knew what she wanted to do," Gehring realized.

So Nüsslein-Volhard prepared to move a hundred miles across the Swiss border to the German-speaking city of Basel. Toward the end of 1974, Gehring's laboratory buzzed with the news: an incredibly talented young man would arrive in January. Then suddenly at the same time a thirty-two-year-old woman appeared too. No one had heard about her coming, though. It was Nüsslein-Volhard.

Gehring had assembled a group of brilliant young people. Some of them found Switzerland provincial and nationalistic; two referendums had just attempted to rid the country of foreigners, and Swiss women had been able to vote in national elections for only three years. A Canadian postdoc's landlord asked her to hang curtains; otherwise, the neighbors might think she was Italian. Two of the men could not rent a television because they were foreign and might abscond with the set. But Nüsslein-Volhard found Basel, the home of Switzerland's oldest university, charming. It was more cosmopolitan than Tübingen, and she liked hearing the local Swiss-German dialect in the streets and speaking English in the laboratory. Gehring's postdocs were like thoroughbred racehorses, prima donnas if you will, eager to show their stuff. He let them loose in his laboratory.

"I immediately loved working with flies," Nüsslein-Volhard recalled. "They fascinated me and followed me around in my dreams.... I knew nothing about embryology and nothing about flies. It was beautiful to see the flies grow and change, but I had no idea what I was looking at. I would just sit and watch, not able to make heads or tails of anything."

Admittedly, she adds, "Maybe there's nothing particular about flies that makes them more attractive than other things, but when you look at them in detail they get very beautiful. Every embryo is beautiful, and it gets more beautiful the more you know about it."

Eric Wieschaus, who had just finished his Ph.D. thesis with Gehring, taught Nüsslein-Volhard fly embryology. Wieschaus (which he pronounced Veesh-house or Wish-house, depending on whether he was speaking with Europeans or Americans) had a self-deprecating, clownish humor that fooled the unwary. He didn't care what others thought. He was a Roman Catholic, Vietnam War protester from Alabama and a frustrated painter. An imaginative thinker about developmental biology, he struck Nüsslein-Volhard as a charming genius. Five years younger, he already knew more about fruit-fly embryos than almost anyone else in the world. He soon left to be a postdoc in Zurich, but on visits to Basel he met Nüsslein-Volhard for dinner and fly talk.

To understand the mother fruit fly's genetic contribution to the embryo, Nüsslein-Volhard began studying a mutated embryo that grows a tail at each end instead of a head and central section. She was fascinated by the fact that the mutated gene responsible for the deformed embryo came from the female's egg long before it was fertilized by the male's sperm. Genes from a normal male have no influence on the defect. Apparently, some of the embryo's structure is determined by the egg in the female's ovary before fertilization.

A beginner in a new field, she struggled to grow enough mutants to study. The work was tedious and difficult; her two-tailed mutant is still not completely understood. Jeanette Holden, a Canadian postdoc and a brilliant geneticist, taught her fruit-fly genetics. Sipping coffee one idyllic afternoon in a restaurant overlooking the Rhine River, Holden was thrilled to share genetics with someone so eager.

With Holden's help, Nüsslein-Volhard eventually developed a series of cheap, low-tech techniques to quickly identify large numbers of *Drosophila* embryos. Her methods enable even small laboratories to study fruit flies today. Instead of collecting the eggs from one fruit fly at a time, she housed each egg-laying female in a small plastic tube and glued twenty tubes together. Then she inverted this "fly condo" over a gel-coated petri dish so that the females laid their eggs in twenty tidy circles on the plate. Nüsslein-Volhard could search for dead and deformed embryos in twenty families at once—a big time-saver considering that she needed to screen thousands of egg clutches a week. Tens of thousands of embryos later, Nüsslein-Volhard could spot every subtle deviation in a cuticle within seconds. The keen eye that gave her an edge with jigsaw puzzles gave her an advantage in the laboratory too.

Gehring's postdocs formed a closely knit group, whether helping each other feed flies or sharing an apartment for an Alpine ski week. Nüsslein-Volhard spent most of her time in the laboratory, but she occasionally visited a good bakery with a laboratory colleague or attended a "party for girls" given by Gehring's popular secretary, Erika Marquardt. Nüsslein-Volhard relished English mysteries, strong coffee, and cigarettes. She liked cooking for crowds too, and after dinner there were poker sessions or music with Nüsslein-Volhard and visiting Wieschaus at the piano.

Yet Nüsslein-Volhard stood apart within this group too. Thanks to the German educational system, she was a few years older than the others. When the young men got rowdy in the lab, she screamed at them to simmer down. She was the only one with a car. She drove a French Deux Chevaux, an underpowered "Two-Horse" car with windows that flapped in the wind; Germans called it a Duck. Moreover, unlike the other postdocs, she cared about esthetics. She rented an apartment "with character" and filled it with antique furniture, her piano, her flute, and a cat, who precipitated a crisis by eating the landlord's pet cricket and acquiring fleas. She drew her own scientific illustrations and hand-lettered her slides to please the eye; she complained to a speaker who scribbled his figures on the blackboard, "That does not look like an egg."

She was also the only Gehring postdoc with a divorce. Already known professionally as Nüsslein, she regretfully decided that she could not abandon her married name. She would have to hyphenate her two last names. Others may call her Janni Nüsslein for short, but she uses Nüsslein-Volhard. She has remained single. "It just happened," she said. She is emphatic about one thing, though: her German male colleagues, all married to full-time housewives, have easier lives. "When they come home, everything is ready.... I'd like to have an invisible butler and some servants whom I wouldn't have to supervise."

Not only did her sophistication and marital status set her apart, but so did her refusal to join sides in the laboratory's hot debates about feminism and male chauvinism. When someone asked Gehring if women performed better than men at any profession, he thought a long time before answering, "Pottery." Hearing that,

Nüsslein-Volhard stalked out of the room. When she sensed a professional slight, though, she bounced right back with a counterattack. She felt that many men "expected less of a woman. The attitude was, 'I'll give her a chance, but I'm sure she won't perform.'" It took her a long time to conclude that the issue was gender.

At the same time, Nüsslein-Volhard associated "women scientists" with amateurism and refused to let their gender sway her opinion of their work. "In this respect, I am very firm with myself, severe with myself if you wish. I try to do things because of their own merit and not because of personal relationships. [I do it to] an extreme. I try to be very objective with people."

Her scientific isolation within Gehring's laboratory bothered her most. Gehring was switching from genetics and embryology to molecular biology; she was doing the reverse. She had come to Basel for his past, not his future. Since no one in the laboratory worked on a project like hers, she had few opportunities to discuss her research. Sometimes she felt like a lonely outcast, struggling to continue.

She was still betting that she could explain embryonic development with genetics, not with molecular biology. Molecular biologists wanted to use the biochemistry and structure of molecules to study heredity. Nüsslein-Volhard's interest in genetics, embryos, and fruit flies seemed old-fashioned to them. But Nüsslein-Volhard was unusual even among *Drosophila* geneticists, few of whom were knowledgeable about molecules. Traditional geneticists had studied relationships between genes, not in terms of their biochemical reactions, but by their proximity to each other on chromosomes. For Nüsslein-Volhard, a gene's position on a chromosome was irrelevant. She thought of genes as units of functionality. And she intended to explore genes by studying their actual effect on offspring. Nüsslein-Volhard knew what she wanted to work on in Basel. She did it, and she did it alone.

She was driven, focused, and systematic, even in the face of incomprehension. She was often impatient with disagreement, and she was not always easy to get along with. She explained points of genetics patiently, but obviously disliked stupid questions. She was intense and private but could not hide her feelings. At rest, her face was serious, even severe; those who belabored their points could literally watch her lose interest. As a result, she often unintentionally controlled the conversation. On bad days, when she felt moody, she disappeared inside her sweatshirt hood and was silent. Working late nights, she drank too much coffee the next day and removed delicate embryos from their cuticles with shaky hands.

She wrote to Jeanette Holden that Gehring grated on her nerves. When she told him about a new discovery, she wanted him to get excited and praise her. Instead, he just suggested more work for her to do. She thought he was pleased with her progress but unable to show it. She, in turn, was impolite to him.

Nüsslein-Volhard published her Basel research alone, without Gehring's name on it. She was the only Gehring postdoc to do so. She wrote Holden that she had sent a small article about *Drosophila* laboratory techniques to a fruit-fly journal without even telling him. As laboratory director, he was entitled to appear on all his postdocs' publications. She said Gehring didn't understand her major article about the fruit-fly mutant with a tail at each end. But she also said he was gracious enough to let her publish it on her own. He said the work was hers alone and didn't need his name. In any case, Nüsslein-Volhard said, it was "the most difficult mutant I ever studied, with unbelievable patience and in retrospect little reward."

She left Basel after only two years; her fellowship had expired, and Gehring did not invite her to stay longer. Jobs were scarce. Germany had few positions for anyone in developmental biology or fruit-fly genetics, she was new in the field, and her publication record was scanty. The thought of traveling abroad, however, unnerved her. Writing Holden, she worried about getting lost in London and collaborating with a stranger thousands of miles away. After months of trying to interest others in her work, she feared competitors. Writing strangers for jobs was an exercise in humiliation. More than twenty years later, she acknowledged that Gehring's laboratory had "a wonderful, a fun and stimulating atmosphere.... I should have stayed longer."

Once again, though, she left a laboratory not on the best of terms with its director. And once again she moved to a small, medieval university city in southwest Germany. Klaus Sander, an expert in insect embryology and the first to describe chemical gradients in an egg, was a zoology professor at the University of Freiburg. Sander had no job or salary for Nüsslein-Volhard, but he did have space in his laboratory for her. So she used a postdoctoral fellowship from the grant-making German Research Society to work there and job hunt.

While in Freiburg, Nüsslein-Volhard discovered another fruit-fly mutation caused by damage to the mother's genes. Dorsal, as she named it, disrupts an embryo's belly-to-back axis. The mutant she had studied in Basel damages the embryo's head-to-tail axis. Putting the two together, she could see how an embryo develops along two coordinate systems that determine whether cells become a head or tail, or a back or belly.

Invited to talk about her new mutant at a meeting of the American Society of Developmental Biology at the University of Wisconsin, she made her first trip to the United States in 1978. She disliked it; the coffee was too weak, and she missed the brusque directness of German scientists like herself. Americans, she thought, are sweeter and more polite but also more insincere. "With Americans you never know what they mean. They're sweet to you in your face, and then you learn that at the same time they are cheating you." Side-trips to visit friends in New York City and San Francisco scared and depressed her. "I was amazed at how different the world could be. I'm a timid person; I'm not courageous about traveling. I was frightened several times." Later, still nervous about public speaking and traveling, Nüsslein-Volhard sometimes dispatched graduate students to give talks in her stead.

In the meantime, she applied for the second time to lead a small research group at a new European Molecular Biology Organization laboratory (called EMBL) under construction in nearby Heidelberg. This time, she got the job—provided that her friend Wieschaus became a joint leader with her. Promised adequate funding and no teaching responsibilities, they could do some of the experiments they had talked about over long dinners in Basel. It was a wonderful opportunity, and Nüsslein-Volhard and Wieschaus were delighted. In any event, Nüsslein-Volhard, at the age of thirty-six, had no other option.

When Nüsslein-Volhard and Wieschaus moved to Heidelberg in 1978, it was her third medieval college town in southwest Germany. Heidelberg's university, founded in 1386, is the oldest in Germany. And EMBO's laboratory sat perched like an ivory tower overlooking the town.

Inside the building, their quarters were tiny. Barely the size of a bedroom, they were crammed with two group leaders, a talented technician, Hildegard Kluding, a fly stock keeper, thousands of fly bottles, microscopes, fly food, and papers.

They got off to a rocky start. "I felt at the time that Eric was much more successful than I.... I also had the impression that I was dependent on him because he had more fly experience and without him I would not have gotten the job." While Wieschaus pondered experiments, Nüsslein-Volhard busily ordered microscopes, bottles, and fly food. When he inquired why she was not doing experiments too, they had "a little discussion" about sharing administrative duties and he immediately pitched in. "From then on we thoroughly enjoyed working in the same lab," Nüsslein-Volhard noted.

Lack of space soon convinced them to collaborate. At the time, most developmental biologists studied individual genes that had been discovered by chance. "It occurred to us that you can't really know what a gene does without knowing how it fits in with other genes," Nüsslein-Volhard explained. "Is it unique or one of many? How many? What is its place in the hierarchy? It seemed essential to find out how complicated the system was." Rejecting the haphazard approach to gene discovery, they launched a massive, systematic search for all the genes that affect early embryonic development.

They decided to mutate virtually every one of the adult fly's genes and observe the effect on its descendants. They could have mutated the genes on just one chromosome and harvested enough data to study for the rest of their lives. But their goal was not just to identify genes; it was to understand how an embryo grows. It was a daring gamble. No one had ever done anything similar, and they might discover too many genes to classify them meaningfully. But they realized that they were the only two people in the world with the knowledge to attempt the project.

To narrow their focus, they planned to track only those genes that make the embryo develop an abnormally shaped cuticle, or skin, during the first few hours after fertilization. During this period, the fruit fly's cells multiply and gradually specialize to become the larva's head, tail, front, and back. They focused on the genes that divide the larva into its fourteen body segments, because they determine the adult insect's overall body plan of a head, central section, and tail.

First they fed adult male flies sugar water laced with a chemical that randomly damaged their DNA. Then, over the course of a year—with the help of Gerd Jürgens, now a professor at the University of Tübingen, and Hildegard Kluding—they established 27,000 inbred lines, each with three generations. By the summer of 1980 they had screened a total of 18,000 mutations and analyzed seven hundred in more detail.

It was a Herculean job requiring finely tuned teamwork. Soon after the eggs were laid on a dish rimmed with food, they hatched and the living larvae crawled toward the food. The dead embryos remained in the center, awaiting examination. A drop of bleach removed their outer membranes; a drop of mineral oil made their inner cuticles transparent and revealed the embryos inside. Nüsslein-Volhard and Wieschaus, perched on either side of a dual microscope, examined dish after dish of dead embryo cuticles. One female fly lays about fifty eggs daily; one person can screen 10,000 flies in five or six weeks. They went through a round of flies in ten-day life cycles, resting until the next onslaught of embryos. Set them up, mutagenize them, mate them, then collect their progeny and examine them—it could take a month, depending on how fast they worked.

One of their major breakthroughs was artistic. They realized that months of examining embryos would be "totally disgusting," in Nüsslein-Volhard's words, unless the larval cuticles were beautifully displayed. As Wieschaus explained, "My work has always had a strong visual component (probably to assuage my suppressed teenage desires to be an artist or painter)." "Preparing them so we could see things in all their beauty hadn't been done before. But if it's not esthetically pleasing, you can't look at them for a long time," Nüsslein-Volhard agreed. Skillful mounting not only revealed the cuticles' delicate pattern with every hair gleaming but also created a library of mutants for reference.

Each evening, Nüsslein-Volhard and Wieschaus flew down from their ivory tower in the Duck, its windows flapping. After a restaurant dinner and talk, they jumped into the Duck again and lurched back up the hill to EMBL for more work. Saturday afternoons, they took a break for fruit tarts from a local bakery.

Eventually, more than six people crammed into the little laboratory: three technicians on an early shift and the scientists later. As word of the excitement spread through the *Drosophila* community, people came to lend a hand for a week or two. When Jürgens arrived, his share of the bench space was not much more than a meter in length. The sheer density of people focused on a single problem sped the project along. The only things that mattered were the laboratory and the experiments. Coming in to see what new kind of mutant had been discovered the night before made every morning exciting.

Emotional, impolitic, and persistent, Nüsslein-Volhard argued for more room and technicians with EMBL's director, the late Sir John C. Kendrew. Kendrew had won a Nobel Prize in Chemistry in 1962, but he could not understand why tiny fruit flies required so much space. A *Drosophila* expert told Kendrew—in front of Wieschaus and Nüsslein-Volhard—that "he needed some really good fly geneticists, not these two." Nüsslein-Volhard thought Kendrew was hedging his bets where their work was concerned: He didn't want to invest too many resources on a gamble. Nüsslein-Volhard saw people as either for or against her personally, and she did not think Kendrew took her seriously as a scientist. EMBL had three other women scientists, but he either dropped them when their contracts expired or tried to team them with men. Kendrew found Nüsslein-Volhard abrasive and preferred to negotiate with Wieschaus.

With Wieschaus billed as the team's diplomat, Nüsslein-Volhard became the lab cop. She wanted the best, and her opinion was written in every gesture and expression. With thousands of bottles and mounted embryos to keep track of, any deviation from the standard procedure threatened to wreck the project. When someone tried a new fly food and the females refused to eat or lay eggs, disaster loomed, and Nüsslein-Volhard was furious: "You see what happens when you do that." But Nüsslein-Volhard confided to others that she felt the emotional strain of being the lab's bad guy.

The pace was unrelenting. Gazing through the microscope, Nüsslein-Volhard and Wieschaus asked whether the embryos of a given stock looked abnormal in the same way. Was there a group of mutants that did something constant during development? As Nüsslein-Volhard, the jigsaw puzzle fan, said, "I have a strong sense of simplifying matters, an urge to always find a general rule out of large diversities."

Together, they tried to classify the mutants, searching for repetitive defects, different periodicities, commonalities, and differences. It was not always easy to agree. Generating overwhelming masses of data, they tried to focus on important features and let details fall by the wayside.

Far from bogging down, they felt exhilarated. "Those years were probably the most exciting, intellectually stimulating ones of my entire scientific career," Wieschaus said. Biologists working on fruit flies had never studied embryos before. And the experiment revealed natural phenomena that confounded everyone's predictions. "Almost every day we could expect to encounter a new phenotype, a phenotype that would force us to re-evaluate some long held assumption about embryonic development." Nüsslein-Volhard found it "extremely exciting—no major disasters, hard work, and great fun."

Gradually, the patterns became clear, and they knew they'd hit the jackpot. One set of mutants developed only half the segments; every other segment was omitted. This was striking, unexpected, and novel. Then they spotted other mutants behaving oddly: some caused major gaps in the embryo's pattern while others duplicated parts as mirror images.

Nüsslein-Volhard and Wieschaus soon narrowed the number of genes that make the embryo's body develop a normal head-to-tail pattern to an astonishingly small number: fifteen out of the fruit fly's twenty-thousand genes. The fifteen genes fall into three distinct categories and progressively subdivide and refine the embryo into segments. The first group broadbrushes the embryo into regions along head-to-tail and front-to-back axes. Next, the second group divides the regions into segments. The third group details the structure of each segment. Surprisingly, most of the genes affect particular areas of the embryo, rather than specific cell types or tissues.

Their evidence was so unexpected that, when it was previewed, one scientist scornfully called it, "A Thousand and One Ways for *Drosophila* Embryos to Die." Later, at a conference in the United States, Nüsslein-Volhard spent hours polishing her talk as she walked in circles at the center of a large, grassy field. When it was time to give her speech, she began nervously, her voice shaky for several minutes.

Their report made the cover of *Nature* in October 1980. Their model was quickly accepted: The mother fly leaves signposts for her embryos to follow. Different concentrations of a maternal gene product let loose hierarchical cascades of genes that oversee the earliest organization of the embryo, progressively subdividing it step by step, telling cells where they are and where they ought to be. They ended the article with the words, "All mutants are available on request." Fruit flies quickly became one of developmental biology's favorite laboratory animals. Her gamble had paid off: Genetics, fruit flies, and embryos were a winning combination. Genetics and pure observation had carried the day.

Three years later, Nüsslein-Volhard and Wieschaus expanded on their original work with a catalogue of 120 well-defined genes that affect the entire development of the fruit fly's embryonic pattern. These papers, published in *Roux Archive* of 1984, became the basis for almost all later research on the fruit fly's development.

Nüsslein-Volhard and Wieschaus had completed their Nobel Prize–winning project, but neither could get a permanent job. In fact, EMBL let both of them go. As Nüsslein-Volhard observed, "Sometimes it struck us how strange it was to discover very exciting things and know at the same time that there was not a single person in the entire institute outside of our lab who would appreciate it. Admittedly, we also did not have great interest in what other people were doing at the EMBL, it was so far from our work and we had so little time...but we enjoyed the international atmosphere...we had very good working conditions,...and we used our great chance—we could not have been more successful, but the people who had given us this chance were unable to realize this."

Wieschaus, eager to pursue related areas in his own laboratory in the United States, had applied for a Princeton University job. He got a junior position, based on work published before the Heidelberg project. He would be an assistant professor with the possibility of tenure. Kendrew said he'd have to watch Nüsslein-Volhard to make sure she didn't take over Wieschaus' half of the lab.

Reluctant to remain under those conditions, Nüsslein-Volhard applied for a job back in Tübingen. She asked for a junior post as a small group leader at the Friedrich-Miescher Laboratory of the Max Planck Society. She was thirty-nine years old in 1981, a bit old for the position, and had published only a few major papers on her own. Moreover, the Heidelberg experiment was a joint collaboration—with a man. The Max Planck Society could not be sure how much of the work had been hers or whether she could lead a group.

The Max Planck Society is Germany's biggest basic research institution. It is actually a series of specialized institutes, founded as the Kaiser Wilhelm Society for the Advancement of Science in 1911 and renamed for the eminent physicist after World War II. Its Miescher laboratory is an incubation center for talented young researchers to show their stuff. For five years, group leaders are freed from teaching and given generous budgets, space, and technical assistance. Afterward, they return to the job market to be judged by the work he—or occasionally she—has accomplished. In 1981, the Max Planck Society decided to give Nüsslein-Volhard her chance.

Returning to Tübingen that May felt like going home. But for the first time, she was the boss. The pressure was on. If she did not measure up, Germany's all-or-nothing system meant that she might never be in charge of her own research again.

Starting from scratch, she chose a team that, for Germany, was unusually and conspicuously dominated by women. Although her later laboratory teams had many more men than women, her Miescher group included two men and, at different times, Kathryn V. Anderson, Ruth Steward, and Ruth Lehmann. All three women later had distinguished academic careers in the United States.

As a manager, Nüsslein-Volhard gave herself mixed grades. "I'm moody. Sometimes I have depressed moods, and then I'm very enthusiastic again.... It's sometimes difficult for people who worked with me.... I would exaggerate in both directions. I'd get very excited about things, and then also very negative about things, and neither would be totally justified."

Nüsslein-Volhard thinks she inherited mild tendencies toward manic-depression, a condition that involves swings between elevated and depressed moods. "Manic-depressive is probably what I am, not in a pathological [clinical] sense, but tendency-wise." Whatever the reason, as Anderson discovered, when Nüsslein-Volhard was in a good mood, everyone else was too. "She could be the most charming person in the universe. When she was in a bad mood, she dragged everyone else down with her."

Nüsslein-Volhard also identified another difficulty. "Women are brought up to be kind and permissive, and as a laboratory director, you have to change your habits. This causes problems because men aren't used to it and you don't want to do it. I wish there would be more women [in German science], and men could find a less obnoxious mode. They're sometimes offensive."

"Men want to become administrators and directors. But what woman wants to be a director?" Nüsslein-Volhard asked later in her career. "They want to have a nice life, and a little group of people doing interesting work. But in the German system, you either have to run the whole show or do little." She wanted the whole show.

As often happened when she moved, her first months in Tübingen were difficult. The local bank would not issue her a credit card. "How much will you earn?" the clerk asked, looking suspiciously at Nüsslein-Volhard's blue jeans and shirt. "We'd better wait to see the first check before we issue a credit card." Nüsslein-Volhard missed the capable Kluding, fired one technician, let another go, and recruited undergraduates to care for the flies instead. The scientists spent the first hot summer transferring thousands of fruit flies from one jar to another every ten days until a constant-temperature room was built.

Focusing again on maternal influences on the embryo, Nüsslein-Volhard planned another large-scale screening of fruit-fly mutants. Even more difficult than the Heidelberg project, it became a nightmare of technical problems, hard work, and long hours. Nüsslein-Volhard could be both motherly and sternly frank. She taught a postdoc to draw "because that's the way you see" and spent hours teaching a graduate student to write. But accustomed to working with peers like Wieschaus and Jürgens, she returned the student's draft manuscript with a single word in the margin: "Logic." The comment was businesslike, but devastating. And when Lehmann thanked Nüsslein-Volhard for helping her get a job, the senior scientist responded gruffly, "Of course you'd get the job. Don't be so proud of it."

As in Heidelberg, the team broke for dinner in a local Greek or Italian restaurant, picnicked, or cooked dinner together at Nüsslein-Volhard's house. Saturday afternoons were reserved for a streussel cake that Nüsslein-Volhard made at home or a fruit tart from a bakery. Everyone was expected to attend; weekends and late nights were for working. "It was an incredibly intense and idealistic way of looking at science," Lehmann thought. But sometimes she sneaked out after 1 A.M. for a date. Not knowing any wives or mothers who ran successful laboratories, Nüsslein-Volhard seemed to think that serious scientists could not have families or children.

Ruth Steward, who had been a graduate student in Gehring's laboratory and a friend of Nüsslein-Volhard there, thought the first summer was magical. But Nüsslein-Volhard, knowing she would be judged by her laboratory's publications, thought that Steward was not working hard or fast enough. She had given the mutant she discovered in Freiburg to Steward to study. In Heidelberg, Nüsslein-Volhard had been horrified that Steward would think of vacationing in Italy before she finished writing her thesis; now she frowned when Steward visited her mother in Basel one weekend each month. Seventeen years later, the rift—the first of several with old friends and colleagues—was still unhealed.

Late one afternoon in the laboratory, when all the preparations for their experiment were complete, Nüsslein-Volhard sat down to look at the first plate of embryos. Peering through the microscope, she instantly spotted four fascinating mutants, one of them especially dramatic. The project would be resoundingly successful. Standing up, she led everyone triumphantly off for a celebratory supper. Later that evening, they came back and started screening the mutants. It was a quintessential Janni moment, Anderson thought.

Within a year, data was emerging to explain the information that the mother fruit fly gives her egg for the embryo to use in organizing its body. Data from Nüsslein-Volhard's group was complemented by Trudi Schüpbach's lab in Princeton. Focusing on the most important issues in embryonic development, Nüsslein-Volhard encouraged researchers in other laboratories and donated mutant stocks for their studies. She gave Markus Noll of the University of Zurich a mutant, for example; then she supplied a diagram, helped edit his manuscript—and removed her name as coauthor. The article made the cover of *Nature*.

As her term at the Miescher laboratory drew to a close in 1984, however, Nüsslein-Volhard was in a foul mood. She was one of the world's best biologists studying the early development of life forms. But the Max Planck Society could not decide whether to give her a permanent position. Four years after her Nobel Prize project, she had one totally unsuitable offer from a German university to head an already-formed laboratory devoted to other subjects; she was its second choice. She wanted her own Max Planck laboratory in Tübingen. "I'm not very courageous, I'm easily homesick, and I'm not someone who easily makes new friends."

Eventually the Max Planck Society decided that she could direct its Developmental Biology Laboratory in Tübingen, starting in April 1985. At age forty-two, she finally had a permanent job. She was only the third woman to direct a Max Planck institute in its seventy-four-year history; the first was Lise Meitner during the 1920s and 1930s. As late as 1998, only five of the Max Planck Society's 234 laboratory directors were women, and of these only two were German.

The first time she attended the Max Planck Society's annual meeting as a director, she was invited to its social program for wives, not its business meeting for scientists. Nüsslein-Volhard was not amused. Later she was asked to join a society committee as its *Quota Frau,* roughly translatable as the "Affirmative Action Female." Nüsslein-Volhard did not let the society forget these faux pas. But she regarded such irritations as psychological, not material, barriers.

Moving her laboratory across a Tübingen street, she decorated the stairwell to her offices with posters from European and American museums and hung a reproduction of Pieter Bruegel the Elder's *Cornfield* near window-boxes of geraniums in her office. Then she hastened to add molecular biology to her bag of tricks. For years, she had argued that genetics was required to explain the embryo. But soon after the Heidelberg project, a number of *Drosophila* laboratories around the world began a concerted effort to clone and analyze the genes identified by the mutations. The combination of molecular biology and genetics revealed that many genes involved in the fruit fly's early development are also found in vertebrates, including human beings.

Having designed a gene, nature uses parts of it frugally, over and over again, in different organisms and in different ways within the same organism. A gene consists of two regions: one contains the information needed to make a particular protein and the other switches the gene on and off at different times and in different cells. Depending on when and where the switch functions, it can perform widely varying jobs. Thus, parts of the genes that develop fruit-fly embryos also operate in plants, insects, worms, mice, chickens, zebrafish, humans, and other organisms. Defective parts in early developmental genes cause many spontaneous abortions and common congenital disorders, including spina bifida and cleft palate. Parts of the same gene control eye growth in both fruit flies and mice, though the eyes are very different. Areas of the gene that Nüsslein-Volhard discovered in Freiburg help develop the immune system of mammals. Regions of most of the genes involved in segmenting an embryo are used subsequently in the nervous system and other tissues. Later in life, portions of these same faulty genes can cause a wide variety of cancerous tumors. Genes encoding the same protein have been implicated in toxic septic shock, arteriosclerosis, cancerous X-ray damage, organ rejection, and AIDS. The universality of the genes that she and Wieschaus discovered clinched the Nobel Prize.

As Nüsslein-Volhard emerged as the world's leading developmental biologist, the prizes showered down. She received memberships in the National Academy of Sciences and the Royal Society and honorary degrees from Yale, Princeton, and Harvard universities. "No other American university; just those three; this is cute," she said. She and her cats moved back to Bebenhausen, the fourteenth-century monastery complex she lived in as a young married woman. She bought an apartment in a half-timbered mill house built during the fourteenth and seventeenth centuries. Using Alfred Sloan Prize money, she bought the monks' fish ponds and planted a garden with wildflowers from around the world: Virginia bluebells from North America, pulmonarias from the Balkans, and the like. She drives a Volkswagon now and travels for science but never for pleasure; evenings find her at home, sitting under a weeping willow and staring at the ponds until dark. "Even today, I do things in a solitary way," she said.

She continues to insist on perfection. Dinner party tablecloths must be ironed properly, and her chocolate cakes are layered with marzipan and red currant jelly and glazed with chocolate. She jigsaws a Bruegel painting, stripping each object and color from its context; then she assembles similar patterns and connects them. When lab workers come to her house to make Christmas cookies, the points on their cinnamon stars must be straight. Walking down a hallway of the Stockholm Opera House, she burst out singing one of the evening's arias in a clear and resounding voice. But background music is forbidden in her laboratory; it is distracting, she said. Arriving late one night to find a radio blaring and a dog barking, she yelled, as she had in Gehring's lab, to silence the noise.

In other areas, Nüsslein-Volhard's views are still evolving. As before, her students and postdocs form an extended family, but they tend to have their own families now too, so they quit work at 7 P.M. and dine out only once a week with her. Nüsslein-Volhard does not coo over their babies, but she helped establish a daycare center at the Max Planck Institute, no small feat in a regulation-bound country.

She wonders what the young women in her laboratory will do once they marry. "I don't know any women who have succeeded when their husbands were less successful," she said. Increasingly, she speaks out about women's problems in science, noting, for example, that a scientist who complained about her hand-drawn diagrams would not have criticized a man. Nevertheless, she thinks pushing feminism in Germany is politically unwise. German feminists, closely allied with the Green environmentalist party, are widely regarded as anti-science. She feels that, far from applauding her success in embryology and gene technology, feminists have been hostile to it.

Many of Nüsslein-Volhard's prizes were given for demonstrating between 1987 and 1992 that actual concentrations of shape-causing chemicals gradually increase to form gradients within the embryo; different concentrations of the chemical make cells build different structures. These prizes ignored her friend and former research partner Eric Wieschaus, by then a Princeton University professor. Thus, she was doubly delighted to share the 1995 Nobel Prize in Medicine with him and with geneticist Edward B. Lewis of Caltech. As she told Lewis, "They finally got it right when they got Eric, too." The three divided one million dollars.

Nüsslein-Volhard, at fifty-two, was the tenth woman to win a Nobel in science and the first in Germany. Headlines crowned her Lady of the Flies and Dame *Drosophila*.

Having a Nobel can be a part-time job and a burden on those without a stock of small talk. "The more famous you get, the more prizes you win, the more humble you have to be," Nüsslein-Volhard admitted. "You have to put others at their ease. You have to assure them you're just like them, and that can be tedious sometimes.... People don't want to hear you complain about traveling to accept your awards. They don't want to hear about how I had to drag a box around with the Lasker award inside, and how that box felt like a rock."

Some Nobel Prize winners cope by cutting back or abandoning research. But by the time Nüsslein-Volhard won the Nobel for fruit flies, she had already doubled her research commitment by growing 100,000 zebrafish. Curious as ever, she wanted to know whether her discoveries about flies apply to more complex creatures too. As versatile as the fruit fly is, it is not a vertebrate. Its eyes, immune system, heart, and circulatory system are far simpler; it has no lungs or liver. Its external skeleton conceals its internal organs, muscles, and digestive tract. Biologists had long sought a cheap, small, prolific, and easily handled vertebrate to study.

Nüsslein-Volhard decided the popular aquarium zebrafish *Danio rerio,* would fit the bill. George Streisinger of the University of

Oregon had suggested the blue-and-yellow striped fish in a 1981 review article. After his death, Chuck Kimmel, David Grinwald, Judith Eisen, and others at the University of Oregon continued to study ways to use the fish in developmental biology. Zebrafish embryos are transparent. Virtually any mutation can be identified easily, as it develops. Buying some small zebrafish from a Tübingen pet store, she grew green slime in jars on her windowsills as possible fish food. Then she wrote a half-page grant proposal to the Max Planck Society and, in 1992, opened her 6,000-cage fish house. No one had ever raised zebrafish on such a massive scale.

Exposing male zebrafish to a chemical that randomly mutates genes, Nüsslein-Volhard's team fertilized egg cells with the mutated sperm, bred three generations of fish, and examined more than two million embryos. Almost three years to the day after her fish house opened, she and her former student Wolfgang Driever, then at the Massachusetts General Hospital in Boston, submitted for publication a tour de force: thirty-seven articles that turned a single issue of the journal *Development* into a 481-page guide to more than one thousand mutations affecting the brain, heart, blood, skin, eyes, ears, and jaws. The mutants produced defects in nearly every aspect of the fish's early growth. It was the first systematic analysis of the genes affecting the development of a vertebrate embryo. Piecing the animal's genetic puzzle together, she was thrilled to see broad patterns emerging once again.

The little fish has proved particularly useful for learning about the development of the cardiovascular and nervous systems at a molecular level. Understanding how genes affect the formation of neurons, the blood, the heart, and other internal organs should provide crucial information about human diseases. Mutants that affect the fish's heart, blood formation, and eyes have counterparts in human genetic disorders, including the widespread and devastating disease thalassemia. Thanks in large part to Nüsslein-Volhard's pioneering efforts, more than 200 research laboratories now study the little blue-and-yellow striped fish.

Turning a common aquarium fish into an important tool for understanding human development earned Nüsslein-Volhard widespread praise again. Her work was "an accomplishment of historic proportions," opening new vistas for discovery akin to Magellan's voyage around the world. As a delighted writer in *Science* punned, "Developmental genetic studies with the zebrafish are far from finished."

Ever on the lookout for people who do not take her seriously,

Nüsslein-Volhard enjoyed quoting the few scientists who called her move from flies to fish "a terrible thing for science."

"What sort of comments are these?" she bristled. "Do they think I'm stupid? Do they think I don't know what I'm doing?"

She ended her first zebrafish article with her usual words, "All mutants are available upon request."

Afterword

ARE WOMEN SCIENTISTS still hiding under furniture and working without pay? No. By and large, overt discrimination has disappeared.

As Mildred S. Dresselhaus, former president of the American Physical Society, observed, "We've made tremendous strides in the last generation. Between the time that Rosalyn Yalow got started to the time I got started in science, there was already a huge, enormous improvement. And from my time to the present is another enormous improvement. But we're still quite a ways from equal opportunity."

Women scientists still do not get promoted as fast as their male counterparts, agrees Marshal Matyas of the American Association for the Advancement of Science. Nevertheless, women fare much better in science than women in other occupations, she emphasizes. Their unemployment rates are lower, their salaries higher, and their working hours more flexible. Furthermore, unlike law and medical students, science graduate students do not pay tuition. They are paid salaries while they study.

Today women, minorities, and the handicapped are recognized as a large, untapped resource of scientific talent that could help solve the nation's grave shortage of scientific and technological personnel.

Are women racing into science? Yes and no. The number of women earning science degrees rose steadily between the 1960s until the late 1980s. Then it stopped growing. Why?

Women's preparation for science remains poor. Teachers and textbooks still downplay the scientific accomplishments of women. Marie Curie remains the only woman scientist mentioned in many classes. Parents and teachers widely believe outdated statistics that purported to show girls as innately unable to learn mathematics as well as boys. That evidence is now known to be either inaccurate or nonexistent. More girls are now completing advanced algebra classes than boys, although fewer high school girls study calculus. In countries like France where all college-preparatory students—male or female—are required to take mathematics and physics, the percentage of women scientists is higher. Their schools keep doors open for young men *and* women to enter science-based careers; more than half of all college majors require calculus. Also in France, 35 percent of the physicists who earn advanced doctoral degrees are women. In comparison, a mere seven percent of employed American physicists and astronomers are women. Only about 14 percent of all bachelor's degrees in engineering go to women in the United States; and fewer women earned degrees in engineering in 1990 than in 1984.

What is the biggest problem? One of them is certainly the fact that women are primarily responsible for bearing and raising children. For Marie Curie as for Maria Goeppert Mayer as for Jocelyn Bell Burnell, combining a scientific career and raising children is difficult. Science moves fast, and women who take time off for child-rearing may need major retraining.

Stephen G. Brush, professor of the history of science at the University of Maryland, summarized the problem in *American Scientist*: "By the time a woman lands an assistant professorship, she is likely to be in her late twenties or early thirties. She then has five or six years to turn out enough first-rate publications to gain tenure. If she has children, she must fulfill her family obligations while competing against other scientists who work at least sixty hours a week. If she postpones childbearing, the biological clock will run out at about the same time as the tenure clock."

So far, no one—not universities, private industry, government, or the public at large—has solved the problem that nature gave to women scientists. Capable as they are, they will probably solve it themselves—as their intellectual progenitors did before them.

* * *

Notes

This is a list of my main sources, published and otherwise. I have not included scientific publications by any of the women, however. Here I also acknowledge the many individuals who helped me with scientific and biographical information and with editorial advice.

Reader's note Profiles of women scientists appear in the following books for young people:

Joan Dash, *Triumph of Discovery: Women Scientists Who Won the Nobel Prize* (Englewood Cliffs: Julian Messner, 1991).

Diana C. Gleasner, *Breakthrough, Women in Science.* (New York: Harcourt Brace Jovanovich, 1979).

Louis Haber, *Women Pioneers of Science* (New York: Harcourt Brace Jovanovich, 1979).

Beatrice S. Levin, *Women and Medicine* (Metuchen, N.J.: Scarecrow Press, 1980).

Iris Noble, *Contemporary Women Scientists of America* (New York: Julian Messner, 1979).

Olga Opfell, *The Lady Laureates* (Metuchen, N.J.: Scarecrow Press, 1976, 1986).

Barbara Shiels, *Women and the Nobel Prize* (Minneapolis: Dillon Press, 1985).

Edna Yost, *Women of Modern Science* (New York: Dodd, Mead, 1959).

Dedication

Biologist Viktor Hamburger, close friend of Gerty Cori and Levi-Montalcini, wrote "Hilde Mangold, Co-Discoverer of the Organizer," *Journal of the History of Biology* 17 (1984 Spring): 1–11.

1. A Passion For Discovery

This section of the book is based on help received from Mildred S. Dresselhaus, Frances F. Ekern, James Hamilton, Christine V. Hampton, Jean Johnson, Robert Loeb, Diana I. Marinez, Marsha Matyas, Lucretia McClure, Margaret W. Rossiter, and Eileen van Tassell. The idea for this book was suggested by a 1988 calendar prepared by the Detroit Area Chapter of the Association for Women in Science.

The single most important book about women in science is:

Margaret W. Rossiter, *Women Scientists in America, Struggles and Strategies to 1940* (Baltimore: Johns Hopkins University Press, 1982).

Other published sources include:

Sandra Harding, *The Science Question in Feminism* (Ithaca: Cornell University Press, 1986).

Harriet Zuckerman, *Scientific Elite, Nobel Laureates in the United States* (New York: Macmillan, 1977).

2. Marie Skłodowska Curie

This chapter is based on interviews with Ève Curie Labouisse; Monique Bordry; Hélène Langevin-Joliot; Pierre Joliot; and Daniel Grinberg.

Among published sources are the following:

Peter Craig, "The Light and Brilliancy of Marie Curie," *New Scientist*, July 26, 1984.

Elisabeth Crawford, *The Beginnings of the Nobel Institution, the Science Prizes 1901–1915* (Cambridge: Cambridge University Press, 1989). Crawford's study of Nobel documents revealed Pierre's insistence that his wife share the Nobel Prize and the reason for Marie's second Nobel.

Ève Curie, *Madame Curie* (New York: Doubleday, 1937) is the classic biography by Marie's daughter and the source for excerpts from Marie Curie's diary after Pierre's death.

Marie Curie, *Pierre Curie* (New York: Macmillan, 1923). I am particularly indebted to Marie Curie's brief autobiography published in this volume. Missy Meloney's story of their first meeting appears in the introduction to the same volume. Other firsthand accounts by Marie Curie appear in letters that she and her daughter Irène Joliot-Curie exchanged: *Correspondance, Choix de Lettres, 1905–1934* (Paris: Les Éditeurs Français Réunis, 1974); her 1911 Nobel Lecture, in *Nobel Lectures 1901–1922* (New York: Elsevier, 1966); and her account of her World War I experiences, *La Radiologie et la Guerre* (Paris: 1921). The French translations are mine.

Pierre Curie, "Nobel Lecture 1903," in *Nobel Lectures 1901–1921,* vol. 3 (New York: Elsevier, 1967).

Norman Davies, *Heart of Europe: A Short History of Poland* (Oxford: Clarendon Press, 1984). Contains the Positivist poem.

Françoise Giroud, *Marie Curie, A Life* (New York: Holmes & Meier, 1986).

Irène Joliot-Curie, "Marie Curie, Ma Mère," *Europe, Revue Littéraire Mensuelle,* (1954): 89–121; and "La Vie et L'Oeuvre de Marie Skłodowska-Curie," *La Pensée,* n.s., 58 (1954): 19–30. Two lengthy articles about Joliot-Curie's mother.

André Langevin, *Paul Langevin, Mon Père* (Paris: Les Éditeurs Français Réunis, 1971).

Camille Marbo, *Souvenirs et Rencontres* (Paris: Éd. Grasset, 1968), in which Marguerite Borel, writing under her pen name, told about the Langevin affair.

L'Oeuvre and *L'Intransigeance* in November 1911, Parisian newspapers recounting the Langevin scandal.

Rosalynd Pflaum, *Grand Obsession: Madame Curie and Her World* (New York: Doubleday, 1989).

Robert Reid, *Marie Curie* (New York: E. P. Dutton, 1974). This is the authoritative biography and includes the most extensive account of the Langevin affair.

Elizabeth Rona, *How It Came About: Radioactivity, Nuclear Physics, Atomic Energy* (Oak Ridge, Tenn.: Oak Ridge Associated University, June 1978). Physicist Rona describes the atmosphere in the Curie laboratory.

Margaret W. Rossiter, *Women Scientists in America, Struggles and Strategies to 1940* (Baltimore: Johns Hopkins University Press, 1982). Rossiter discusses the Curie tours of America.

Emilio Segrè, *From X-Rays to Quarks, Modern Physicists and Their Discoveries* (San Francisco: W. H. Freeman, 1980).

Spencer Weart, *Scientists in Power* (Harvard University Press: Cambridge 1979) discusses the politics of the Curie circle.

David Wilson, *Rutherford: Simple Genius* (Cambridge: MIT Press, 1983).

3. Lise Meitner

This chapter is based on interviews with Günter Herrmann, Peter A. Brix, Ruth L. Sime, and Wolfgang Paul.

I also thank for their help Hans Bethe, Kerstin Borysowicz, Ulla Frisch, Ikuko Hamamoto, Edwin N. Hiebert, Paul Kienle, Charlotte Kerner, Charlotte Keyes, Shubrick Kothe, David Marwell, Evelies Mayer, Ulla Nothhacksberger, Rudolf Peierls, Günter Siegert, Barbara Jaeckel, and Chien-Shiung Wu.

I am particularly indebted to Ruth L. Sime's articles (see below) for accounts of the discovery of protactinium, Meitner's escape from Germany, and her relationship with Otto Hahn after World War II. Sime, who is preparing a biography of Meitner, is also the source for Meitner's doubts and Hahn's certainty about their early, incorrect results regarding fission; Strassmann's remark that Meitner was the team's intellectual leader; Hahn's postwar press release; Meitner's comment about Hahn's suppressing the past; and Meitner's description of herself as a "wind-up doll."

Especially useful publications include the following:

Peter Brix, "The Discovery of Uranium Fission: Its Intricate History and Far-Reaching Consequences," *Interdisciplinary Science Reviews* 15 (1990):4. A helpful description of the fission search.

Sigvard Eklund, "Lise Meitner och Otto Robert Frisch," unpublished speech, April 13, 1989, tells about his friend Meitner's hiding behind furniture during chemistry lectures; about her "lost years"; and about her conviction that remaining in Berlin had supported Hitlerism. I am indebted to Kerstin Borysowicz for translating from the Swedish.

Renate Feyl, *Der lautlose Aufbuch Frauen in der Wissenschaft* (Berlin: Luchterhand, 1983).

James Franck, taped interviews, July 9–11, 1962, Oral History Interviews, Archive for the History of Quantum Physics, American Philosophical Society Library, Philadelphia.

Otto Robert Frisch, Meitner's physicist nephew, wrote extensively about his aunt. He told the stories of the iridescent puddle; her siblings' teasing; her reply to questions about marriage; the Hertz alcohol story; and her fateful New Year's visit. Frisch also provided the translation of Hahn-Strassmann's fission publication. Frisch's publications about Meitner include *Working With Atoms* (New York: Basic Books, 1965); "Lise Meitner," *Biographical Memoirs of Fellows of the Royal Society of London,* vol. 16 (London: Royal Society of London, 1970–1971); "Lise Meitner, Nuclear Pioneer," *New Scientist,* Nov. 9, 1978; *What Little I Remember* (Cambridge: Cambridge University Press, 1979); and "The Discovery of Fission," *Physics Today* 43 (Nov. 1967).

Hans G. Graetzer and David L. Anderson, *The Discovery of Nuclear Fission: A Documentary History* (New York: Van Nostrand Reinhold, 1971).

Otto Hahn, *A Scientific Autobiography,* trans. by Willy Ley (New York: Scribner's Sons, 1966); and *Otto Hahn: My Life* (New York: Herder and Herder, 1970). *My Life* tells about his formal relationship with Meitner; her shopping with Mrs. Rutherford; his worry that she endangered the institute.

Günter Herrmann, "Five Decades Ago: From the 'Transuranics' to Nuclear Fission," *Angewandte Chemie*, International Edition in English 29 (May 5, 1990): 481–508. This is an authoritative account of the fission experiment by a former student of Fritz Strassmann. Herrmann also discusses Ida Noddack's contribution.

Paul Kienle, "Lise Meitner: An der Wege der Kernphysik", *Lise Meitner: Ausstellung über Leben und Werk einer Kernphysikerin* (Darmstadt: Gesellschaft für Schwerionenforschung mbH, 1988). Hahn's Dec. 19, 1938, letter and Meitner's reply appear here. Hahn's letter also appears in Herrmann (see above). George F. Bertsch provided translations.

Charlotte Kerner, *Lise Atomphysikerin* (Weinheim: Beltz & Gelberg, 1986). This biography, in German, relates Meitner's reaction to Christmas shopping with Mrs. Rutherford.

Fritz Krafft, *Im Schatten der Sensation: Leben und Wirken von Fritz Strassmann* (Weinheim: Verlag Chemie, 1981). Krafft's authoritative biography contains much information about Meitner, including the story about Strassmann's late-night experiment; the Hahn-Meitner argument about her endangering the institute; her complaint that Hahn lost no sleep over the Nazis; the government minister's refusal to let Meitner leave Germany; and her dismay at being called Hahn's assistant.

Evelies Mayer, "Lise Meitner: Ein Leben im doppelten Exil," *Lise Meitner: Ausstellung Über Leben und Werk einer Kernphysikerin* (Darmstadt: Gesellschaft für Schwerionenforschung mbH, 1988). Mayer includes Meitner's anguished letter to Hahn, written after the concentration camps were publicized. The translation is by George F. Bertsch.

Lise Meitner, "Looking Back," *Bulletin of the Atomic Scientists*, November 1964: 2–7; "The Status of Women in the Professions," *Physics Today*, 13 (August 1960): 17–21. In the first, Meitner talks about her adolescent dream for a full life; Boltzmann's classroom; her interviews with Planck; her first meetings with Einstein, Bohr, and Rutherford; the bigwig-free dinner; the joy of friendship; Fischer's assistants' greeting only Hahn; her enthusiasm for Planck and for Berlin's colloquia; the Nazi-era atmosphere of cooperation in her laboratory; and other stories. In the second, Meitner names the books that opposed educating women in her youth.

Lise Meitner, Taped Oral History interview, 1963, American Institute of Physics, New York.

David Nachmansohn, *German-Jewish Pioneers in Science, 1900–1933* (New York: Springer-Verlag, 1979).

Robert Olby, *Path to the Double Helix* (Seattle: University of Washington Press, 1974) describes Delbruck's informal, Nazi-era seminars.

Rudolf Peierls, *Bird of Passage* (Princeton: Princeton University Press, 1985).

Richard Rhodes, *The Making of the Atomic Bomb* (New York: Simon & Schuster, 1986).

Patricia Rife, *Lise Meitner: Ein Leben für die Wissenschaft,* a German-language biography by an American. It provided Meitner's remark about her inability to do research in Sweden.

Rockefeller Foundation Archives, Record Group 1.1, Series 713D, Box 5, Folder 57. These documents tell about Meitner's agreement to stay in Berlin after 1935 at Planck's request.

Saturday Evening Post, "Interview With Lise Meitner," Jan. 5, 1946.

Ruth L. Sime, "Lise Meitner's Escape from Germany," *American Association of Physics Teachers* 58 (3) (March 1990): 262–67; Talk, International Congress on the History of Science, Munich, Aug. 7, 1989; "Lise Meitner and the Discovery of Fission," *Journal of Chemical Education* 66 (5) (May 1989): 273–376; "The Discovery of Protactinium," *Journal of Chemical Education* 63 (August 1986): 653–57; "Belated Recognition: Lise Meitner's Role in the Discovery of Fission," *Journal of Radioanalytical and Nuclear Chemistry* 142 (1) (1990): 13–26.

Sallie A. Watkins, "Lise Meitner and the Beta-Ray Energy Controversy: An Historical Perspective," *American Association of Physics Teachers* (June 1983); and "The Making of a Physicist," *The Physics Teacher* (Jan. 1984), pp. 12–15. In the latter, Watkins tells about Planck's responses to the poll on education for women.

David Wilson, *Rutherford: Simple Genius* (Cambridge: MIT Press, 1983). Wilson tells about Christmas shopping with Mrs. Rutherford.

C. S. Wu, "History of Beta Decay," *Beiträge zur Physik und Chemie des 20. Jahrhunderts* (Braunschweig: Verlag Vieweg).

4. Emmy Noether

This chapter is based on interviews with Elizabeth Monroe Boggs, Clark Kimberling, Ruth Stauffer McKee, Emiliana Pasca Noether, Herman Noether, Richard E. Phillips, Martha K. Smith, and Olga Taussky-Todd, and on correspondence with Auguste Dick. I am indebted to Hartmut Schulz for translations from German.

I would also like to thank for their assistance: J. Southerland Frame,

Hanna Lifson, Caroline Rittenhouse, Paul R. Sweet, Robert C. Ward, and J. Werner.

Important published sources are:

"Anna Pell Wheeler," *Bryn Mawr Alumnae Bulletin*, Summer 1966.

Alan D. Beyerchen, *Scientists under Hitler: Politics and the Physics Community in the Third Reich* (New Haven: Yale University Press, 1977).

P. S. Alexandrof, "In Memory of Emmy Noether," in Brewer and Smith (see below).

James W. Brewer and Martha K. Smith, eds., *Emmy Noether, A Tribute to Her Life and Work* (New York: Marcel Dekker, 1981). This contains Hermann Weyl's famous eulogy of Noether, including his description of her as "warm" bread. It is also the source for Clark Kimberling's biography, S. Mac Lane's remark, and Taussky-Todd's reminiscences.

Robert P. Crease and Charles C. Mann, *The Second Creation, Makers of the Revolution in Twentieth-Century Physics* (New York: Macmillan, 1986). Discussion of Noether's theorem.

Hans-Joachim Dahms, Cornelia Weger, eds., *Die Universität Göttingen unter dem Nationalsozialismus* (Munich: KG Saur, 1987). For university documents.

Auguste Dick, *Emmy Noether, 1882–1935,* trans. by H. I. Blocher (Boston: Birkhauser, 1981). The groundbreaking biography of Noether, based on Dick's interviews with Noether's associates. It includes Hermann Weyl's eulogy, as well as Noether's characterization of her thesis; Hilbert and Klein's comments about Noether's relativity work; Hilbert's bathhouse comment and course catalogue; the ministry's definition of her lowly legal position; Noether's comment about her hypercomplex number class; descriptions of her teaching style; Noether's letter to Hasse; and her characterization of Princeton University.

Richard J. Evans, *The Feminist Movement in Germany 1894–1933* (London: Sage, 1976); and *Comrades and Sisters: Feminism, Socialism and Pacifism in Europe 1870–1945* (New York: St. Martin's Press, 1987). For background regarding the position of women in Germany.

Walter Feit, "Richard D. Brauer," *Bulletin of the American Mathematical Society* n.s. 1 (Jan. 1979): 1–38.

James Franck, taped interviews July 9–11, 1962, Oral History Interviews, Archive for the History of Quantum Physics, American Philosophical Society Library, Philadelphia. This contains Hilbert's bathhouse story.

Louise S. Grinstein, "Anna Johnson Pell Wheeler," Association of Women in Mathematics *Newsletter* 1978.

Clark Kimberling, "Emmy Noether," in James W. Brewer and Martha K. Smith, eds., *Emmy Noether, A Tribute to Her Life and Work* (New York: Marcel Dekker, 1981). Other articles are "Emmy Noether," *American Mathematical Monthly* 79 (1972): 136–49, and "Emmy Noether," *The Mathematics Teacher* (March 1982), pp. 246–49. Kimberling offers the most extensive English-language accounts of Noether's life. I am indebted to him for the story about Ernst Witt; Noether's characterization of her doctoral thesis; Noether's "pig-in-the-poke" remark; her swimming; the Natasha Artin story; and Noether's remark about her happiness in the United States.

David Nachmansohn, *German-Jewish Pioneers in Science, 1900–1933* (New York: Springer-Verlag, 1979).
Emmy Noether, *Gesammelte Abhandlungen* (Collected Papers), ed. Nathan Jacobson (New York: Springer Verlag, 1983).
Emmy Noether, *Gesammelte Abhandlungen* (Collected Papers), edited by Nathan Jacobson. (New York: Springer-Verlag, 1983).
Gottfried E. Noether, "Emmy Noether," in *Women of Mathematics, A Bio-bibliographic Sourcebook*, eds. Louise S. Grinstein and Paul J. Campbell (New York: Greenwood Press, 1987).
Lynn M. Osen, *Women in Mathematics* (Cambridge: MIT Press, 1974).
Constance Reid, *Hilbert* and *Courant in Göttingen and New York* (New York: Springer, 1970, 1976). Reid's biography of Hilbert is the source for several Noether stories, including the faculty discussion of her promotion and Hilbert's bathhouse rejoinder; Hilbert's "zero" remark and response to the education minister; Noether's "another foreigner" comment, and the Noether Guard uniform. Reid's Courant biography contains the account of Weber's boycott.
Fritz K. Ringer, *The Decline of the German Mandarins: The German Academic Community 1890–1933* (Cambridge: Harvard University Press, 1969). For the position of university professors in German society.
Rockefeller Foundation Collection: Record Group 1.1, Series 200; Box 128, folder 1580. For letters of reference from Lefschetz, Wiener, and Birkhoff and the attempt to find her jobs.
S. L. Segal, "Helmut Hasse in 1934," *Historia Mathematica* 7 (1980): 46–56. For his relationship to the Nazi party.
Bhama Srinivasan and Judith Sally, eds., *Emmy Noether in Bryn Mawr* (New York: Springer-Verlag, 1983). Reports presented at a symposium held at Bryn Mawr to honor Noether.
Olga Taussky-Todd, in *Mathematical People: Profiles and Interviews*, eds. Donald J. Albers and G. L. Alexanderson (Boston: Birkhauser, 1985). Other memories appear in Brewer and Smith's book (see above).
Von Cordula Tollmien, "Emmy Noether 1882–1935," *Göttingen Jahrbuch* 38 (1990): 153–219. A lengthy and authoritative biography. It is the source of stories about the 1915 faculty debate before Noether gave her Habilitation; Frankfurt's job offer; the intercession of Einstein and Klein in 1918–1919; and Einstein's full comment to Hilbert about Noether's paper on invariant forms. I am indebted to Hartmut Schulz for translations.
Hermann Weyl, "Emma Noether," appears in Dick (see above).

5. Gerty Cori

This chapter is based on interviews with Barbara Illingworth Brown, David Brown, Mildred Cohn, Ann Fitz-Gerald Jones Cori, Marvin Cornblath, William Daughaday, Luis Glaser, Viktor Hamburger, David M. Kipnis, Arthur Kornberg, Edwin G. Krebs, Joseph Larner, Neil Madsen, Charles Rollo Park, Jane H. Park, Sidney Velick. I am indebted to Viktor Hamburger and Ann Fitz-Gerald Jones Cori for information about Gerty Cori's religious background.

I am also grateful for help from Paul G. Anderson, Hugh Blaschko,

Philip Randle, and Susan Killenberg. I am particularly grateful to Joseph Larner, who permitted me to read his *Biographical Memoir* of Gerty Cori before publication by the National Academy of Sciences.

For additional reading, see:

Carl Cori, "The Call of Science," *Annual Review of Biochemistry* 38 (1969): 1–19. This autobiographical essay is the source for the opening story about his job offer (which his widow Ann Fitz-Gerald Jones Cori identified as the University of Rochester). The essay also contains his description of Gerty as a student; the cancer director's intravenous cancer cure; working with Gerty; his anatomy examination at St. Louis; the limitations of working with laboratory animals; and the excitement of biochemistry.

Gerty Cori, "This I Believe," ed. Edward R. Murrow, *Columbia Records*; "Some Thoughts on Science and Society," Society of Sigma Xi panel discussion, October 1954; and "Biochemistry, the Science of Life Processes," speech, Smith College, undated. All courtesy of the Cori Papers, Archives, Washington University School of Medicine. The Columbia Record includes her statements about art and science as "the glories of the human mind" and about the benefits of a European education, as well as the closing quotation. Her Sigma Xi talk includes the comment about salaries and prestige associated with basic research.

Herman M. Kalckar, "The Isolation of Cori-ester," in *Selected Topics in the History of Biochemistry: Personal Recollections (Comprehensive Biochemistry)* 35 (1983) ed. G. Semenza; and "Gerty T. Cori," *Science* 126 (July 4, 1958): 16.

Arthur Kornberg, *For the Love of Enzymes: The Odyssey of a Biochemist* (Cambridge: Harvard University Press, 1989).

Joseph Larner and Carlos Villar-Palasi, "Commentary," *Biochemica et Biophysica Acta* 1000 (1989): 311–13.

Severo Ochoa, "Gerty T. Cori, Biochemist," *Science* 126 (July 4, 1958): 16.

Olga Opfell, *The Lady Laureates* (Metuchen, N.J.: Scarecrow Press, 1976, 1986). This contains Gerty Cori's comments about her high school examination and the state of United States biochemistry and Carl Cori's Nobel remarks.

Philip Randle, "Carl Ferdinand Cori," *Biographical Memoirs of Fellows of the Royal Society* 32 (1986): 67–95.

Rockefeller Foundation Archives, Record Group 1.1, Series 228, Box 4, Folder 48. The documents include Gerty Cori's reason for abandoning animal research and her revised report to the foundation.

6. Irène Joliot-Curie

This chapter is based in part on interviews with Hélène Langevin-Joliot, Pierre Joliot, Ève Curie Labouisse, and Monique Bordry.

Important published sources include:

Pierre Biquard, *Frédéric Joliot-Curie, The Man and His Theories* (London: Souvenir Press, 1965) contains Joliot's description of his wife.

P. M. S. Blackett, "Jean Frédéric Joliot," *Biographical Memoirs of the Fellows of the Royal Society (of London)* (London: Royal Society, 1960).

James Chadwick, "Some Personal Notes on the Search for the Neutron," *Proceedings, 10th International Congress of the History of Science, New York 1962* (Paris: Hermann, 1964).

Eugénie Cotton, *Les Curies* (Paris: Seghers 1963). For childhood stories about Irène's shyness and her reaction to dinosaurs and art.

Marie Curie, *Correspondance, Choix de Lettres, 1905–1934* (Paris: Les Éditeurs Français Réunis, 1974). Most published sources about Irène and Frédéric Joliot-Curie concentrate on Frédéric, so I have quoted extensively from the letters in this volume. The French translations are mine. *La Radiologie et la Guerre* (Paris: 1921) and *Pierre Curie* (New York: Macmillan, 1923) contain, respectively, her version of World War I and her comparison of her two daughters.

Bertrand Goldschmidt, *Pionniers de L'Atome* (Paris: Stock, 1987).

Maurice Goldsmith's *Frédéric Joliot-Curie, A Biography* (London: Lawrence and Wishart, 1976).

Irène Joliot-Curie, "Marie Curie, Ma Mère," *Europe, Revue Littéraire Mensuelle* (1954): 89–121. In this lengthy article, Joliot relates her World War I experiences; views on religion, books, childbearing and her parents; and Marie Curie's attitude toward Frédéric Joliot-Curie. The translation is mine.

Irène and Frédéric Joliot-Curie, *Nobel Lectures 1922–41* (New York: Elsevier, 1966).

Lew Kowarski, taped interview, October 1969, Niels Bohr Library, American Institute of Physics, New York. Kowarski uses the Italian word *coglione* to describe Irene's use of the French word *testicule*, best translated into English vernacular today as *asshole.*

Noelle Loriot, *Irène Joliot-Curie* (Paris: Presses de la Renaissance, 1991). This authoritative biography in French contains stories about Irène as government minister. Translations are mine.

Camille Marbo, *Souvenirs et Rencontres* (Paris: Grasset, 1968) for accounts of Joliot-Curie's childhood shyness and the Langevin affair.

Newsweek March 29, 1948. Joliot-Curie vilified by the American press.

Rosalynd Pflaum, *Grand Obsession: Madame Curie and Her World* (New York: Doubleday, 1989) is about the Joliot-Curies as well as Marie Curie. It is the source for several of Frédéric's remarks about his marriage and his wife's remarks about working women's rights.

Angèle Pompéi, "Irène Joliot-Curie," *Europe*, (May 1961), pp. 230–41.

Robert Reid, *Marie Curie* (New York: E. P. Dutton, 1974) is useful for Irène's childhood and upbringing. He tells about how Marie Curie informed Irène of her father's death and about Irène's post-doctoral press interview. Reid also gives the fullest account of the Langevin scandal.

Margaret W. Rossiter's *Women Scientists in America: Struggles and Strategies to 1940* (Baltimore: Johns Hopkins University Press, 1982) discusses Curie's American tours.

Time, Mar. 29, p. 28, and Apr. 17, 1950. Examples of how Joliot-Curie was treated by the American press after World War II.

Spencer R. Weart, "Scientists in Power: France and the Origins of Nuclear Energy, 1900–1950," *Bulletin of the Atomic Scientists* March 1979; and *Scientists in Power* (Cambridge: Harvard University Press, 1979). For the political activities of the Joliot-Curies, their campaign to increase French funding for science, and Frédéric's protestations of affection for his wife.

7. Barbara McClintock

This chapter is based on interviews with Barbara McClintock, Ernest Abbe, Lucy Boothroyd Abbe, Bruce M. Alberts, Guenter Albrecht-Buehler, Charles Burnham, Harriet Creighton, Helen Crouse, Nina V. Fedoroff, Barbara Sears, James A. Shapiro, and Evelyn Witkin.

I also want to thank for their help Marjorie Bhavnani, Susan Cooper, Judith R. Goodstein, and Howard Green.

Important publications include:

Jeremy Cherfas and Steve Connor, "How Restless DNA Was Tamed," *New Scientist*, Oct. 13, 1983.

Stanley N. Cohen and James A. Shapiro, "Transposable Genetic Elements," *Scientific American* 242 (Feb. 1980): 40–49.

E. L. Konigsburg, "Barbara McClintock Retrospective," *The Nobel Prize Annual 1988*, pp. 15–27.

Nina V. Fedoroff, "The Restless Gene," *The Sciences* 31 (Jan. 1991): 22–27 "Transposable Genetic Elements in Maize," *Scientific American* 250 (June 1984): 84–98.

Nina Fedoroff and David Botstein, ed., *The Dynamic Genomes: Barbara McClintock's Ideas in the Century of Genetics* (Plainview, N.Y.: Cold Spring Harbor Laboratory Press, 1992). This festschrift, prepared for McClintock's ninetieth birthday by her friends, contains the flood story and Sturtevant's remark.

Evelyn Fox Keller, *A Feeling for the Organism: The Life and Work of Barbara McClintock* (San Francisco: W. H. Freeman, 1983). Keller is the source for Rhoades's remark about McClintock's brilliance; McClintock's comment about getting down into the cell; her letter about discrimination; and the insults of two biologists.

Gina Maranto, "At Long Last," *Discover,* December 1983, pp. 26–32.

Barbara McClintock, "The Significance of Responses of the Genome to Challenge," *Science* 226 (Nov. 16, 1984): 792–801; "Expanding Opportunities for Women in Science," American Association of University Women Achievement Award speech, 1947; "Chromosome Organization and Genic Expression," *Cold Spring Harbor Symposia on Quantitative Biology*, 1951; "Introduction," *The Discovery and Characterization of Transposable Elements: The Collected Papers of Barbara McClintock* (New York: Garland Press, 1987), vii–xi; and "Remarks at a Press Conference," Cold Spring Harbor Laboratory, Oct. 10, 1983. McClintock discussed the importance of fellowships and growing opportunities for women with the AAUW and the significance of chromosome repairs in her Nobel speech. Her prophesy that transposans had wide implications appeared in Cold Spring Harbor reports.

Thomas Hunt Morgan, "Chromosomes and Heredity," *American Naturalist* 44 (Aug. 19, 1910): 496ff.

New Scientist, Oct. 13, 1983, for Shapiro's praise of the experiment with Creighton.

Newsday, Oct. 11, 1983, contains Baltimore's observation.

Rockefeller Foundation Archives, Record Group 1.1, Series 200, Box 136, Folder 1679, and Record Group 1.1, Series 205, Folder 72, for Weaver's characterization of the Cornell botany department and Morgan's appraisal of McClintock's work and resentment.

Barbara Shiels, *Women and the Nobel Prize* (Minneapolis: Dillon Press, 1985), for McClintock's Statue of Liberty remark.

Jamie Talan, "Organisms 'Speak' for Nobel Winner," *New York Times*, Long Island edition, Oct. 16, 1985.

Winifred Veronda, "James Bonner Recalls," *Caltech News*, Feb. 18, 1984, p. 3.

John Noble Wilford, "A Brilliant Loner in Love with Genetics," *New York Times*, Oct. 11, 1983.

8. Maria Goeppert Mayer

This chapter is based on interviews with Elizabeth Urey Barenger, Hans Bethe, Jacob Bigeleisen, Robert G. Sachs, Hans Suess, Viktor F. Weisskopf, and Marianne Mayer Wentzel.

I also want to thank for their help Marie B. Kuhn, Martin Levitt, James Stimpert, the Maryland Department of the Enoch Pratt Free Library, and Hartmut Schulz for German translations.

Significant publications include:

Elizabeth Urey Baranger, "The Present Status of the Nuclear Shell Model," *Physics Today* 26 (June 1973): 34–42. By the physicist daughter of two of Mayer's close friends.

Max Born, Selections from *Recollections*. American Philosophical Society Library, Philadelphia, Pa. This is the source of Born's accounts about Mayer's university years, her explanation of why she switched from mathematics to physics, and her descriptions of Göttingen's exciting atmosphere.

Joan Dash, *A Life of One's Own: Three Gifted Women and the Men They Married* (New York: Harper & Row, 1973). Based on interviews with Mayer and letters, this book contains a lengthy chapter about Mayer. It is the source for the opening paragraph of my chapter and for Mayer's remarks about associating only with men; gaining her independence from Joe; the onion shell model; Jensen's eyeglass prescription; and for much of our knowledge of Mayer.

Laura Fermi, *Atoms in the Family* (Chicago: University of Chicago Press, 1954).

James Franck, transcript of taped interview with Maria Goeppert Mayer, 1962, "Sources for the History of Quantum Physics," American Philosophical Society. Here she tells stories about her high school examination, Hilbert's anemia and lecture invitation, Ehrenfest's house, why she switched from mathematics to physics, and her descriptions of Göttingen's exciting atmosphere..

Mary Harrington Hall, "Maria Mayer: The Marie Curie of the Atom," *McCall's* 91 (July 1964): 38+; and "The Nobel Genius," *San Diego Magazine* (August 1964), pp. 64ff. The San Diego article is a fuller version of McCall's. They contain Mayer quotations about her social status in Göttingen; visiting Ehrenfest; cooking Christmas dinner; guilt about leaving her children; sensing resentment; Joe's firing; opera with Teller; filing systems; kaffeeklatsches; Sarah Lawrence interview; telling Joe everything; not contributing to the bomb; not rushing her friends; the Swedish palace; and her love of doing physics. Also, an American's confession that "everyone" loved her.

Hans Jensen letters to Maria Goeppert Mayer, University of California, San Diego Mandeville Department of Special Collections, Mss. 20, Box 1, Folder 16–29. Especially letters dated June 14, 1952; February 1953; July 8, 1953; and March 15, 1961. I am indebted to Hartmut Schulz for translations.

Karen E. Johnson, "Maria Goeppert Mayer: Atoms, Molecules and Nuclear Shells," *Physics Today* 39 (Sept. 1986): 44–49; and Ph.D. Thesis, University of Minnesota. The former contains Mayer's characterization of Sarah Lawrence as "rather swell."

Correspondence and notes regarding Joseph Mayer's firing, Record Group 02.001, Records of the Office of the President, series 1, file numbers 47 (Physics Department) and 48 (Chemistry Department); Karl Herzfeld correspondence Record Group 02.001, series 1, file 47 (Physics Department). Ferdinand Hamburger Jr., Archives, The Johns Hopkins University.

Maria Goeppert Mayer, "The Shell Model," *Science*, Sept. 1964: 999–1006; "The Structure of the Nucleus," *Scientific American*, Dec. 1948; "Changing Status of Women as Seen by a Scientist," manuscript of speech to Japanese women, Tokyo, in University of California at San Diego, Mandeville Dept. of Special Collections. Mss. 20, Box 5, Folder 11. Mayer's comments about her parents' expectation that she would earn her own living, the emotional strain on working mothers, universities' refusal to pay Depression-era wives, and working for fun and insurance are in the Tokyo speech. In the *Science* article, which is her Nobel lecture, she calls her Columbia research "nice, clean physics" and describes Fermi's contribution to the shell model and her excitement afterwards. Her *Scientific American* article lists the "mysteries of nuclear physics."

Joseph Mayer, "My Wife's Secret," University of California at San Diego, Mandeville Dept. of Special Collections. This contains Joseph Mayer's stories about meeting and dating Mayer.

Julia B. Morgan, *Women at the Johns Hopkins University* (Baltimore: Johns Hopkins University, 1986).

Robert G. Sachs, "Maria Goeppert Mayer," *Biographical Memoirs of the National Academy of Sciences* 50 (1979): 311ff; "Maria Goeppert Mayer, Two-Fold Pioneer," *Physics Today*, Feb. 1982; and "Maria Goeppert Mayer" *Remembering the University of Chicago* (Chicago: University of Chicago Press, 1992). In the last, Sachs tells about hiring Mayer.

Science Digest, February 1964, pp. 30–36.

Eugene Wigner's comment about the excitement of nuclear physics during the 1930s comes from *Nuclear Physics in Retrospect, Proceedings of a Symposium on the 1930s,* ed. Roger H. Stuewer (Minneapolis: University of Minnesota Press, 1979).

Edward Teller letters in the the University of California, San Diego, Mandeville Department of Special Collections. Mss. 20, Box 3, Folder 1–16.
Sam B. Treiman, "On Physics Graduate Students," *American Journal of Physics* 53 (Sept. 1985): 817–18.
Hans A. Weidenmüller, "Why the Shell Model Came as a Surprise," *Nuclear Physics* A507 (1990): 5c–14c.
Viktor F. Weisskopf, *Privilege of Being a Physicist* (New York: W. H. Freeman, 1989).
Harriet Zuckerman, *Scientific Elite, Nobel Laureates in the United States* (New York: Macmillan, 1977). Mayer says she was dismayed by Jensen's paper.

9. Rita Levi-Montalcini

This chapter is based on interviews with Rita Levi-Montalcini, Luigi Aloe, Ruth Hogue Angeletti, Ralph Bradshaw, Stanley Cohen, Renato Dulbecco, Lloyd A. Greene, Viktor Hamburger, Ronald Oppenheim, Robert R. Provine, and Dale Purves.

I would also like to thank for their help Mamar Blosser, Verena Brink, and Mildred Cohn.

Other, published sources include:

Rita Levi-Montalcini, *In Praise of Imperfection: My Life and Work,* transl. Luigi Attardi (New York: Basic Books, 1988); "The Nerve Growth Factor Thirty-Five Years Later," *Science* 237 (Sept. 4, 1987): 1154–62; "NGF: An Uncharted Course," in *The Neurosciences: Paths of Discovery,* Frederic G. Worden, Judith P. Swazey, George Adelman, eds. (Cambridge: MIT Press, 1975); "Reflections on a Scientific Adventure," in *Women Scientists: The Road to Liberation*, ed. Derek Richter (New York: Macmillan, 1982). In "Uncharted Course," she told about G. Levi's visit in Florence and said her St. Louis years were her happiest. *In Praise of Imperfection* is the source for quotations about roosters in the coops; childhood memories of religion; kissing her mother; her father's permission to study; test scores; a friend's advice to do wartime research; her mother's guarding the laboratory; why she did research despite the war; the importance of optimism in research; the need to continue working; and Dante's verse.
Rita Levi-Montalcini and Pietro Calissano, "The Nerve-Growth Factor," *Scientific American* 240 (June 1979): 68–77.
Omni, "Interview," 10 (March 1988): 70–74f. Here she comments that she succeeded during wartime despite primitive instrumentation; she did not respect G. Levi's ideas; and why Hamburger did not win a Nobel.
Dale Purves, *Body and Brain: A Trophic Theory of Neural Connections* (Cambridge: Harvard University Press, 1988).
Dale Purves and Joshua R. Sanes, "The 1986 Nobel Prize in Physiology or Medicine," *Trends in NeuroSciences* 10 (June 1987): 231–35.
Frederika Randall, "The Heart and Mind of a Genius," *Vogue*, March 1987. This is the source for Levi-Montalcini's comment about America's cordiality.

10. Dorothy Crowfoot Hodgkin

This chapter is based on interviews with Dorothy Crowfoot Hodgkin, Thomas L. Blundell, Louise Johnson, Aaron Klug, Barbara Rogers Low, Jenny Pickworth Glusker, Max Perutz, Anne Sayre, David Sayre, Kenneth N. Trueblood, and Alexander Tulinsky.

I would also like to thank for their help Pauline Adams, David Brink, Verena Brink, Jack D. Dunitz, Judith Howard, Mandy A. Mackenzie, Fernanda Perrone, and Keith Prout.

In addition, the following publications were especially useful:

Robert C. Brasted and Peter Farago, "Interview with Dorothy Crowfoot Hodgkin," *Journal of Chemical Education* 54 (April 1977): 214–15. Here Hodgkin explains why she continued working during World War II, child care, and a university salary.

Guy Dodson, Jenny P. Glusker, David Sayre, ed., *Structural Studies on Molecules of Biological Interest, a Volume in Honor of Professor Dorothy Hodgkin* (Oxford: Clarendon Press, 1981). This festschrift includes a particularly useful article by Max Perutz and is the source of quotations by Dennis Parker Riley, John H. Robertson, John G. White. In it, also, Hodgkin talks about phoning her friends to see the penicillin model; telegraphing UCLA; and the Order of Merit.

P. P. Ewald, *Fifty Years of X-Ray Diffraction* (Utrecht: 1962). This is the source for W. L. Bragg's "sound barrier" quotation.

Maurice Goldsmith, *Sage: A life of J. D. Bernal* (London: Hutchinson, 1980).

Dorothy Crowfoot Hodgkin, "John Desmond Bernal," *Biographical Memoirs of the Fellows of the Royal Society* 41 (1980); "It's Up to Us!" *The Bulletin of the Atomic Scientists*, Jan. 1981, pp. 38–39; "The X-Ray Analysis of Complicated Molecules," *Science* 150 (Nov. 19, 1965): 979–88; "Some Ancient History of Protein X-Ray Analysis," in *Structural Chemistry and Molecular Biology: A Volume Dedicated to Linus Pauling*, eds. Alexander Rich and Norman Davidson (San Francisco: W. H. Freeman, 1968). The *Science* article, which is Hodgkin's Nobel talk, includes her quotation about "lurking questions" and spending time failing to solve problems. Her biography of Bernal includes her comment about his "colored" life and the wiring in his laboratory.

Maureen M. Julian, "Profiles in Chemistry," *Journal of Chemical Education* 59 (Feb. 1982): 124–25.

Patricia Phillips, *The Scientific Lady: A Social History of Women's Scientific Interests 1520–1918* (London: Weidenfeld & Nicholson, 1990).

Rockefeller Archive Center, Record Group 1.1, Series 401, Box 38, Folders 491 and 487. These documents contain the Rockefeller official's glowing reports about Hodgkin's laboratory.

"Tribute: Thomas Hodgkin," *London Times* March 26, 1982, p. 10f.

Lewis Wolpert and Alison Richards, *A Passion for Science* (Oxford: Oxford University Press, 1988). Based on BBC interviews, this book is the source for Hodgkin's stories about identifying ilmenite, why she liked archaeology, and collusion with White on B_{12}.

11. Chien-Shiung Wu

This chapter is based on interviews with Chien-Shiung Wu, Ernest Ambler, Felix Boehm, William Fowler, Joel Groves, William Havens, Raymond Hayward, David G. Hitlin, Evelyn Hu, Noemie Koller, Ursula Schaefer Lamb, Leon J. Lidofsky, Rudolf Peierls, Melba Phillips, Robert R. Wilson, C. N. Yang, Luke Yuan, Vincent Yuan.

I also want to thank for their help Steven Averill, Gloria Blatt, Patricia Cianciolo, Linda Cooke Johnson, Jane Repko, and Margaret Steneck.

Among important publications about Wu are:

Henry A. Boorse and Lloyd Motz, *The World of the Atom* (New York: Basic Books, 1966).

Robert P. Crease and Charles C. Mann, *The Second Creation, Makers of the Revolution in Twentieth-Century Physics* (New York: Macmillan, 1986).

Lynn Gilbert and Gaylen Moore, *Particular Passions, Talks with Women Who Have Shaped Our Times* (New York: Crown Publishers, 1981). The Wu chapter is the source for the story about her father's gift of textbooks.

Gloria Lubkin, "Chien-Shiung Wu, the First Lady of Physics Research," *Smithsonian*, Jan. 1971.

Jacquelyn A. Mattfeld and Carol G. Van Aken, ed. *Women and the Scientific Professions: The MIT Symposium on American Women in Science and Engineering* (Cambridge: MIT Press, 1965). This contains Wu's concluding comments about women in science.

New York Post profile, Jan. 22, 1959, cited in *Current Biography*, 1959: 492.

Rudolf Peierls, *Bird of Passage* (Princeton: Princeton University Press, 1985).

Emilio Segrè, *From X Rays to Quarks: Modern Physicists and Their Discoveries* (San Francisco: W. H. Freeman, 1980).

Viktor F. Weisskopf, *Privilege of Being a Physicist* (New York: W. H. Freeman, 1989). Weisskopf tells the story about Pauli's bet.

C. S. Wu, "Recent Investigations of the Shapes of Beta-Ray Spectra," *Reviews of Modern Physics* 22 (Oct. 1950); "One Researcher's Personal Account," *Adventures in Experimental Physics* 1973; "Subtleties and Surprises: The Contribution of Beta Decay to an Understanding of the Weak Interaction," *Annals of the New York Academy of Sciences*, Nov. 8, 1977; "The Discovery of the Parity Violation in Weak Interactions and Its Recent Developments," Nishina Memorial Foundation, April 1983. Both the New York Academy and the Nishina Foundation publications recount the parity experiment for general audiences; I relied on them extensively. The former article also includes the story about her airplane flight with Pauli.

Edna Yost, *Women of Modern Science* (New York: Dodd, Mead, 1959).

12. Gertrude Elion

This chapter is based on interviews with Gertrude B. Elion, James Burchall, Herbert Elion, Jonathan Elion, Elvira Falco, Cora Himadi, George H. Hitchings, Howard J. Schaeffer, Thomas A. Krenitsky.

The main published sources are:

Katherine Bouton, "The Nobel Pair" *New York Times Magazine*, Jan. 29, 1989.
George H. Hitchings and Gertrude B. Elion, "Layer on Layer," *Cancer Research* 45 (June 1985): 2415–20.

13. Rosalind Franklin

This chapter is based on interviews with Dorothy Hodgkin, Donald L. D. Caspar, Francis Crick, John T. Finch, Jenifer Glynn, Raymond Gosling, Steve Harrison, Kenneth C. Holmes, Aaron Klug, Vittorio Luzzati, Anne Sayre, David Sayre, James Watson. I am particularly grateful to Anne Sayre for the story of Ellis Franklin's opposition to his daughter's entering a university. I also would like to thank Pauline Adams.

Among the numerous publications featuring Franklin are:

J. D. Bernal, "The Department of Crystallography," *The Lodestone* (Birkbeck College, University of London) 55 (1965): 37–44.

Francis Crick, *What Mad Pursuit: A Personal View of Scientific Discovery* (New York: Basic Books, 1988).

Miriam Franklin, "Rosalind," unpublished manuscript. This brief biography, written by Rosalind Franklin's mother, is the source of quotations about Rosalind's childhood.

Horace Freeland Judson, "Annals of Science: The Legend of Rosalind Franklin," *Science Digest*, Jan. 1986; "DNA," *New Yorker* Nov. 27, 1978, Dec. 4, 1978, Dec. 11, 1978; and *The Eighth Day of Creation* (New York: Touchstone, 1979). Crick's "patronizing attitude" remark appears in *The Eighth Day of Creation.*

Aaron Klug, "Rosalind Franklin and the Discovery of the Structure of DNA," *Nature* 219 (Aug. 24, 1968): 808–44; "Rosalind Elsie Franklin," *Dictionary of National Biography*, and "Rosalind Franklin and the Double Helix," *Nature* 248 (Apr. 6, 1974): 787–880.

Robert Olby, *Path to the Double Helix* (Seattle: University of Washington Press, 1974). This is the source for Crick's views on collaboration.

Peter Pauling, "DNA—The Race That Never Was?" *New Scientist* 58 (May 31, 1973): 558–60.

Rockefeller Foundation Archives, Record Group 1.2; Series 401D; Box B18; Folder F167.

Anne Sayre, *Rosalind Franklin and DNA* (New York: W. W. Norton, 1975). This book, which countered Watson's portrayal of Franklin and created the Franklin legend, is the source of Wilkins's remark that Watson and Crick scooped King's College.

Anthony Serafini, *Linus Pauling: A Man and His Science* (New York: Paragon House, 1989).

Gunther S. Stent, ed., *The Double Helix: Text, Commentary Reviews, Original Papers* (New York: W. W. Norton, 1980). This contains comments by André Lwoff and Robert L. Sinsheimer.

James D. Watson, *The Double Helix* (New York: Atheneum, 1968). This is the source for quotations about Franklin's lecture; Watson's and Crick's faulty model; Wilkins's comment about Franklin's bark and her so-called refusal to see a helix; and Watson's fear that Franklin might attack him.

14. Rosalyn S. Yalow

This chapter is based on interviews with Rosalyn S. Yalow, Maurice Goldhaber, Stanley J. Goldsmith, Joseph Meites, Johanna Pallotta, J. Edward Rall, Ira Rosenthal, John T. Potts, Jr., Eugene Strauss, Benjamin Yalow, and others.

I would like to thank also for their help Jesse Roth and A. Rees Midgley.

Significant publications about Rosalyn Yalow include:

Fred A. Bernstein, *The Jewish Mothers' Hall of Fame* (Garden City: Doubleday, 1986). Clara Zipper Sussman's quotations appeared here, as well as Yalow's remark about packing her own valise.

Lynn Gilbert and Gaylen Moore, *Particular Passions, Talks with Women Who Have Shaped Our Times* (New York: Crown Publishers, 1981). This is the source for Yalow's remarks about library rules, "discriminators," nuclear physics in the 1930s, Greer Garson, waking up early, luck and creativity, and playing tennis.

Diana C. Gleasner, *Breakthrough: Women in Science* (New York: Walker, 1983). Stories about baseball games, braces, and bright Jewish girls appear here.

Carol Kahn, "She Cooks, She Cleans, She Wins the Nobel Prize," *Family Health* 10 (June 1978): 24–27.

Eileen Keerdoja and William Slate, "A Nobel Woman's Hectic Pace," *Newsweek*, Oct. 29, 1979. Yalow jokes about being a public figure after winning a Nobel.

Leticia Kent, "Winner Woman!" *Vogue* 168 (Jan. 1978): 131 ff. Yalow's comment about women needing to work harder than men appeared here.

"*Festschrift* for Rosalyn S. Yalow: Hormones, Metabolism and Society," *Mount Sinai Journal of Medicine* 59 (March 1992): 95–185. See this for Seymour Glick's quotation; making Berson's lunch; Yale professor; posh suburb; and live-in help.

J. Edward Rall, "Solomon A. Berson," *Biographical Memoirs of the National Academy of Sciences* (Washington, D.C.: National Academy Press, 1990).

William P. Rayner, *Wise Women* (New York: St. Martin's Press, 1983). Yalow talks of her husband's support here.

Elizabeth Stone, "A Mme. Curie from the Bronx," *New York Times Magazine*, April 9, 1978, pp. 29ff. I am indebted to Stone for her description of Yalow as "earth mother and aggressor" and for Yalow's quotations about her love of logic; Purdue University; and her graduate examination. In addition, Aaron and Clara Yalow talk about his contribution and Strauss describes lab workers as "flying around."

Rosalyn S. Yalow, "Radioimmunoassay," *Nuclear Medicine*, ed. Henry N.

Wagner, Jr., (New York: HP Publishing, 1975): 225–32; Nobel Banquet Speech and autobiography, *Les Prix Nobel 1977* (Stockholm: The Nobel Foundation, 1978); "Radioimmunoassay: A Probe for the Fine Structure of Biologic Systems," *Science* 200 (June 1978): 1236–45; "A Physicist in Biomedical Investigation," *Physics Today* 32 (Oct. 1979): 25–29; "Presidential Address: Reflections of a Non-Establishmentarian," *Endocrinology* 106(1) (1980): 412–14; "Radioactivity in the Service of Man," her Nobel lecture, *BioScience* 31 (Jan. 1981): 23–28; "Need for Scientific Literacy in a Modern Society," text of unpublished speech, 1991; and "Women in Science," unpublished manuscript of speech, 1991. Yalow's Nobel speeches contain quotations about her as a stubborn child and student; her acceptance and teaching skills at Illinois; and Dr. Failla.

15. Jocelyn Bell Burnell

This chapter is based on interviews with Jocelyn Bell Burnell, George Greenstein, Jeremiah P. Ostriker, and Joseph Taylor.

I want to thank H. Allison Bell, David Brink, Verena Brink, and the Franklin Institute for their help.

For the account of Burnell's pulsar discovery, I have relied extensively on Burnell's own account, "Little Green Men, White Dwarfs, or What?" *Sky and Telescope*, March 1978, pp. 218–21. Other useful sources by Burnell include a speech, "Female Scientists—Feat or Freak?" *Wise Week 1984* (Edinburgh: Royal Observatory); and her booklet, *Broken for Life*, published under the name of S. Jocelyn Burnell (London: Quaker Home Service, 1989). *Broken for Life* covers her son's illness and her thoughts on religion.

I used the *Sky and Telescope* to describe the press conference, sensitivity of radio telescopes; Hewish's comments and her reactions as the pulsar appeared and disappeared; her reaction to the LGM theory; the appearance of more "scruff" before Christmas; and her return from Christmas. Her "Feat or Freak" talk covers criticism of science teaching; "freaks"; physics in astronomy; her comments on women's attitudes towards physics; sexist language; sex differences; and part-time jobs.

Other important sources are:

George Greenstein, *Frozen Star* (New York: Freundlich Books, 1983).

Anthony Hewish, "Pulsars and High Density Physics" *Science* 188 (June 13, 1975): 1079–83. Hewish's Nobel speech discussed LGM and secrecy.

Paul and Lesley Murdin, *Supernovae* (Cambridge: Cambridge University Press, 1985).

Vera Rubin, "Women's Work," *Science* 58 (July-August 1986): 58–65.

N. Wade, "Discovery of Pulsars: A Graduate Student's Story," *Science* (1975). Burnell's description of discovering the first "bit of scruff": her "conundrum" remark; and the Hoyle controversy appear here.

16. Christiane Nüsslein-Volhard

This chapter is based on interviews with Kathryn Anderson, Spyros Artavanis-Tsakonas, Lorraine Daston, Wolfgang Driever, Walter Gehring, Jeanette Holden, Nancy Hopkins, David Ish-Horowicz, Gerd Jürgens, Judith Kimble, Ruth Lehmann, Maria Leptin, Edward B. Lewis, Mary Mullins, Markus Noll, Christiane Nüsslein-Volhard, Vincenzo Pirrotta, Heinz Schaller, Paul Schedl, Trudi Schüpbach, Leslie Stevens, Ruth Steward, and Eric Wieschaus. I am particularly grateful to Jeanette Holden for letting me read correspondence about the Gehring laboratory in the 1970s.

I also want to thank Anthony Capitos of the American Association of University Women, Hellmut Ammerlahn, George Bertsch, the Commission on Professionals in Science and Technology, Antje Hoering, Jörn Knoll, Wolfgang Nörenberg, Susan Parkhurst, David W. Raible, Elsbeth Rass, Margaret Rossiter, and Barbara Wakimoto for their help. I am particularly indebted to the Gesellschaft für Schwerionenforschung, Darmstadt, for its hospitality while I wrote this chapter.

Among the significant publications by and about Nüsslein-Volhard are:

Jennifer Ackerman, "Journey to the Center of the Egg," *New York Times Magazine* (Oct 12, 1997): p. 42. This is the source of passages about "Intelligence didn't matter," and loneliness; staring at ponds, jigsaw puzzles; "I knew nothing about flies"; gender issues and lower expectations; Magellan.

Natalie Angier, "'The Lady of the Flies Dives Into a New Pond." *New York Times* (Dec. 5, 1995) p. B5 (N) p. C1 (L) col. 1: This is the source of passages about her parents attitude toward her interest in science; the burdens of prize winning; other scientists' complaints about zebrafish.

David Brown, "Two Americans, German Share Nobel Prize for Genetics Research," *Washington Post* (Oct. 10, 1995): p. A3.

Kenneth M. Brown, *Women, Minorities, and Persons With Disabilities in Science and Engineering: 1996* (National Science Foundation: Division of Science Resources Studies, 1996): This is the source for figures on women's participation in American science faculties.

Michael Dean, "Polarity, Proliferation, and the Hedgehog Pathway," *Nature Genetics* 14 (Nov. 1996): 245–47.

Judith S. Eisen, "Zebrafish Make a Big Splash: Review," *Cell 87* (Dec. 13, 1996): 969–77.

Saskia Esser and Carla Fandrey, *Bebenhausen: Kloster, Schule, Schloss, Landtagssitz, Gemeinde* [n.p.,n.d.].

Meg Gordon, "See How We Grow," *New Scientist* 155 (Sept. 6, 1997): 30–33.

David Jonah Grunwald, "A Fin-De-Siecle Achievement: Charting New Waters in Vertebrate Biology," *Science* 274 (Dec. 6, 1996): 1634–35. This is the source of the "fin-ished" pun.

Pascal Haffter et al., "The Identification of Genes With Unique and Essential Functions in the Development of the Zebrafish, *Danio Rerio*," *Development* 123 (Dec. 1996): 1–36.

Nigel Holder and Andrew McMahon, "Genes From Zebrafish Screens," *Nature* 384 (Dec. 12, 1996): 515–16.

P. W. Ingham, "Zebrafish Genetics and Its Implications for Understanding Vertebrate Development," *Human Molecular Genetics* 6 (no. 10) 1755–60.

Patricia Kahn, "Zebrafish Hit the Big Time," *Science* 164 (May 13, 1994): 904–905.

Patricia Kahn, "Germany Warily Maps Genome Project," *Science* 268 (June 16, 1995): 156–58.

Gertrud Lehnert, "Women in the Academic World," *AvH Magasin* 68 (1996): This is a source for statistics on women students in German science.

Edward B. Lewis, "Nobel Address," *Les Prix Nobel* (Stockholm: Almqvistand Wiksell International, 1996): pp. 233–60.

Max Planck Society Archives, private communication (May 6, 1998): This is the source for the number of directorships.

Christiane Nüsslein-Volhard, "Axis Determination in the *Drosophila* Embryo," *The Harvey Lectures* series 86, (Wiley-Liss Inc., 1992): p. 129-48.

Christiane Nüsslein-Volhard, "Gradients That Organize Embryo Development," *Scientific American* 297 (August 1996): 54–61.

Christiane Nüsslein-Volhard, Letters to Jeanette Holden: Sept. 15, 1975; Nov. 10, 1975; April 5, 1976; April 29, 1976; June 18, 1978.

Christiane Nüsslein-Volhard and Eric Wieschaus, "Mutations Affecting Segment Number and Polarity in *Drosophila*," *Nature* 287 (Oct. 30, 1980): 795–801.

Christiane Nüsslein-Volhard, "Nobel Address," *Les Prix Nobel* (Stockholm: Almqvist and Wiksell International, 1996): pp. 263–94. This is the source of passages about her personal background, especially girlhood; quotations about her dreams; her initial problems and excitement at EMBL.

Christiane Nüsslein-Volhard, "Von Fliegen und Menschen," *Bild der Wissenschaft* (Jan. 1996): This is a source for the good overall description of her work.

Judith Rauch, "Verstehen, Wie das leben Funktioniert," *Madame Curie und ihre Schwestern: Frauen, die den Nobelpreis Bekamen,* ed., Charlotte Kerner (Weinheim, Germany: Beltz and Gelberg, 1997). This is the source of passages about childhood foods and singing; responsiblity for siblings; frog; anonymity in Frankfurt; changing her thesis topic.

Wade Roush, "Nine Make the Nobel Grade," *Science* 270 (Oct. 20, 1995): 380–81.

J. Travis, "Nobel Prize for Genes That Shape Embryos," *Science News* 158 (Oct. 14, 1995) *Statistisches Jahrbuch 1997 fuer die Bundesrepublik Deutschland,* (Wiesbaden: Metzler Poeschel, 1997). This is the source for statistics on German women in science.

U.S. Department of Education, National Center for Educational Statistics, Integrated Postsecondary Education Data Systems (IPEDS), "Fall Staff Survey, 1995." Table B–7b is the source for the percentage of tenured U.S. faculty members who are women.

Bjoern Vennstroem, "Introduction," *Les Prix Nobel* (Stockholm: Almqvist and Wiksell International, 1996): 22–23.

Eric Wieschaus, "Nobel Address," *Les Prix Nobel* (Stockholm: Almqvistand Wiksell International, 1996): 297–314. This is the source for passages about Wieschaus's personal and art background; work in Heidelberg. It is an excellent overall description of the Heidelberg experiment.

Afterword

Stephen G. Brush, "Women in Science and Engineering," *American Scientist* 79 (Sept.–Oct. 1991): 404 ff.

* * *

The author wishes to gratefully acknowledge the use of photos and illustrations on the following pages:

13, 14, 15, 119, 120, 121: Archives Curie et Joliot-Curie, Musée de Laboratoire Curie
39, 40: Mrs. Otto Frisch and Churchill Archives
40 (bottom): Historical Archives of the Max Planck Society
65: Emiliana P. Noether and Journal für de Reine und Angewandte Mathematik
66 (top): Dr. Gottfried Noether
66 (bottom): Emiliana Pasca Noether
95: Joseph Larner
96 (top): Arthur Kornberg
96 (bottom): Washington University
97: Edwin G. Krebs, M.D.
145: Cold Spring Laboratory Research Library Archives, David Miklos, photographer
146, 147: Marjorie Bhavnani
177: Louise Baker, AIP Niels Bohn Library
178, 179: Stein Collection, AIP Niels Bohr Library
203: Viktor Hamburger
204 (top left, and bottom): Rite Levi-Montalcini
204 (top right): Maria Levi-Montalcini
215: Laboratory of Cell Biology, Rome, and *Science*
227: Anita Corbin, The British Council
228: Dorothy Crowfoot Hodgkin
229: The Royal Society
241: Reprinted from Robertson, J. Monteith, *Organic Crystals and Molecules.* (Ithaca, New York: Cornell University Press, 1953), page 283.
242: ibid., page 285
251: Reprinted from Blundell, Tom L. et al. 1972. "Crystal Structure of rhombohedral 2 Zinc Insulin," *Structure and Forum of Proteins at the Three-dimensional Level* 36:238.
257, 259: AIP Niels Bohr Library
258 (top): Dr. C. S. Wu
258 (bottom): Segré Collection, AIP Niels Bohr Library
281, 283: Burroughs Wellcome
282: Gertrude Belle Elion
293: *The Chemical Educator.*
305, 306 (bottom): Donald L. D. Caspar
306: (top): Jenifer Glynn
317: Aaron Klug and *Nature*
329: Kenneth C. Holmes
335: AIP Meggers Gallery of Nobel Laureates
361, 362 (bottom): Jocelyn Bell Burnell
362 (top left and right): M. Allison Bell
381: Max-Planck-Gesellschaft
382, 383 (top): Christiane Nüsslein-Volhard
383 (bottom): Judith Kimble

Index